MODERN
SHOP
PROCEDURES

MODERN SHOP PROCEDURES

EDWARD R. KRATFEL

Scientific Research Instruments
A Subsidiary of G. D. Searle
Baltimore, Md.

GEORGE R. DRAKE

A.A.I. Corporation
Baltimore, Md.

RESTON PUBLISHING COMPANY, INC., Reston, Virginia 22090
A Prentice-Hall Company

Library of Congress Cataloging in Publication Data

Kratfel, Edward R. 1938-
   Modern shop procedures.

   1. Machine-shop practice. I. Drake, George R.,
1938- joint author. II. Title.
TJ1160.K678   621.7'5   73-19644 ✓
ISBN 0-87909-506-7
ISBN 0-87909-505-9 (pbk.)

© 1974 by
Reston Publishing Company, Inc.
A Prentice-Hall Company
Box 547
Reston, Virginia 22090

10 9 8 7 6 5 4 3 2 1

Printed in the United States of America.

DEDICATED TO

Rose, Michael, and Sandra

the wife and children of Edward R. Kratfel

and

Mary, Natalie, Jeffrey, and Paul

the wife and children of George R. Drake

CONTENTS

PREFACE

Modern Shop Procedures, as its name implies, is a modern text of shop procedures for the beginning student in mechanical shop training programs in high schools, vocational schools, technical insitutes, and junior colleges. The text contains both the English and metric systems and their measuring tools, math, sketching and drawing evaluation, use and care of general hand tools, layout procedures, drills, taps and dies, power tools, and fasteners.

The text is presented in sequence such that the student progressively builds his knowledge as he completes each chapter. Review questions and problems at the end of each chapter allow the instructor to assign specific exercises to the students as home assignments.

Hobbyists, electrical technicians, and mechanics can also benefit by the use of this text as a guide in the solution of their particular mechanical application.

In Chapter 1, the student is taught to use and read the very accurate measuring devices used in the shop. This chapter provides the first practical work encountered by the student. Chapter 2 provides the math essentials used in the shop—plain geometry, solid geometry, and trigonometry. Over 80 equations are provided. Practical construction of figures such as parallel lines, hexagons, etc., and solutions to practical problems such as cost estimation, weights, and volumes are presented.

Chapter 3 teaches the student how to read engineering drawings and how to make sketches; it also discusses projections, drawing formats, lines, dimensions, tolerances, symbols, and abbreviations. Chapters 4, 5, 7, 8, and 9 discuss hand tools, tools used in layout, drills, taps and dies, and power tools. Use, application, and care are discussed for each tool. The procedures of laying out a pattern onto a workpiece in two and three dimensions are discussed in depth in Chapter 6. Chapter 10 discusses material fasteners.

The authors have endeavored to give the student a solid theoretical foundation upon which he can build his knowledge; no text can teach or demonstrate adequate practical applications. The student must take the authors' suggestions and work with an instructor or supervisor to gain the practical experience which is important in his development as a technician.

Edward R. Kratfel   *George R. Drake*

1

MEASURING TOOLS

Measuring tools are the instruments used by the mechanical technician to determine the dimensions or size of a workpiece. If the workpiece is very large, a flexible steel scale or steel tape may be used for the measurement; if the workpiece is to be very accurately measured, a micrometer is used. A different type of micrometer is used for outside measurements, inside measurements, and depth measurements. For a measurement range that is greater than the micrometer, a vernier caliper is used—for both inside and outside measurements. A vernier height gauge is used for measuring and marking heights of workpieces. And finally, the last tool we'll discuss is one you've probably used—a thickness gauge. It's used to gauge the width of an opening, such as the gap in spark plugs.

This chapter discusses steel scales and tapes, outside micrometer calipers, inside micrometers, depth micrometers, vernier calipers, vernier height gauges, and thickness gauges. You learn what they are, how they are used, how the measurements of the workpieces are read, and how the tools are cared for. Both English and metric measuring tools are covered.

The authors have chosen the measuring tools that are most often used by the beginning mechanical technician. By learning the information present-ed in this chapter, the technician can make most of the measurements that will be required of him in his first years of work. The chapter does not cover every measuring tool nor every variation of a tool because of the large number of various designs by the many manufacturers.

STEEL SCALES AND TAPES

1-1. Introduction

A steel scale is a flat strip of metal with accurately calibrated graduations marked on its face for use in measuring linear distances, and as an aid in drawing straight lines. A steel tape is a strip of flexible metal with accurately calibrated graduations marked on its face for use in measuring larger linear distances than the steel scale and for measuring large curved surfaces such as the perimeter of a cylinder. Between the scale and the tape, the scale is the more accurate measuring device.

Steel Scales

The steel scale is available in numerous models with lengths varying from $\frac{1}{4}$ inch to 12 feet (typical lengths: 1, 2, 3, 4, 6, 18, 24, 36, 48, 60, 72, 96, 120, and 144 inches) and 5 centimeters to a meter (typical lengths: 15, 30, 50, and 100 centimeters). The scales are made of spring tempered, semi-flexible, or full flexible steel.

Scales are graduated in either English inches, metric millimeters, or in both English and metric. Depending upon the model of scale chosen, the graduations may appear on the front top and bottom edges or both the front and the back top and bottom edges of the scale. English graduations may be in $\frac{1}{4}$, $\frac{1}{8}$, $\frac{1}{16}$, $\frac{1}{32}$, or $\frac{1}{64}$ inches and each scale edge may use different graduations. English graduations may also be in the decimal system, graduated in $\frac{1}{50}$ (0.02) and $\frac{1}{100}$ (0.01) inches. The metric graduations may be decimeters, centimeters, millimeters, and half millimeters.

An example of a scale with four different graduated edges is:

$$\text{front} \begin{cases} \text{first edge:} & \tfrac{1}{10}, \tfrac{1}{20}, \tfrac{1}{50}, \text{ and } \tfrac{1}{100} \text{ inch} \\ \text{second edge:} & \tfrac{1}{12}, \tfrac{1}{24}, \text{ and } \tfrac{1}{48} \text{ inch} \end{cases}$$

$$\text{back} \begin{cases} \text{third edge:} & \tfrac{1}{14} \text{ and } \tfrac{1}{28} \text{ inch} \\ \text{fourth edge:} & \tfrac{1}{16}, \tfrac{1}{32}, \text{ and } \tfrac{1}{64} \text{ inch} \end{cases}$$

The graduations of $\frac{1}{10}$, $\frac{1}{20}$, $\frac{1}{50}$, and $\frac{1}{100}$ inch on one edge should raise a question. How are all of these graduations on one edge? The answer is, different graduations are along different inch segments; for example, the first inch is divided in $\frac{1}{100}$ inch graduations, the second in $\frac{1}{50}$ inch, the third in $\frac{1}{20}$ inch, and the remainder of the scale in $\frac{1}{10}$ inch graduations.

On metric scales, three edges may be graduated in millimeters and the fourth edge in half millimeters. Combined English–metric scales may have edges graduated in $\frac{1}{32}$ and $\frac{1}{64}$ inches and in millimeters and half-millimeters.

A shrink scale is a special oversize scale that compensates (over the entire length of the scale) in measurement for shrinkage of a cast metal. The shrink scale is used in laying out wood and metal patterns and core boxes. For example, shrinkage per foot of lead is $\frac{5}{16}$ inch and aluminum and brass shrinkage is $\frac{3}{16}$ inch per foot. For casting aluminum, you would purchase a $\frac{3}{16}$ inch shrink scale, either 12 or 24 inches long. Graduations are in $\frac{1}{50}$ (0.02) and $\frac{1}{10}$ (0.1), or on another scale, $\frac{1}{8}$, $\frac{1}{16}$, $\frac{1}{32}$, and $\frac{1}{64}$ inch.

An adjustable hook steel scale is used to make measurements from a "blind" corner or edge where it is not possible to be sure that the end of a standard scale (without a hook) is even with the end of the work. The hook simply slips over the edge surface and holds the scale. The hook scale is also used to set the leg openings of calipers and dividers (Chap. 5) to any graduated length.

A set of tempered steel scales of $\frac{1}{4}$, $\frac{3}{8}$, $\frac{1}{2}$, $\frac{3}{4}$, and 1 inch lengths is available for making special measurements. Each scale is graduated in $\frac{1}{32}$ and $\frac{1}{64}$ inch. A holder is used to grip the small scales before they are placed into areas which are not easily accessible. Another type of special scale has a tapered end (from $\frac{1}{2}$ down to $\frac{1}{8}$ inch) which is used in small holes, narrow slots, grooves, and recesses.

A flexible steel scale with a pocket clip attached is a very versatile scale used by the technician to make linear measurements on workpieces and as an aid in sketching (Sec. 3-1). The pocket scale is graduated on the front side only in $\frac{1}{32}$ and $\frac{1}{64}$ inch graduations.

Steel Tapes

Pocket steel tapes are used for measuring linear distances or large curved distances such as cylinder circumferences. The tapes are about $\frac{3}{8}$ inch wide and are usually 72 inches long. Graduations are in inches and $\frac{1}{16}$ inch. A spring is used as a rewind mechanism; rewinding occurs on some models when a push button is depressed. The pocket tape is convenient for making large measurements in the field. It is also conveniently stored.

Large steel tapes measure long linear and curved distances. They are available in lengths of 25, 33, 50, 66, 75, and 100 feet and 10, 15, 20, 25, and 30 meters. Combination English/metric tapes are also available. The English tapes are graduated in feet, inches, and $\frac{1}{8}$ inches. Each foot is marked in a color different than the inch markings. Marks are set at each 16 inches for standard stud (upright posts in a wall) centers. Special tapes are available with graduations in $\frac{1}{10}$ and $\frac{1}{100}$ foot. Steel tapes are used extensively in the construction and surveying fields where large dimensions are often encountered.

1-2. Using and Reading Steel Scales and Tapes

The steel scale is placed on the workpiece along the line to be measured. The zero inch (or millimeter) end is placed against the left edge of the workpiece or on the mark from which the measurement is to be made. A reading is then taken at the end of the workpiece or the end of the area to be measured.

In making the reading, look in a perpendicular direction at the mark to be measured and at the corresponding graduation on the scale. This ensures that parallax does not cause the reading to be inaccurate. Parallax is the apparent displacement of the reading mark or graduation on the scale due to a change of direction in the position of the observer. Therefore, to prevent parallax and a subsequent inaccurate reading, the observer must be directly in front of the workpiece mark and scale when the reading is made. This procedure also holds true when reading all other measuring tools and gauges discussed in this book.

If a hook steel scale is used, the hook may be placed over the edge of the surface to be measured.

With a scale marked in $\frac{1}{64}$ inch, the scale can be accurately read to $\frac{1}{64}$ of an inch (0.0156). If the mark to be measured falls between $\frac{1}{64}$ inch graduations, the reading is rounded down to the lower $\frac{1}{64}$ if the mark appears to be less than half of $\frac{1}{64}$ or is rounded up to the next $\frac{1}{64}$ if the mark appears to be half or greater than half of $\frac{1}{64}$ inch. As the technician becomes more proficient in using the scale, he will develop the ability of measuring with a scale to within 0.005 inch of the actual size. This is done by mentally and visually "thirding" a $\frac{1}{64}$ inch increment ($\frac{1}{64}$ inch equals 0.0156 inch).

Metric scales can be read accurately to one half of one millimeter. Metric scales graduated in millimeters are rounded to the next millimeter the same as described above for $\frac{1}{64}$ inch readings.

In using the steel tape, the end of the pull ring is placed against the beginning measurement mark. The pull ring is a part of the length of the tape and is included in every measurement. In using the tape for a measurement, the technician must also pull the tape taut and ensure that the tape is along the line to be measured.

As with the steel scale, make a reading of the tape from an observation point perpendicular to the mark to be read. This eliminates parallax. Measurements are made accurately to $\frac{1}{16}$ inch (0.0625) by rounding off.

1-3. Care of Steel Scales and Tapes

As with all tools, avoid accidental scratches or nicks to the surfaces and especially the edges of the steel scales and tapes. Wipe scales and tapes clean to remove dirt and fingerprints (moisture placed on metal surfaces from the skin can cause corrosion). Keep the scales and tapes separated from other tools to prevent damage. Wipe metal scales and tapes occasionally with a light oil applied with a soft lint-free cloth. Many tapes are coated with baked enamel; these tapes should not have oil applied to them.

OUTSIDE MICROMETER CALIPERS

1-4. Introduction

Micrometer calipers, more often called micrometers or "mikes," are used to make a variety of accurate measurements with little possibility of error through misreading. Outside micrometers (Fig. 1-1) are used to measure the outsides of a workpiece (i.e., the workpiece is placed inside of the measuring faces on the anvil and spindle of the micrometer). Measurements may be made accurately in the English inch units to thousandths (0.001) or ten-thousandths (0.0001) of an inch and in the metric system to hundredths of a millimeter (0.01 mm, which equals 0.00039 inch). Note that all micrometer measurements are in decimal notation: not fractional notation.

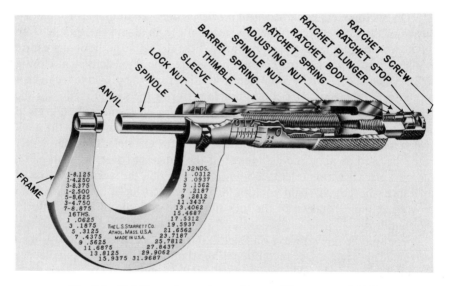

Fig. 1-1. Outside micrometer calipers. (Courtesy of L.S. Starrett Co.)

Outside micrometers are made in a variety of frame sizes to accommodate workpieces up to 24 inches or 300 millimeters. It is interesting to note though, that the working parts of the micrometer do not move through more than 1 inch (English) or 25 millimeters (metric). Instead, various sized frames and interchangeable anvils are used to adapt the micrometer to the workpiece size.

Various special purpose outside micrometers are available to adapt to measuring or checking a wide variety of workpieces or parts of workpieces such as: thicknesses of closely spaced sections; out of roundness, odd fluted taps, milling cutters, reamers, etc.; diameters and depths of narrow grooves, slots, keyways, recesses, and shallow depths between lands and fins; wire diameters; screw threads; sheet metal; and paper thicknesses. Tungsten carbide faces are also available for use in unusual applications where severe abrasive conditions exist.

Figure 1-1 illustrates the external and internal parts of outside micrometer calipers. The technician should become familiar with the nomenclature of the external parts: *frame, anvil, spindle, lock nut, sleeve, thimble,* and *ratchet stop.* In making a measurement, the workpiece to be measured is placed between the faces of the anvil and spindle. The thimble is rotated until the faces of the anvil and spindle close lightly on the workpiece. The dimension of the workpiece is then read from the graduations on the sleeve and thimble.

On micrometers equipped with a *ratchet stop*, the thimble is rotated by means of the ratchet stop. When the faces of the anvil and the spindle close on the workpiece, the ratchet stop slips or clicks so that no additional pressure is placed on the workpiece. This ratchet assures the same consistent closing tightness on successive measurements, whether performed by one technician or another technician. A friction thimble performs the same function as the ratchet stop except that the spindle will not turn when more than a given amount of turning pressure is applied. The lock nut is used to lock the anvil and spindle faces to a fixed dimension; thus, for example, the micrometer could be used as a "go/no-go" gauge (if the gauge fits over the workpiece, the workpiece is within tolerance—go; if it doesn't fit, the workpiece is out of tolerance—no-go). A slight rotation of the lock nut locks the spindle from moving; a slight rotation of the lock nut in the opposite direction unlocks the spindle.

The spindle screw thread on English micrometers has a pitch of $\frac{1}{40}$ inch (40 threads per inch). One revolution of the thimble advances the spindle face toward the anvil $\frac{1}{40}$ inch, which equals 0.025 inch.

Note the divisions along the longitudinal line on the sleeve of the micrometer (Fig. 1-2). These divisions, called the major sleeve divisions, are marked from 0 (zero) to 0 (ten). This distance is exactly 1.000 inch. The distance between 0 and 1 is $\frac{1}{10}$ inch or 0.100 inch. Note that between the major sleeve divisions with numbers there are four smaller divisions, known as the minor sleeve divisions; each minor sleeve division is 0.025 inch. Therefore, the divisions (counting major and minor) along the scale are 0.000, 0.025, 0.050, 0.075, 0.100, 0.125, . . . , 0.900, 0.925, 0.950, 0.975, and 1.000 inch. As previously described, one revolution of the thimble moves the spindle 0.025 inch, which is indicated on the sleeve by minor sleeve divisions.

The beveled scale of the thimble is graduated from 1 to 25 and represents a total of 0.025 inch; the divisions are called the thimble divisions. The space

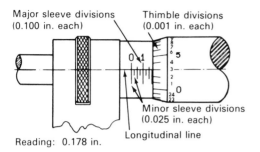

Major sleeve divisions
(0.100 in. each)

Thimble divisions
(0.001 in. each)

Minor sleeve divisions
(0.025 in. each)

Longitudinal line

Reading: 0.178 in.

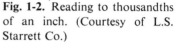

Fig. 1-2. Reading to thousandths of an inch. (Courtesy of L.S. Starrett Co.)

between each thimble division corresponds to an advance of the spindle of 0.001 inch. The longitudinal line of the sleeve (previously discussed) is the reading line for the thimble.

1-5. Reading to Thousandths of an Inch

With the graduations discussed thus far, it is now possible for the technician to read all micrometers that are graduated in thousandths of an inch. Refer to Fig. 1-2. The sleeve indicates one major sleeve division and three minor sleeve divisions. The thimble divisions indicate a 3 opposite the longitudinal line. Thus:

$$
\begin{array}{lll}
\text{Major sleeve divisions:} & 1 \times 0.100 = & 0.100 \\
\text{Minor sleeve divisions:} & 3 \times 0.025 = & 0.075 \\
\text{Thimble divisions:} & 3 \times 0.001 = & \underline{0.003} \\
& \text{by addition} & 0.178 \text{ inch}
\end{array}
$$

The dimension of the workpiece measured is equal to 0.178 inch.

1-6. Reading to Ten Thousandths of an Inch

A micrometer that is graduated in ten thousandths of an inch has an additional group of divisions called the vernier divisions [Fig. 1-3(A)]. There are ten vernier divisions numbered 0 (zero) through 0 (ten); each division equals 0.0001 inch. The spacing of the vernier divisions is such that the total length of the ten divisions fits exactly into the space of nine of the thimble divisions. This always causes one of the vernier division lines to fall exactly opposite one of the thimble divisions. The correct vernier reading is the reading (on the vernier divisions) where a vernier division line aligns exactly opposite a thimble division line.

With the graduations discussed, the technician now has the capability of reading a micrometer graduated in ten thousandths of an inch. Refer to Fig. 1-3(B) and read the micrometer distance measured to ten thousandths of an inch.

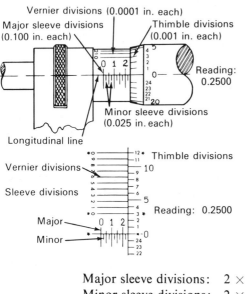

Fig. 1-3. Reading to ten thousandths of an inch. (Courtesy of L.S. Starrett Co.)

Major sleeve divisions:	2×0.100	$= 0.200$
Minor sleeve divisions:	2×0.025	$= 0.050$
Thimble divisions:	0×0.001	$= 0.000$
Vernier divisions:	0×0.0001	$= 0.0000$
	by addition	0.2500

In the above problem, the 0 thimble division aligns with the longitudinal line. Thus, there are no 0.001 divisions. The vernier division line (and the 0—10 line) align with a thimble division line (the 3). Therefore, there are no 0.0001 divisions.

Refer to Fig. 1-3(C) and read the micrometer distance measured to one ten thousandths inches. How many major sleeve divisions are there? Minor sleeve divisions? Which thimble division line most nearly aligns with the longitudinal line (the lower line is read)? Which vernier division aligns with a thimble division?

Major sleeve divisions:	2×0.100	$= 0.200$
Minor sleeve divisions:	2×0.025	$= 0.050$
Thimble divisions:	0×0.001	$= 0.000$
Vernier divisions:	7×0.0001	$= 0.0007$
	by addition	0.2507

1-7. Reading the Metric Micrometer

Physically, the metric micrometer caliper (Fig. 1-4) is the same as the English micrometer caliper except that the graduations are based on the metric millimeter rather than the English inch. The spindle screw thread has a pitch of $\frac{1}{2}$ millimeter; two revolutions of the thimble advances the spindle

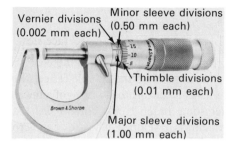

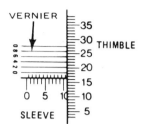
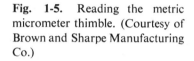

Fig. 1-4. The metric outside micrometer caliper. (Courtesy of Brown and Sharpe Manufacturing Co.)

Fig. 1-5. Reading the metric
micrometer thimble. (Courtesy of
Brown and Sharpe Manufacturing
Co.)

face toward the anvil 1 millimeter. The maximum opening between the anvil
and the spindle face is 25 millimeters. Thus, there are 50 threads per 25 millimeters on the metric micrometer spindle screw.

On the metric micrometer, the major sleeve divisions each represent 1.00
millimeters (mm), the minor sleeve divisions 0.50 mm, the thimble divisions
0.01 mm and the vernier divisions 0.002 mm. The method of reading the
metric micrometer is basically the same as for the English micrometer (Sec.
1-6). This is illustrated in Fig. 1-5.

Major sleeve divisions:	10×1.00	$= 10.00$
Minor sleeve divisions:	1×0.50	$= 0.50$
Thimble divisions:	16×0.01	$= 0.16$
Vernier divisions (read directly or third line)	$3 \times 0.002 =$	0.006
	by addition	10.666 mm

Figure 1-6 illustrates another type of readout of micrometer calipers.
Although a metric version is shown, English versions are also available.
The metric micrometer in Fig. 1-6 reads 7.266 mm.

Fig. 1-6. Metric micrometer with a digital readout. (Courtesy
of Brown and Sharpe Manufacturing Co.)

1-8. Use of the Micrometer Calipers

The micrometer is held in the right hand. The workpiece is held in the left hand against the anvil face. The right hand rotates the thimble or the friction thimble or ratchet stop (if available) to close the spindle face slowly against the workpiece. The measurement is not forced—a light contact pressure assures correct reading. After some practice, the correct feel that develops accuracy in repeated measurements is developed. The use of a micrometer with a ratchet stop or a friction thimble allows for consistent readings, even from a number of technicians using the same micrometer.

The workpiece should not be removed from the micrometer until the reading is taken. If it is necessary to remove the workpiece in order to read the micrometer, then rotate the lock nut to secure the spindle before removing the workpiece. This procedure should be avoided, if possible, to prevent any possible scratching to the micrometer faces or to the workpiece.

When measuring the diameter of a round or cylindrical workpiece, the following procedure is recommended. Hold the anvil of the micrometer firmly to the workpiece at some point. With this point as a pivot point, swing the micrometer back and forth slowly and at the same time, rotate the thimble in toward the workpiece. As contact of the spindle against the workpiece is felt, decrease the swinging action until a smooth firm feel is developed between the anvil and the spindle. Read the micrometer to determine the diameter of the workpiece.

1-9. Care of the Micrometer Calipers

The micrometer should be kept clean at all times. Wipe the micrometer with a soft lint-free cloth; dampen with a solvent, if necessary, to dispose of hard-to-remove dirt. Ensure that the faces of the anvil and spindle are clean. An occasional drop of instrument oil on the spindle and spindle threads assures good performance. Never use an air hose for cleaning as the pressure could force dirt into the spindle threads. Store the micrometer in a case or box to protect it and to keep it clean.

On occasion, the micrometer may develop play or the longitudinal line may not align with the thimble 0 (zero) division when the micrometer faces are closed. To adjust for play, back off the thimble until the adjusting nut (refer to Fig. 1-1 for location) is exposed. Insert the spanner wrench (included with the micrometer) into the adjusting nut slot and tighten just enough to eliminate play. It is better to make several small adjustments with the spanner wrench rather than one large adjustment that could damage the threads.

To align the thimble 0 division on the longitudinal line, first ensure that the anvil and spindle faces are clean, then close the spindle against the face. Insert the spanner wrench in the small slot in the sleeve (Fig. 1-7) and turn the sleeve until the longitudinal line aligns with the thimble 0 division.

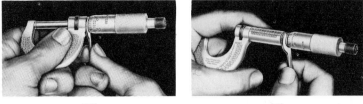

(A)   (B)

Fig. 1-7. Adjusting the micrometer: (A) for play; (B) to align the thimble 0 division on the longitudinal line. (Courtesy of L.S. Starrett Co.)

It is a good habit to store the micrometer with a space of at least $\frac{1}{8}$ inch between the anvil and spindle faces. This practice prevents thermal expansion damage, which could distort the frame or the threads of the micrometer.

THE INSIDE MICROMETER

1-10. Introduction

The inside micrometer caliper is used by technicians for measuring inside diameters of cylinders and rings, for setting calipers, comparing gauges, and for measuring between parallel surfaces. Inside micrometers come in sets with typical measuring ranges of 2–8, 2–12, and 2–32 inches in the English system and 50–200, 50–300, and 50–800 mm in metric. Typical large tubular inside micrometers are 300 inches in English and 1000 mm in metric. Measurements are made by a moveable screw head, which is similar to an outside micrometer, to which one or more rods and spacers are attached.

The inside micrometer set (Fig. 1-8) consists of a 0.000–0.500 inch micro-

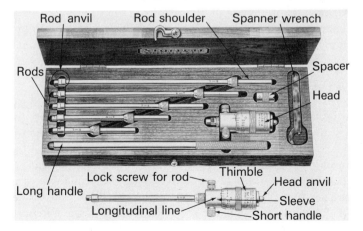

Fig. 1-8. The inside micrometer. (Courtesy of L.S. Starrett Co.)

meter head, spacer, a set of rods, and a handle. The head operates on the screw principle and when the thimble is rotated, the head extends or retracts up to $\frac{1}{2}$ inch or 1 inch, depending upon the size of the micrometer head. The head has an anvil on each end; the anvils contact the workpiece when the measurement is made. Thus, measurements as small as the distance between the head anvils plus one-half or one inch can be made with the head alone. The rods extend the capability of the head. The spacer is a one-half inch collar, which extends the rod length by one-half inch. A spanner wrench is included for making adjustments for wear.

1-11. Reading the Inside Micrometer

The inside micrometer is read in the exact same manner as the outside micrometer caliper described in Sec. 1-6 except that the length of the head, the gauge(s), and the rod(s), all as applicable, are added to the micrometer reading. The graduations are in thousandths (0.001) of an inch and typical micrometer screws move $\frac{1}{2}$ or 1 inch.

For example, if the head in Fig. 1-8 is used alone, it can measure from 2.000 inches (micrometer reading of 0.000) to 2.500 inches (micrometer screw fully extended, reading 0.500). By placing the 5-6 rod into the head, as shown in Fig. 1-8, the micrometer can measure from 5.000 to 5.500 inches. If the 5-6 rod is then removed, and the spacer collar is placed over the rod shaft and the rod is reinserted into the head, the micrometer would then read between 5.500 and 6.000 inches.

Metric inside micrometers are read the same as the metric outside micrometer discussed in Sec. 1-7 except that the length of the head, the spacer(s), and the rod(s), all as applicable, are added to the micrometer reading. The graduations are in hundredths (0.01) of a millimeter (0.00039 inch). The micrometer screw moves 13 mm or 25 mm on a typical manufacturers micrometer head. Metric spacers are typically 12, 25, or 50 mm.

1-12. Use of the Inside Micrometer

The inside dimensions of the workpiece should first be approximated using a scale. This approximation determines which rod, and spacer, if required, should be assembled to the head.

Knowing the approximate measurement, assemble the head, gauge (if required), and the appropriate rod by following the manufacturer's instructions. Be sure to clean any dirt from the anvils, head, and rod shoulder. Rotate the micrometer thimble to the 0.000 reading. If the micrometer is to be inserted deep into a hole, the long handle may be screwed into the head in place of the short handle.

Insert the micrometer into the workpiece opening, holding it as shown in Fig. 1-9. One anvil is held against one of the workpiece sides. The micrometer head is extended by rotating the thimble until the opposite anvil comes into contact with the opposite side of the workpiece. If there is a lock nut on the thimble, it is locked tight, preventing thimble movement. The micrometer is then carefully removed from the workpiece and a reading is made (to 0.001 inch).

Fig. 1-9. Use of the inside micrometer. (Courtesy of L.S. Starrett Co.)

On internal cylindrical surfaced workpieces, more caution and practice are required to make accurate readings. One anvil is held against one contact point of the workpiece; the other anvil is "swung" in different sweeping directions, as in arcs or circles, while the thimble is rotated, extending or retracting the micrometer length until the true dimension is attained. This method ensures accurate and repeatable measurements. It takes practice to acquire the "feel" for making accurate inside measurements of cylindrical surfaced workpieces.

1-13. Care of the Inside Micrometer

The inside micrometer is cared for in the same manner as the outside micrometer (Sec. 1-9). The head is adjustable for wear on its screw. Individual rods are also adjustable with specially furnished wrenches to compensate for wear. Make the adjustments in accordance with the manufacturer's recommendations.

THE DEPTH MICROMETER GAUGE

1-14. Introduction

The technician uses the depth micrometer gauge to measure the depth of holes, slots, recesses, keyways, projections, and similar measurements. The gauge is sold in sets with typical measuring depths of 0–3, 0–6, and 0–9 inches in the English system and 0–75, 0–150, and 0–225 millimeters in the metric system.

Measurements are made by rotation of a micrometer screw that extends a rod into the recess until it contacts the workpiece. A measurement reading is then taken from the micrometer scale and added to the length of the rod, if applicable.

The depth micrometer gauge (Fig. 1-10) consists of a $2\frac{1}{2}$–6 inch $\times$ $\frac{1}{2}$ inch base and 0.000–1.000 inch micrometer head, and a rod which is inserted through the micrometer screw. The head operates on the screw principle and increases the length of the rod from the base until the rod contacts the bottom of the workpiece hole, recess, etc. to be measured. Various sizes of rods increase the depth capability of the depth micrometer gauge. Three, six, or nine round or flat rods increasing in length by one inch (English) or 25 mm (metric) are usually contained in a set. The spanner wrench is used to adjust the micrometer head for screw wear.

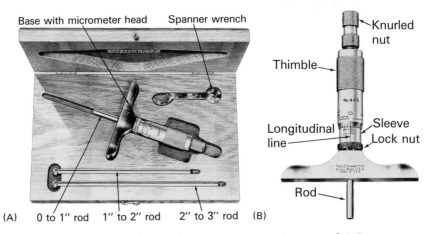

Fig. 1-10. The depth micrometer gauge. (Courtesy of L.S. Starrett Co.)

A depth micrometer gauge with only one-half of a base is also available. This style permits gauging depths close to shoulders and between obstructions.

Some models feature a lock nut which locks the micrometer tightly until a measurement reading is taken. Some models also feature a ratchet stop that is used in turning the micrometer to increase the length of the gauge. When the rod contacts the bottom surface, the ratchet slips. This permits consistent measurements regardless of the technician making the measurement.

1-15. Reading the Depth Micrometer Gauge

The depth micrometer gauge is read in a manner similar to that of the outside micrometer described in Sec. 1-5; however, the depth micrometer gauge is read from right to left. The depth micrometer gauge is read to thousandths (0.001) of an inch. This reading is added to the rod length of the rod inserted.

The illustration of the depth micrometer gauge in Fig. 1-10 indicates a depth reading. Before reading on, can you determine the reading? The reading is obtained as follows:

Length of rod used: 0–1.000   = 0.000
Major sleeve divisions: 8 × 0.100 = 0.800
Minor sleeve divisions: 2 × 0.025 = 0.050
Thimble divisions: 0 × 0.001 = 0.000
                                  by addition   0.850 inch

If the 2–3 inch rod is installed into the depth micrometer gauge shown in Fig. 1-10, then the reading would be: rod length 2.000 plus micrometer reading of 0.850 inch, which equals 2.850 inches. It is evident then that the technician must remember to add the appropriate rod length to the micrometer reading to obtain the correct depth reading.

1-16. Use of the Depth Micrometer Gauge

If the hole, slot, or projection is not too small in cross-section, insert a pocket scale and approximate the depth. This determines which rod is to be used in the depth micrometer gauge.

Rotate the thimble to the approximate 0.000 setting. Remove the knurled nut (cap) and insert the appropriate length rod. Replace the knurled nut. Be sure that the base, rod end, and workpiece are clean.

Fig. 1-11. Use of the depth micrometer gauge. (Courtesy of L.S. Starrett Co.)

Hold the base of the depth micrometer gauge firmly against the workpiece (Fig. 1-11). Rotate the thimble, knurled nut, or ratchet stop until you "feel" the rod touch the bottom of the workpiece or, if using a micrometer with a ratchet stop, until the ratchet slips. If available, rotate the lock nut until tight. Remove the depth micrometer gauge from the hole, slot, or projection and read the micrometer. Add on the applicable rod length.

1-17. Care of the Depth Micrometer Gauge

The depth micrometer gauge is cared for in the same manner as the outside micrometer (Sec. 1-9). The micrometer screw length is adjustable for a 0.000 inch setting. The rod lengths are also adjustable to compensate for wear. Ensure that the collars on the ends of the rods and the micrometer mating face are clear of dirt and burrs. Make length adjustments to the rods in accordance with the manufacturer's recommendations.

VERNIER CALIPER

1-18. Introduction

The vernier caliper, most often called "vernier," is used by toolmakers, machinists, layout men, mechanics, and inspectors to measure outside and inside measurements of a wide variety of tools and parts. Measurements are made to an accuracy of 0.001 inch.

The vernier was invented by Pierre Vernier in 1631. It has been developed through the years to the modern-day vernier caliper shown in Fig. 1-12.

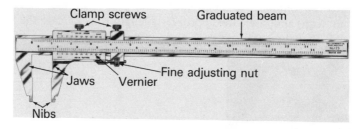

Fig. 1-12. Vernier caliper. (Courtesy of L.S. Starrett Co.)

The vernier consists of a graduated beam with a fixed jaw. A vernier with a second jaw slides along the graduated beam. Clamp screws hold the fine adjustment and vernier assemblies to the graduated beam. The fine adjusting nut is used for the final closing of the jaws upon the workpiece being measured.

In measuring a workpiece, the jaws are closed upon the workpiece (or the nibs opened against a workpiece for an inside measurement) and a reading is made at the vernier. Two graduated scales are available: one for outside measurements and one for inside measurements. Some verniers have both the outside and inside scales on one side of the vernier caliper (Fig. 1-12 and 1-13) while others have the outside scale on one side and the inside scale on the opposite side (Fig. 1-14). In the case of the latter, the outside scale is read left to right; the inside scale is read right to left. Some verniers are 50-division (Fig. 1-13), while others are 25-division (Fig. 1-14). Measurements can be made accurately to 0.001 inches in the English system and 0.02 millimeters in the metric system.

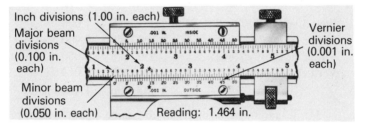

Fig. 1-13. Reading the 50-division vernier. (Courtesy of L.S. Starrett Co.)

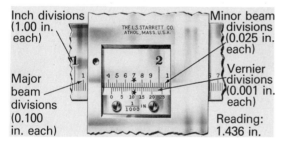

Fig. 1-14. Reading the 25-division vernier. (Courtesy of L.S. Starrett Co.)

Vernier calipers are made in a variety of sizes with graduated beams of 6, 12, 24, 36, and 48 inches (60 and 72 or other sizes on special orders) and to 600 millimeters metric. Jaw depths are from approximately $1\frac{1}{2}$ inches for the 6-inch vernier to $4\frac{1}{2}$ inches for the 72-inch vernier. Nib widths, with the jaws closed, are from $\frac{1}{4}$ to $\frac{3}{4}$ inches.

1-19. Reading the 50-Division Vernier

Figure 1-13 illustrates the 50-division vernier across a portion of the graduated beam. Note the inch divisions shown as 1, 2, 3, etc. Major beam divisions are numbered 1–9 and each represents 0.100 inch. Minor beam divisions, located between major divisions, are each 0.050 inch. Vernier divisions, numbered each five divisions, are 0.001 inch each.

The reading is begun at the 0 (zero) line of the vernier and is the sum of the inch divisions, major beam divisions, minor beam divisions, and the vernier divisions. The vernier division line that is read is the line which is exactly opposite one of the division lines of the graduated beam; only one of the 50-division lines of the vernier will be exactly opposite one of the graduated beam divisions. This is because the vernier is constructed so that the space of 50 divisions of the vernier fits exactly into the space of 49 divisions of the graduated beam.

The dimension of the width of the opening of the jaws (an outside measurement) of the vernier caliper shown in Fig. 1-13 is:

Inch divisions:	1×1.00 inch $= 1.000$
Major beam divisions:	4×0.100 inch $= 0.400$
Minor beam divisions:	1×0.050 inch $= 0.050$
Vernier divisions:	14×0.001 inch $= \underline{0.014}$
	by addition 1.464 inches

An inside measurement is made in the same manner as for outside measurements; the difference in the position of the numbers on the graduated beam for the inside measurements is because of the measurement of the nibs, which is automatically compensated for on this type of vernier caliper.

1-20. Reading the 25-Division Vernier

The 25-division vernier (Fig. 1-14) is similar to the 50-division vernier except that the minor beam divisions are each 0.025 inch. The reading of the 25-division vernier is:

Inch divisions:	1×1.00 inch $= 1.00$
Major beam divisions:	4×0.100 inch $= 0.400$
Minor beam divisions:	1×0.025 inch $= 0.025$
Vernier divisions:	11×0.001 inch $= \underline{0.011}$
	by addition 1.436 inches

1-21. Reading the Metric Vernier

The metric vernier (Fig. 1-15) is read in a similar manner to the English vernier. Centimeter divisions each equal 1 centimeter, which equals 10 millimeters. The major beam divisions are 1 mm each; the minor beam divisions are 0.50 mm each; and the vernier divisions are 0.02 mm each.

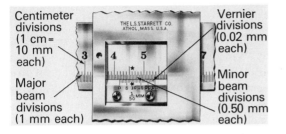

Centimeter divisions (1 cm = 10 mm each)

Vernier divisions (0.02 mm each)

Major beam divisions (1 mm each)

Minor beam divisions (0.50 mm each)

Fig. 1-15. Reading the metric vernier. (Courtesy of L.S. Starrett Co.)

The dimension of the width of the opening of the jaws of the metric vernier caliper shown in Fig. 1-15 is:

Centimeter divisions:	4×10	mm =	40.00
Major beam divisions:	1×1	mm =	1.00
Minor beam divisions:	1×0.50	mm =	0.50
Vernier divisions:	9×0.02	mm =	0.18
		by addition	41.68 mm

1-22. Use of the Vernier Caliper

The vernier caliper is held for a measurement as shown in Fig. 1-16, which illustrates the use of the vernier in making an inside measurement of a workpiece. Prior to making the measurement, the clamp screws should both be loosened and the vernier slid along the graduated beam to the approximate dimension of the workpiece. The vernier nib (or jaw) opening is then compared to the workpiece and is slid to a final approximated nib (jaw) opening. The clamp screw over the fine adjusting nut is then tightened with the fingers against the graduated beam. The fine adjusting nut is rotated until the nibs (or jaws) firmly (but not too tightly) contact the workpiece. Then carefully tighten the vernier clamp screw and remove the vernier from the workpiece. Take the reading.

Fig. 1-16. Use of the vernier caliper. (Courtesy of L.S. Starrett Co.)

1-23. Care of the Vernier Caliper

The vernier caliper should be kept clean at all times to prevent inaccuracies and for ease of operation. Wipe the vernier after each use with a clean dry lint-free cloth. Dampen the cloth with a solvent to remove stubborn dirt. An occasional drop of instrument oil rubbed onto the vernier and the moving jaw assembly will ensure free-running performance. Store the vernier in a case or box to protect it; the clamp screws should be loose during storage.

Handle the vernier gently, but firmly. Never force it to obtain a measurement. When you are not using the caliper, lay the vernier flat on the work surface, away from the edges of the work surface.

Check the vernier periodically for accuracy of the zero line. If the 0 (zero) line of the vernier is no longer accurate or it shows signs of wear with the jaws closed, return it to the factory for adjustment or repair.

VERNIER HEIGHT GAUGE

1-24. Introduction

The vernier height gauge, normally called height gauge, is used by layout men, toolmakers and inspectors for layout, jig, and fixture work where vertical distances must be marked off or measured. Height measurements are read on a 50-division vernier the same as for the outside vernier caliper previously described in Sec. 1-19. Measurements are accurate to 0.001 inch.

The vernier height gauge (Fig. 1-17) consists of a graduated beam mounted to a heavy base. The graduated beam contains 1-inch, 0.1-inch, and 0.05-

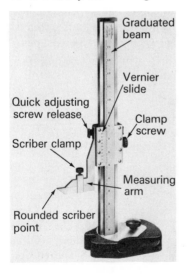

Fig. 1-17. Vernier height gauge. (Courtesy of L.S. Starrett Co.)

inch graduations. Typical beam lengths are 12, 18, 24, 36, and 48 inches, with longer lengths available on special order. Beam lengths to 600 millimeters are available in metric. Some height gauges are graduated in English on one side and in metric on the opposite side.

The vernier slide is held in place along the graduated beam by means of the clamp screw. A spring-loaded quick-adjusting screw release allows the slide to be moved up or down the beam when the release is squeezed. The fine adjustment knob is rotated to move the vernier slide along the beam until the rounded scriber point is at the required height. When heights are being measured (and the scriber point is not needed), the scriber point is removed by loosening and removing the scriber clamp.

Various attachments for the height gauge extend its capability. A depth gauge attachment adapts the height gauge so that depths of holes, recesses, and inside measurements of jigs, fixtures, etc. can be accurately determined. An offset scriber attachment increases the height gauge range and permits measurements from the base and in shallow recesses. A universal bevel protractor attachment is used to check angles (the angle is read on the bevel protractor—a similar bevel protractor is discussed in Sec. 5-7). A direct-reading dial test indicator can also be attached for use as a precision inspection gauge for comparing heights from a plane surface (such as a layout table).

1-25. Reading the Vernier Height Gauge

The graduated beam is marked with 1-inch, 0.1-inch, and 0.050-inch divisions. The vernier is marked with 50 divisions (total readings to 0.001 inch). The vernier height gauge is marked the same and thus is read the same as the 50-division outside vernier caliper described in Sec. 1-19. Reading of the metric vernier height gauge is the same as that described for the metric vernier caliper in Sec. 1-21. Metric readings are to $\frac{1}{50}$ millimeter (0.02 mm equals 0.00079 inch).

1-26. Use of the Vernier Height Gauge

When the height gauge is used as an inspection gauge, the rounded scriber point (Fig. 1-17) is removed by loosening the scriber clamp screw and removing the scriber clamp. The height gauge is placed in close proximity to the workpiece and the vernier slide is set to an approximate height by squeezing the quick-adjusting screw release while moving the vernier up or down the graduated beam to the desired height. The fine-adjustment knob is then rotated until the height gauge measuring point is located exactly at the layout line or point to be checked. A reading is then made.

Figure 1-18 illustrates the use of a similar type of height gauge. Here, the height gauge is checking the proper center distances (vertical) of the holes in the plate.

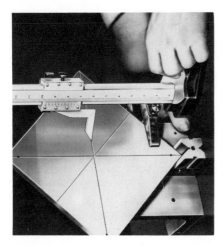

Fig. 1-18. Use of the vernier height gauge. (Courtesy of L.S. Starrett Co.)

The height of the measuring point of the height gauge in Fig. 1-18 is measured in the same manner as distance is measured with the vernier caliper in Fig. 1-12. Both clamp screws are loosened; an approximate height is set by moving the vernier with the hand. The clamp screw opposite the fine adjusting nut is then tightened. The fine adjusting nut is rotated until the height gauge measuring point is opposite the point or line to be checked.

To scribe a line, the rounded scriber point (available in tungsten carbide) is clamped to the measuring arm with the scriber clamp. The scriber point is placed lightly, but firmly, against the workpiece which is clamped on a layout table. The height gauge is slid carefully across the layout table, making a scribed mark on the workpiece. (Layout procedures are fully described in Chap. 5).

When an offset scriber, universal bevel protractor, depth gauge, or dial test indicator attachments are used with the height gauge, they replace the rounded scriber point that is clamped to the measuring arm.

1-27. Care of the Vernier Height Gauge

The vernier height gauge is cared for the same as the vernier caliper described in Sec. 1-23. The vernier height gauge should always be placed in a vertical position when it is not stored in its case. Wooden cases are made for storing the height gauge in a horizontal position.

THICKNESS GAUGES

1-28. Introduction

Thickness gauges, also called feeler gauges, are used by technicians to check bearing clearances, check gear play, fit pistons, rings, and pins, and gauge narrow slots. Thickness gauges are also used in the automotive and aircraft industries in setting spark plug gaps, valve tappets, and distributor points to the proper openings.

Thickness gauges come as a set of from 6 to 26 metal leaves from 0.0015 to 0.200 inch; they are also available in metric. Each leaf is marked with its thickness. A locking device permits one or more leaves to be locked into any desired combination.

1-29. Use and Care of Thickness Gauges

Choose a gauge leaf that you think is less than the thickness of the opening you're planning to measure. Place the leaf into the opening. If it fits loosely, try the next thicker leaf or combination of leaves until the leaves fit with just a little tightness. This is the correct thickness of the opening. Do not force a leaf into the opening.

Keep the thickness gauge leaves clean from dirt. Occasionally apply a thin coat of oil to leaf surfaces with a cloth.

SUMMARY

Accurate measurements are required throughout industry to ensure that manufactured parts will mate with other parts during assembly and to ensure interchangeability of parts when a worn part must be replaced. Accuracy is achieved and maintained by the use of the measuring tools described in this chapter: outside micrometer caliper, inside micrometer, depth micrometer, vernier caliper, vernier height gauge, and thickness gauges. Steel scales and tapes are also useful in making measurements, but are not as accurate as the other tools discussed.

This chapter should not be considered as only theoretical; it is the practical aspect of this chapter which is very important. The technician should endeavor to complete each of the review questions in this chapter and should extend his practice with the tools with which he is having difficulty, until finally he can use every tool described to make rapid and accurate measurements.

REVIEW QUESTIONS

NOTE

Some of the following problems request you to measure width, depth, *and* height. *The accompanying illustration defines these measurements.*

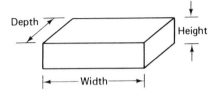

1. Describe the difference between a shrink scale and a steel tape. How does the steel tape differ from the steel rule?

2. What is parallax? How can you prevent parallax from affecting the accuracy of your measurements?

3. How accurately can you read the English steel scale? The metric steel scale?

4. List the functions of the external parts of the outside micrometer caliper.

5. If you are reading an outside micrometer caliper and the resultant answer is 0.253 inch, what are the number of major, minor, and thimble divisions read on the micrometer?

6. What part of an inch does each of the following divisions on an English ten-thousandth micrometer represent?
   a. a major sleeve division
   b. a minor sleeve division
   c. a thimble division
   d. a vernier division

7. How many millimeters are there in each of the following metric micrometer divisions?
   a. a major sleeve division
   b. a minor sleeve division
   c. a thimble division
   d. a vernier division

8. Describe the procedure of using an outside micrometer to measure the thickness of a steel plate.

9. An inside micrometer consisting of a 2-inch head, a $\frac{1}{2}$-inch spacer (gauge), and a 4–5 rod is used to measure the diameter of a hole. The micrometer head reads 0.273 inch. What is the diameter of the hole?

Describe the procedure you would use to make an inside measurement with an inside micrometer.

10. How is a rod inserted into the depth micrometer gauge? Describe how a measurement is made. What are the values of the major sleeve, minor sleeve, and thimble divisions on the head?

11. What are the values of the following divisions on the 50-division vernier caliper?
    a. an inch division
    b. a major beam division
    c. a minor beam division
    d. a vernier division

12. Describe how the vernier height gauge is used.

13. How would you use a thickness gauge to determine the gap of an engine spark plug?

14. Using a steel tape, measure the width and depth of one of your rooms.

15. Using a steel scale, measure the width, depth, and height of this book.

16. Using an outside micrometer caliper, measure the thickness of the following (either an English or a metric micrometer may be used): pencil lead, pencil, steel scale, human hair, thickness of a page of this book, and thickness of the cover of this book.

17. Using an inside micrometer, measure the inside diameter of a gallon can and a quart can.

18. Using a depth gauge, measure the depth of the following: a slot in a car or house key, and a cigar box.

19. Using a vernier caliper, measure the following: the width, depth, and thickness of this book (compare your answers with those from Question 15); the length of your pencil; the outside width, depth, and thickness of a cigar box; and the inside and outside diameters of a can.

20. Place a book flat on a table. Using a vernier height gauge, measure the height (thickness) of the book. Place a second book on top of the first and measure the total height.

21. Using a thickness gauge, measure the gap in a spark plug or the opening in a pair of fingernail clippers.

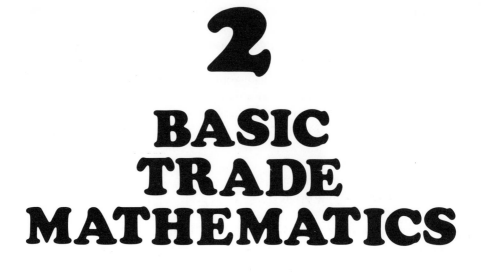

BASIC
TRADE
MATHEMATICS

Today's world of modern machinery is growing more and more complex. However through all of the complexity, some old and basic geometric forms still serve as design shapes of tools and instruments. These shapes are the circle, square, triangle, and parts or variations of these shapes.

The first part of this chapter breaks each of these shapes into their basic parts. The knowledge of these basic figures will impart a sense of understanding when the technician is faced with a complex problem or piece of equipment at a later time.

The mechanical technician of today must be able to look at a drawing or a workpiece and instinctively reduce it to its basic design shapes. After exposing its basic form, he must apply his knowledge of form through mathematics to arrive at an accurate answer to an unknown part.

Problem solving is a logical step-by-step approach of fact gathering, with a precise application of these facts to achieve an answer. Steps to aid in problem solving are:

1. Isolate and define the problem area.
2. Determine the basic geometric shape of the problem.
3. Sketch the part or parts on paper.
4. Determine what unknown or unknowns are to be found and in what units (inches, feet, meters, etc.). Never mix units of measure in one

problem, e.g. don't mix inches and feet if the answer is to be in inches. Change all dimensions to the same unit.

5. Know what units the answer should be in. Collect all available known information about the problem.
6. Mark the known facts on the sketch along with any known angles that can help describe the workpiece shape.
7. Break complex geometric shapes into simpler shapes.
8. Using knowledge of these particular figures or shapes, proceed to set up a mathematical condition for its solution.

NOTE

Sometimes three or more preliminary solutions must be found before the final needed solution can be found.

The use of these eight steps in problem solving is shown in problem solutions in this chapter. As the technician becomes proficient in observing a problem and in breaking a problem into its component parts, he will begin to solve many problems in a one- or two-step solution. The real key to problem solving is the isolation of the problem from its sometimes complex surroundings.

This chapter is divided into three main sections: plane geometry, solid geometry, and trigonometry. Plane geometry deals with dimensions of width and depth—two dimensions only. Solid geometry deals with figures that enclose a space or volume. Thus, solid geometric figures are three dimension: width, depth, and height. Trigonometry deals with the relationships between the sides and angles of triangles and the performance of the various calculations based on the sides and angles of triangles.

The word *geometry* is taken from the Greek language, and translated means "earth measurements." Its original use was in the laying out and dividing of land tracts for farmlands and for cities. The word *trigonometry* when translated means "triangle measure." This form of mathematics was developed to solve the many problems of computing angles and sides of triangles.

The technician should become familiar with the essentials of each section in this chapter because he will find applications of the material in his everyday work.

PLANE GEOMETRY

To understand the make-up of figures and shapes such as squares, circles, triangles, etc., the technician must be familiar with both their component parts and the terminology used in describing these figures and shapes.

Once this is accomplished, the more complex geometric problems that are encountered are merely variations and combinations of these basic parts. Therefore, every effort should be made to fully understand the following discussion of geometry.

2-1. Some Basics in Geometry

To begin our discussion of geometry, let us look at the parts of the object shown in Fig. 2-1. The object shown is a six-sided solid and is made up of a series of connected parts. The object has width, depth, and height. The object can be further described as a figure made up of six planes or surfaces sharing common edges. A *surface* or *plane* is a closed form or shape having dimensions of only width and depth and is not considered to have a thickness. Surfaces do not have to be square, but can exist in any shape such as the surface form of a pond (forming curved lines).

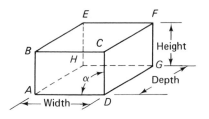

Fig. 2-1. Solid figure.

Let us take the sample plane *ABCD* from Fig. 2-1 and break it into its component parts. This plane is composed of connected straight lines (of defined length) and defined angles (α) alpha. Starting at one of the simplest parts of this plane, we will inspect a line (e.g. line *AD*). This line has a defined dimension as noted by the end points *A* and *D*. There are an infinite number of paths that may be drawn from point *A* to point *D*. But, there is only one path which if measured, is the shortest distance. A path which goes from point *A* to point *D* without any deviation or wandering is thus defined as a *straight line*.

A curved line path or a random line path are two of the infinite number of methods of going from point *A* to point *D*.

Surface *ABCD* (Fig. 2-1) is thus bound by four straight lines with length *AB* equal to *CD* and length *BC* equal to *AD*. When two straight lines meet at a common point, such as line *AD* and line *CD*, they form a vertex of some angle α. Lines *AD* and *CD* are thus called the sides of the angle.

The measurement of angles is noted by degrees and/or portions of degrees (minutes and seconds). To understand degrees and what they are, consider a circle with two points *A* and *B* on the circle's periphery (Fig. 2-2). To accurately describe or locate these two points on the circle only two facts are necessary: the value of α (the angle) and the radius of the circle. The angle is therefore a rotational displacement of one point or line from another point or line. The total range in measuring angles is based on a standard that states that a full circle is composed of 360 one- (1) degree increments, or a total of 360 degrees in a circle. Therefore, if point A went around the periphery of the circle a full *360* degrees in either direction, it would return to its original starting point.

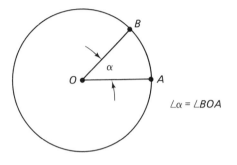

$$\angle \alpha = \angle BOA$$

Fig. 2-2. Location of two points on a circle, as defined by radius and central angle α.

$$1 \text{ degree} = \tfrac{1}{360} \text{ of a circle} \tag{2-1}$$

A degree is broken into smaller parts called minutes and seconds so that greater accuracy of measurement may be defined. A minute is $\tfrac{1}{60}$ of a degree and a second is $\tfrac{1}{60}$ of a minute. Therefore,

$$1 \text{ degree} = 60 \text{ minutes} \tag{2-2}$$

$$1 \text{ minute} = \frac{\text{degree}}{60} \tag{2-3}$$

$$1 \text{ minute} = 60 \text{ seconds} \tag{2-4}$$

$$1 \text{ second} = \frac{\text{minute}}{60} \tag{2-5}$$

General notation used in problem work makes use of the symbol ° for degrees, the symbol ′ for minutes, and the symbol ″ for seconds, e.g.

$$10° \quad 15′ \quad 45″$$

This reads 10 degrees, 15 minutes, and 45 seconds.

By use of the above values in Eqs. (2-2) through (2-5) for minutes and seconds, it is possible to express the angle α in decimal parts of a degree only. To perform the conversion of minutes and seconds to degrees, we must find what part of a degree the minutes and the seconds are.

$$\alpha = 10° \quad 15′ \quad 45″$$

Using Eq. 2-4, we first convert seconds to minutes:

$$45″ \times \frac{1′}{60″} = \frac{45}{60} = 0.750$$

Adding this to the previous 15 minutes, we then convert minutes to degrees: to 15′ we add 0.750′ = 15.750′.

$$\therefore \quad 15.750' \times \frac{1°}{60'} = 0.2625°$$

NOTE

The symbol $\therefore$ *is defined as* therefore.

Since we have converted both seconds and minutes to degrees, we can now express the original angle as:

$$\alpha = 10° \quad 15' \quad 45''$$
$$\alpha = 10.2625°$$

Angles can be specified by a lower case letter placed at the vertex of the angle or by calling out the capital letters forming the angle.

The angle formed by two intersecting lines 90° apart is called a right angle. Two other classifications of angles are an *acute angle* and an *obtuse angle*. An angle between 0° and 90° is an acute angle. An angle between 90° and 180° is an obtuse angle.

When working with acute or obtuse angles, it is often necessary to know its parent or adjacent angle. In the case of an acute angle, its adjacent angle is called the *complementary angle* and when added to the acute angle the total is 90°:

$$\text{complementary } \angle = 90° - \text{acute } \angle \qquad (2\text{-}6)$$

The adjacent angle in the case of an obtuse angle is called the *supplementary angle*. When the supplementary angle is added to the obtuse angle, the total equals 180°:

$$\text{supplementary } \angle = 180° - \text{obtuse } \angle \qquad (2\text{-}7)$$

One final discussion will be made in defining the plane surface *ABCD* (Fig. 2-1) and that is a discussion of parallel lines, or surfaces. When looking at the cover and pages of a closed book, a person observes a series of lines or pages. The average thickness of a page is approximately 0.005 inch. If the page content of the book were 100 pages, this means that the front and back cover are exactly 0.500, or $\frac{1}{2}$ inch apart from each other. The covers of the book are therefore parallel to each other. For the covers to be parallel, all pages must be parallel to one another. The book, therefore, is made up of a series of parallel pages or planes, each with a thickness of 0.005 inch. Parallel lines or planes therefore are defined as being equidistant from one another at any point of perpendicular measurement [Fig. 2-3(A)].

If a line *AE* is passed through another line *BD* as shown in Fig. 2-3(B), the following angle relationship is true:

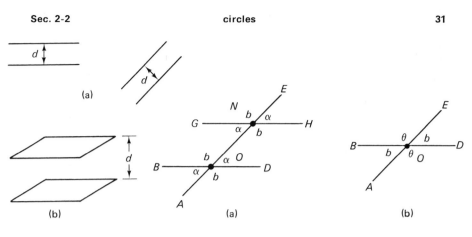

Fig. 2-3. (A) Parallel (a) lines and (b) planes; (B) angle relationships of (a) a line intersecting two parallel lines and (b) two intersecting lines.

$$\angle AOB \ (b) \text{ will be equal to } \angle DOE \ (b)$$

$$\text{and } \angle DOA \ (\theta) \text{ will be equal to } \angle EOB \ (\theta)$$

$$\angle \theta + \angle b = 180°$$

If we add all the angles about point O, we find that they equal 360°:

$$2\angle \theta + 2\angle a = 2(180°)$$

$$= 360°$$

If the line AE were now cut through parallel lines BD and GH, the following angle relationship would be true.

$$\angle GNA = \angle EOD \quad \text{(opposite interior angles)}$$

$$\angle HNA = \angle EOB \quad \text{(opposite interior angles)}$$

A line cutting through parallel lines will form alternate equal interior angles [Fig. 2-3(C)] where

$$\angle a = \angle a, \qquad \angle b = \angle b, \qquad \text{and } \angle \theta = \angle \theta$$

This same theory of cutting lines is extended to parallel planes cut by another plane.

2-2. Circles

The circular shape is used in today's mechanized world for more design applications than is any other geometric shape. Wheels, shafts, gears, and bearings are typical examples of applications of the circle.

By definition, a circle is the path left by a point that moves around a fixed point through a full rotation and always the same distance from the fixed point. The measured total path length is called the *circumference* of the circle. The fixed point is called the *center of the circle.* The straight line distance from the center to any point on the circular path is called the *radius* of the circle. The full turn in our definition is equal to 360°, the total number of degrees in a circle (Fig. 2-4).

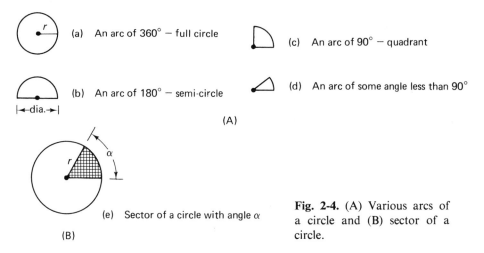

(a)   An arc of 360° − full circle

(c)   An arc of 90° − quadrant

(b)   An arc of 180° − semi-circle

|←dia.→|

(d)   An arc of some angle less than 90°

(A)

(e)   Sector of a circle with angle α

(B)

Fig. 2-4. (A) Various arcs of a circle and (B) sector of a circle.

In drawing a line from a point on the circle through the center and on to the other side of the circle, two half circles called *semicircles* are made. This bisecting line is called the *diameter* (*D*) of the circle and is equal to two times the radius (*r*). Thus:

$$2r = D \tag{2-8}$$

Semicircles are only one half of full circles and therefore contain 180° ($\frac{1}{2}$ of 360° = 180°). This angle of 180°, which is defined at the center, is called the *central angle.*

The semicircle can also be divided by drawing a line perpendicular to the diameter at the midpoint of the diameter (this is the center of the circle). This new line divides the two semicircles into halves so that there are now four sections to the circle. Each section is called a *quadrant.* Thus, there are four quadrants within the circle, labeled 1, 2, 3, and 4. The central angle of each quadrant is one-fourth of the circle: $\frac{1}{4}$ of 360°, which equals 90°.

A full circle, or the parts thereof, are all defined by two basic units, the radius and the angular rotation of the radius. For example, a circle is a given radius rotated in a plane about one end of the radius through 360°. A quadrant is thus defined as a given radius rotated through 90°.

Arcs (Fig. 2-4) are another part of a circle. Arcs are small portions of the circular path of a circle and can be defined in an infinite number of angular sizes including degrees, minutes, and seconds.

Subdivisions of the circle are called *sectors* [Fig. 2-4(E)]. Sectors are defined by their central angle and by the two radii of the circle that the sector is bound by.

Before further discussion, let us study a unique relationship found within the circle. If a circle is drawn with any radius (r) and if the circle's total path length (circumference) and diameter (two times the radius) are measured, the following constant is found by dividing:

$$\frac{\text{circumference}}{\text{diameter}} = 3\tfrac{1}{7} \tag{2-9}$$

This number $3\tfrac{1}{7}$, or approximately 3.1416, is represented by the Greek letter π (pi). Since π is a constant for all circles, the following equation is formed for computing the circumference of a circle:

$$C = \pi D \text{ or } \pi 2r \tag{2-10}$$

also

$$C = 2\pi r \tag{2-11}$$

where

$$C = \text{circumference}$$
$$\pi = 3.1416 \text{ (constant)}$$
$$d = \text{diameter} = 2r$$
$$r = \text{radius}$$

The area of a circle is found to be equal to the following equation:

$$\text{area} = \pi r^2 \tag{2-12}$$

where

$$\pi = 3.1416$$
$$\text{radius } r^2 = r \times r$$

If the diameter is used, the above equation becomes

$$\text{area} = \frac{\pi D^2}{4} \tag{2-13}$$

Very often, the area of a ring section (Fig. 2-5) is to be calculated. The solution to this problem at first glance is the area of the large circle with radius R, minus the area of the small circle with radius r. This solution is

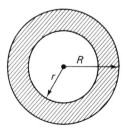

Fig. 2-5. Ring section.

perfectly accurate. To help save time and lessen the chance for error, however, the following equation may be used for ring area problems.

$$\text{Area (ring)} = \pi(R^2 - r^2) \qquad (2\text{-}14)$$

where

$$R = \text{radius of large circle}$$
$$r = \text{radius of small circle}$$

Since a sector is a portion of a whole circle, the computation of its area and circumference (arc) is based on its fractional part of the whole circle. For example, if a 60° sector with a 1-inch radius is formed, find its area and length of arc.

Solution: Since this sector is a part of a whole circle, compute area and circumference of the whole circle.

$$\text{Radius} = 1 \text{ inch}$$
$$\text{Area} = \pi r^2 \text{ (from Eq. 2-12)}$$
$$\text{Area} = 3.1416 \text{ sq. inches}$$
$$C = \pi D, \qquad D = 2 \text{ inches}$$
$$C = 6.2832 \text{ inches}$$

Looking at a 60° sector [Fig. 2-4(E)], compute the area of the sector in proportion to the area of the circle. This is done as follows:

$$\text{Sector portion} = \frac{\text{sector angle}}{360°} \qquad (2\text{-}15)$$

$$\frac{60°}{360°} = \frac{1}{6}$$

This sector, which is $\frac{1}{6}$ of the whole circle, is also $\frac{1}{6}$ of its area and its circumference,

$$\therefore \quad \text{the area of a sector} = \frac{\text{sector angle}}{360°} \times \pi r^2 \qquad (2\text{-}16)$$

or

$$\tfrac{1}{6}(3.1416) \text{ sq. inches} = 0.5236 \text{ sq. inch}$$

$$\text{the arc length of a sector} = \frac{\text{sector angle}}{360°} \times \pi D \qquad (2\text{-}17)$$

or

$$\tfrac{1}{6}(6.2832) = 1.0472 \text{ inches}$$

A *segment* of a circle is that figure enclosed by an arc and a straight line joining the ends of the arc (Fig. 2-6). The line AB joining the ends of the arc is called a *chord*. If a line is drawn from the center of the circle O to the midpoint of the chord AB and is then continued on to intersect the arc of the segment, the line would both bisect the central angle θ of the sector into two equal angles ($\theta/2$) and would also intercept the chord at a 90° angle (perpendicular). The usefulness of this will become apparent at a later time. The line DC is called the *segment height* (h).

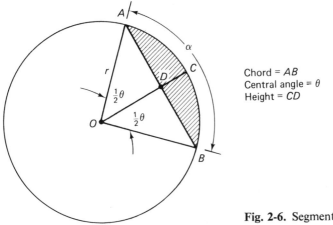

Chord = AB
Central angle = θ
Height = CD

Fig. 2-6. Segment of a circle.

Generally, three facts can be deduced from a segment: its area, its chord length, and its height. Computation of these three measurements is quite easily performed by use of trigonometry which is subsequently discussed.

The area of a circular segment, the length of a chord, and segment height are calculated by the following equations:

$$\text{Area } A = \frac{2}{3}hC + \frac{h^3}{2C} \tag{2-18}$$

or

$$A = \left(\frac{4}{3}\right)h^2\sqrt{\frac{2r}{h} - 0.608} \tag{2-19}$$

where

A = area of segment
h = segment height
C = chord length
r = radius

NOTE

These are approximate area solutions and give an error of 0.1% or less.

$$\text{Length of chord } C = 2\sqrt{h(2r - h)} \tag{2-20}$$

where

C = chord length
h = segment height
r = radius of arc

$$\text{Segment height } h = r - \frac{1}{2}\sqrt{4r^2 - C^2} \tag{2-21}$$

where

$$h = \text{segment height}$$
$$r = \text{radius of arc}$$
$$C = \text{chord length}$$

2-3. Triangles

The word *polygon* is a general term used to describe a closed geometric figure consisting of three or more connecting straight lines. To be more definative:

A polygon having three sides is a *triangle*.
A polygon having four sides is a *quadrilateral*.
A polygon having five sides is a *pentagon*.
A polygon having six sides is a *hexagon*.

This section discusses the three sided polygon—the triangle.

Figure 2-7 illustrates a triangle *ABC*. Its construction consists of three lines: *AB*, *BC*, and *CA* connected in such a fashion that they form a closed plane with three vertices and three angles noted as *a*, *b*, and *c*. These angles are called the interior angles of the triangle. A very important fact concerning these interior angles is that the sum of the angles equals 180°. Therefore, in triangle *ABC*, $\angle a + \angle b + \angle c = 180°$. That is, the sum of the interior angles of a triangle equals 180°.

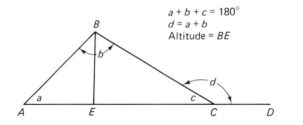

$a + b + c = 180°$
$d = a + b$
Altitude = BE

Fig. 2-7. Exterior angle relationship of any triangle.

If side *AC* of the triangle in Fig. 2-7 was extended to some point *D*, the extension would form an *exterior angle* (*d*). An important relationship of this exterior angle (*d*) to the two opposite interior angles of the triangle (*a*) and (*b*) is formed:

$$\angle a + \angle b + \angle c = 180° \qquad (2\text{-}22)$$

The angle *ACD* also = 180°

or stated differently

$$\angle c + \angle d = 180° \qquad (2\text{-}23)$$

Solving for angle *c* in Eq. 2-22

$$\angle c = 180° - (\angle a + \angle b) \qquad (2\text{-}24)$$

If we substitute this value of $\angle c$ into Eq. (2-23) we have

$$180° - (\angle a + \angle b) + \angle d = 180°$$

or

$$\angle d = \angle a + \angle b \qquad (2\text{-}25)$$

That is, *the exterior angle of a triangle is equal to the sum of the opposite two interior angles.*

If a line starting at, and bisecting the vertex angle of an interior angle continues to the opposite side, it becomes known as a *median*. Since a triangle contains three angles, three medians can be constructed. Three medians constructed in a triangle meet at a common point (O). This point, if used as a center, allows a circle to be drawn within the confines of the triangle's sides.

Another term used in triangles is *altitude*. The altitude or height (Fig. 2-7) of a triangle is the perpendicular distance from the base to the opposite angle. Depending upon which side of the triangle is chosen as the base, there could be three altitudes to a triangle.

Of the numerous types and shapes of triangles possible, three have been given names which best describe their geometric shapes. These are the *isosceles triangle*, the *equilateral triangle* and the *right triangle*.

The isosceles triangle (Fig. 2-8) is so named because it has two of its sides equal in length. It also has two of its interior angles equal to each other. These equal angles are the angles that are opposite the equal sides.

The equilateral triangle is unique in that all three sides are of equal length. Likewise, the interior angles of all equilateral triangles are also equal to each other and therefore each equals 60° (180°/3).

The right triangle (Fig. 2-9) is any triangle that has one of its interior angles equal to 90°. This means that the two sides of the triangle that form this 90° angle are perpendicular to each other. The third side of the right triangle is called the *hypotenuse* and is always the longest side of the triangle; it is also always opposite the 90° angle.

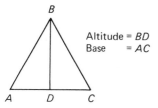 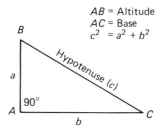

Fig. 2-8. Isosceles triangle having side AB equal to side BC.

Fig. 2-9. Pythagorean solution for right triangles.

Two forms of right triangles that are most often used are the 45–45–90° right triangle, and the 30–60–90° right triangle. The 45–45–90° right triangle is in reality an isosceles triangle with a 90° vertex angle formed by the two equal sides. The other two angles of this triangle are equal to each other, and using Eq. 2-21, are found to be 45°.

The 30–60–90° right triangle can be formed from an equilateral triangle by the following method. If any one of the 60° interior angles of an equilateral triangle is bisected (cut in half) by a line that intersects the opposite side, the following occurs: first the line bisecting the angle meets the opposite side and forms a 90° angle to it. Then the line bisects or halves the intersected side.

If we isolate this 30–60–90° triangle, we can see an important relationship develop. The side A, which is opposite the 90° angle, is now the hypotenuse of the right triangle. The base of the triangle, the side opposite the 30° angle, is equal to $\frac{1}{2}$ of A, or $A/2$. This relationship of side to angle for a 30–60–90° triangle is of extreme importance and should be fully understood before proceeding further. Thus, *in any 30–60–90° triangle, the side that is opposite the 30° angle is equal to one half the value of the hypotenuse.*

In many problems concerning right triangles, the need to compute the value of the hypotenuse is of the utmost importance. Pythagoras, a Greek mathematician who lived about 500 B.C., developed a formula to handle this problem. Consider the right triangle in Fig. 2-9 with sides a, b, and hypotenuse c. The Pythagorean theorem states mathematically that

$$c^2 = a^2 + b^2 \tag{2-26}$$

where

$$c = \text{value of the hypotenuse}$$
$$a = \text{value of the altitude}$$
$$b = \text{value of the base}$$

Equation (2-26) states that given the value of any two sides of a right triangle, the third side can be calculated. By algebra, Pythagorean's theorem can be rewritten as

$$c \text{ (hypotenuse)} = \sqrt{a^2 + b^2} \tag{2-27}$$

$$a \text{ (altitude)} = \sqrt{c^2 - b^2} \tag{2-28}$$

$$b \text{ (base)} = \sqrt{c^2 - a^2} \tag{2-29}$$

When this equation is applied to the 45–45–90° and 30–60–90° right triangles, we find the value for the hypotenuse of the 45° triangle to be $(\sqrt{2})$ a; thus, the value of the sides of the 45–45–90° triangle are in the ratio of 1, 1, and $\sqrt{2}$ (Fig. 2-10). The $\sqrt{2}$ has the value of 1.41421.

The hypotenuse of a 30–60–90° triangle is found to be $(\sqrt{3})$ a. Thus, the values of the sides of the 30–60–90° triangle are in the ratio of 1, 2, $\sqrt{3}$

(Fig. 2-10). These unit values for the sides of the 45–45–90° triangles are of great value for quick problem solution.

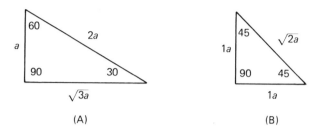

Fig. 2-10. Basic relationship for (A) 30-60-90 degree triangle and (B) 45-45-90 degree triangle.

Example:

Given: A 45° right triangle with a hypotenuse of 5 inches.

Find: The value of its sides.

Solution: The value of the hypotenuse of the unit 45° triangle is $a\sqrt{2}$. where

$$\sqrt{2} = 1.414$$

In the problem, the value to be found is (a). The hypothenuse is equal to 5 inches. Comparing this value to our unit triangle where the hypotenuse is $a\sqrt{2}$, we can set up a relationship between the two 45° triangles and say that the hypot (unit) = hypot (real) or some number $a \times \sqrt{2} = 5$.

$$a \times \sqrt{2} = 5$$

solving for a $1.414a = 5$

$$a = 3.536 \text{ inches}$$

This value $a = 3.536$ inches is also equal to the value of each of the sides originally asked for.

Given: A 30–60–90° triangle with the side opposite the 60° angle equal to 7 inches.

Find: The value of the other two sides.

Solution: Set up a general 30–60–90° triangle with its notations. Set up and label the problem triangle.

By looking at a comparison of the two triangles, we can immediately set up the following relationship.

$$a\sqrt{3} = 7$$

where

$$\sqrt{3} = 1.732$$
$$1.732a = 7$$
$$\therefore \quad a = 4.041 \text{ inches}$$

Having found the value of a, we can now apply it to find the value of the other two sides.

Side (1) is equal to a, or 4.041,

$$\text{side (1)} = 4.041 \text{ inches}$$

The hypotenuse, side (2), is equal to $2a$

or

$$\text{side (2)} = 8.082 \text{ inches}$$

2-4. Areas of Polygons

With some understanding of polygons and their basic parts and forms, we now consider the whole figure and some of its applications. It is often necessary to find the areas of polygons.

The first polygons to be discussed are parallelograms. *Parallelograms* are four sided figures whose opposite sides are parallel. Figure 2-11 shows several forms which include the square and rectangle.

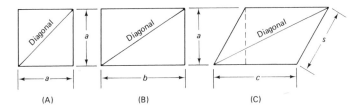

Fig. 2-11. Parallelograms: (A) square, (B) rectangle, and (C) parallelogram.

For ease of understanding, all of the figures have a common height, or altitude, equal to a. A square is shown in Fig. 2-11(A). Therefore, the altitude a and base b are equal. The rectangle [Fig. 2-11(B)] has a base equal to b. Figure 2-11(C) is a parallelogram with a base equal to c, a side equal to s, and an altitude equal to a. It should be noted that squares and rectangles are special forms of parallelograms that have right angles formed at each of their internal corners.

The area of all parallelograms is equal to the product of the base times its altitude. It can be seen that in the case of a square of side a, the area becomes equal to $a \cdot a$, or a^2. The general formula for the area of a parallelogram is

$$A = a \cdot b \qquad (2\text{-}30)$$

where

$$A = \text{area}$$
$$a = \text{altitude}$$
$$b = \text{base}$$

2-5. Areas of Triangles

The area of a triangle equals one half the product of the altitude times its base (see Fig. 2-8). Written in mathematical notation, the equation is

$$A = \tfrac{1}{2}ab \qquad (2\text{-}31)$$

where

$$A = \text{area}$$
$$a = \text{altitude}$$
$$b = \text{base}$$

Since a triangle can have any one of three bases on which it can rest, the proper altitude (a) is the one which is perpendicular to the chosen base (b).

To illustrate the equation for the area of a triangle, let us return briefly to a discussion of a parallelogram. The area of the parallelogram [Fig. 2-11(B)] with sides a and b is equal to ($a \times b$). Since the diagonal bisects the area of the parallelogram into two triangles, the area of one of them is equal to one half the area of the parallelogram. This is the same equation as previously discussed (2-31).

Example

Given: Isosceles triangle ABC with two sides equal to 5 inches and obtuse angle $BCA = 120°$.

Find: Area of ABC.

Solution: Use Eq. (2-31)

$$A = \tfrac{1}{2}ab$$

where

$$b = 5 \text{ inches}$$
$$a = \text{unknown}$$

Since the altitude (a) is unknown, its value must be found first. From B, extend a line to the base line AC to meet and form a perpendicular line formed by BD. A right triangle BCD is now formed.

$$\angle BDC = 90°$$

$$\angle DBC = \frac{120°}{2} \text{ or } 60°$$

$$\therefore \quad \angle BCD = 180° - (90° + 60°) = 30°$$

Side BD is the altitude of the large triangle ACB and also of the small triangle CDB. Triangle CDB is therefore a 30–60–90° triangle with a hypotenuse equal to 5 inches. From our previous problem (Fig. 2-10), we know the side relationship for a 30–60–90° triangle is $1a$, $2a$, $a\sqrt{3}$

$$\therefore \quad 5 \text{ inches} = 2a$$

$$a = 2.5 \text{ inches}$$

This is CD in the problem.

The altitude BD is $a\sqrt{3} = 2.5\sqrt{3} = 4.33$.

Having found the common altitude BD, this value can now be substituted into the original triangle area Eq. (2-31):

$$A = \tfrac{1}{2}ab$$

$$A = (\tfrac{1}{2})(4.33 \times 5)$$

$$A = (\tfrac{1}{2})(21.65)$$

$$A = 10.825 \text{ sq. inches}$$

This problem was chosen so that it would utilize the 30–60–90° triangle relationship. Other problems will be encountered where trigonometry must be used to find the altitude. Trigonometric solutions are covered in the third section of this chapter.

A method of finding the area of a triangle when only the three sides are known is solved by Hero's equation. This equation written in mathematical form is

$$\text{Area} = \sqrt{S(S - a)(S - b)(S - c)} \tag{2-32}$$

where $S = (a + b + c)/2$ and a, b, c are the sides of the triangle.

This equation may be stated in the following words:

> *To find the area of a triangle when its three sides are given, first find one half the sum of the three sides. From this value subtract each side of the triangle. Find the product of the half sum and each of the side differences. The square root of this product is the area of the triangle.*

To illustrate this equation, consider the problem.

Given: A triangle with sides equal to 7, 9, and 10 inches.

Find: Area of the triangle ABC and the altitude with a base equal to 10 inches.

Using Hero's formula (Eq. 2-32):

$$A = \sqrt{S(S - a)(S - b)(S - c)}$$

$$S = \frac{(a + b + c)}{2} = \frac{(7 + 9 + 10)}{2}$$

$$S = 13 \text{ inches}$$
$$S - a = 13 - 9 = 4 \text{ inches}$$
$$S - b = 13 - 10 = 3 \text{ inches}$$
$$S - c = 13 - 7 = 6 \text{ inches}$$
$$\therefore \quad \text{area} = \sqrt{(13)(4)(3)(6)}$$
$$= \sqrt{936} = 30.6 \text{ sq. inches}$$

To find the altitude a, with a base equal to 10 inches, we use Eq. (2-31) for the area of a triangle (area equal to 30.6 sq. inches):

$$A = \tfrac{1}{2}ab$$
$$30.6 = \frac{a(10)}{2}$$
$$30.6 = 5a$$
$$a = 6.11 \text{ inches}$$

2-6. Trapezoid

The trapezoid is a quadrilateral that has only two of its sides parallel. These parallel sides are in all cases called the bases (B, b) of the trapezoid. The altitude (a) is the perpendicular distance between these two bases.

Consider the trapezoid $ABCD$ in Fig. 2-12. If a diagonal AC were drawn, it would divide the figure into two triangles with a common altitude (a).

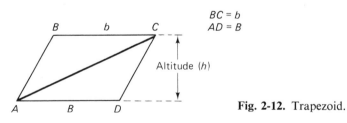

Fig. 2-12. Trapezoid.

To find the area of this figure, the area of the two triangles ABC and ACD can be added together:

$$\text{area } ABC = \tfrac{1}{2}a(BC)$$
$$\text{area } ACD = \tfrac{1}{2}a(AD)$$

But BC and AD are the bases of the trapezoid.

$$\therefore \quad \text{area trapezoid} = \tfrac{1}{2}a(BC + AD)$$

or

$$\text{Area trapezoid} = \tfrac{1}{2}a(B + b) \tag{2-33}$$

where

$$B = \text{large base (lower base)}$$
$$b = \text{small base (upper base)}$$
$$a = \text{altitude}$$

This formula can be written as: *the area of any trapezoid is equal to one-half the sum of the two bases times the altitude.*

Given: Trapezoid with an upper base equal to 20 cm, lower base equal to 36 cm, and altitude equal to 6 cm.

Find: Area of trapezoid.

Solution: From Eq. (2-33), area $= \frac{1}{2}a(B + b)$ where

$$b = 20 \text{ cm}$$
$$B = 36 \text{ cm}$$
$$a = 6 \text{ cm}$$
$$\text{Area} = \tfrac{1}{2}(6)(36 + 20)$$
$$\text{Area} = \tfrac{1}{2}(6)(56)$$
$$\text{Area} = 168 \text{ sq cm}$$

The following equations are provided for use in solving geometry problems.

Formula

Area (A) of Square

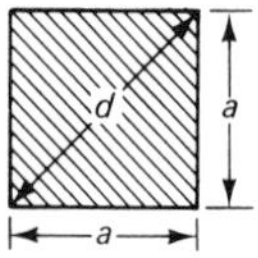

where a = side
d = diagonal

$A = a^2$	(2-34)
$A = \frac{1}{2}d^2$	(2-35)
$d = 1.414a$	(2-36)

Area (A) of Rectangle

where a = side
b = base
d = diagonal

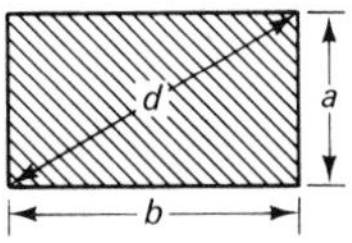

$A = ab$	(2-37)
$A = a\sqrt{d^2 - a^2}$	(2-38)
$A = b\sqrt{d^2 - b^2}$	(2-39)
$d = \sqrt{a^2 + b^2}$	(2-40)

Area (A) of Parallelogram

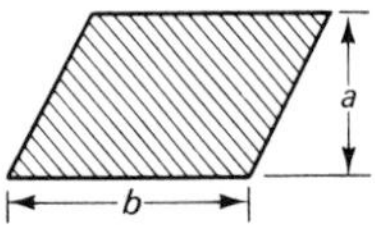

where a = altitude
b = base

$A = ab$	(2-41)
$a = \dfrac{A}{b}$	(2-42)
$b = \dfrac{A}{a}$	(2-43)

Formula

Area (A) of Right Triangle

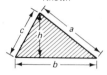

$$A = \tfrac{1}{2}(ab) \tag{2-44}$$
$$c = \sqrt{a^2 + b^2} \tag{2-45}$$
$$b = \sqrt{c^2 - a^2} \tag{2-46}$$
$$a = \sqrt{c^2 - b^2} \tag{2-47}$$

where a = altitude

b = base

c = hypotenuse

Area (A) of Triangle with 3 Sides Known

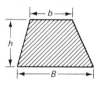

$$A = \sqrt{S(S - a)(S - b)(S - c)} \tag{2-48}$$

where $S = \tfrac{1}{2}(a + b + c)$

Area (A) of Trapezoid

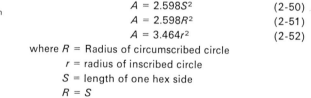

$$A = \frac{(b + B)h}{2} \tag{2-49}$$

where b = small base

B = large base

h = height

NOTE: In England this figure is called the trapezium.

Area (A) of Regular Hexagon

$$A = 2.598S^2 \tag{2-50}$$
$$A = 2.598R^2 \tag{2-51}$$
$$A = 3.464r^2 \tag{2-52}$$

where R = Radius of circumscribed circle

r = radius of inscribed circle

S = length of one hex side

$R = S$

Area (A) of Circle

$$\text{Area} = \pi r^2 = \pi \frac{d^2}{4} \tag{2-53}$$
$$\text{Cir} = \pi d = 2\pi r \tag{2-54}$$

where π = 3.1416

r = radius

d = diameter

Area (A) of Circular Sector

$$A = \tfrac{1}{2}rl \tag{2-55}$$
$$l = \frac{r \times \alpha \times 3.1416}{180°} \tag{2-56}$$

where l = length of arc

α = angle in degrees

r = radius

Formula

Area (A) of Circular Segment	$A = \frac{1}{2}[rl - c(r - h)]$	(2-57)
	$l = 0.0174r$	(2-58)
	$h = r - \frac{1}{2}\sqrt{4r^2 - c^2}$	(2-59)
	$r = \dfrac{c^2 + 4h^2}{8h}$	(2-60)

where *c* = chord
     h = height
     l = length of arc
     r = radius

Area (A) of Circular Ring	$A = \pi(R^2 - r^2)$	(2-61)
	$A = 0.7854(D^2 - d^2)$	(2-62)

where π = 3.1416
     D = diameter of large ring
     d = diameter of small ring
     R = radius of large ring
     r = radius of small ring

2-7. Geometric Construction

This section is included as an aid to the technician in the field. Construction problems, particularly in the art of layout, which are discussed in Chap. 6, are graphically shown. It is recommended that the technician perform these exercises so as to become familiar with the procedures. Paper, pencil, scale and a pencil compass are the only tools required to perform these constructions.

To start our discussion, let us begin with a simple line *AB* of length *a*. Three important constructions will be performed on this line: construction of a parallel line, a perpendicular bisector, and a perpendicular line constructed at an endpoint.

Construction of a line or lines parallel to a given line is very often required. This procedure is used in the construction of squares, rectangles or just merely in producing a line parallel to a given line.

Construction of a Line or Lines Parallel to a Given Line (Arc Method).

Figure 2-13 shows a given line *AB*. For purposes of better understanding, let us assume that a parallel line $\frac{3}{8}$ inch from line *AB* is required. First, set the distance required ($\frac{3}{8}$ inch) between the pivot point and pencil point of the compass. Choose two pivot points close to or even use the end points of line *AB*. Resting the pivot point on the line *AB*, draw arcs as shown at both ends. Lay a scale or straight edge so that its edge is tangent to both arcs, and scribe a line. This line is both parallel to and $\frac{3}{8}$ inch from line *AB*. (Tangent means touching. It is usually the intersection of a curved and straight line.)

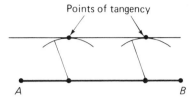

Points of tangency

Fig. 2-13. Construction of a line parallel to a given line (AB).

Naturally, the second method of producing this parallel line is to measure two $\frac{3}{8}$ inch points perpendicular to the line AB. It should be pointed out that the arc method of producing parallel lines is a far more superior method due to both speed and accuracy of construction.

Perpendicular Bisector

To construct a perpendicular bisector to a given line AB, the following procedure is used. Set the compass at some distance that is at least greater than the distance from an end point to the midpoint of the line AB. Set the compass pivot point at one end point of the line and scribe arcs on either side of the line. Using the same compass setting, perform the same using the other line end point as the compass pivot point. The arcs on either side of the line will meet and cross. Draw a line connecting the points of intersection. This line will be both perpendicular to line AB and also will bisect or split the line AB. This procedure is often used in dividing a circle into four parts. In this case, the end points of the diameter are used to scribe the arcs.

Erect a Perpendicular at an End Point or Selected Point of a Line AB

Assume that the need to erect a line perpendicular to line AB at point A is required (Fig. 2-14). The following procedure is recommended: extend line AB at the end where the construction is to be performed. Using a compass and point A as the center, construct a circle of some convenient size. This scribed circle will intersect the line and its extension at points M and N. Point A is now midway between points M and N. The procedure used for construction of the perpendicular bisector using points M and N as centers for the arcs can now be used.

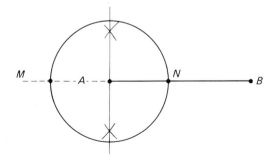

Fig. 2-14. Construction of a line perpendicular to and at end point of a line (AB).

If more lines perpendicular to line AB are required, we merely use the parallel line construction method of Fig. 2-13 to construct them.

Angle Construction (Chord Method)

Very accurate angle construction for angles from 0 to 89.9° is achieved by a simple relationship of the chord length of a circular arc to its central angle. Table 2-1 is provided for the chords of a circle with a radius equal to 1 (unit).

TABLE 2-1. CHORDS OF ANGLES (UNIT CIRCLE OF RADIUS 1)

Lengths of Chords for Whole Degrees

Angle	0°	1°	2°	3°	4°	5°	6°	7°	8°	9°
0°	0.000	0.017	0.035	0.052	0.070	0.087	0.105	0.122	0.140	0.157
10°	0.174	0.191	0.209	0.226	0.243	0.261	0.278	0.296	0.313	0.330
20°	0.347	0.364	0.382	0.399	0.416	0.433	0.450	0.467	0.484	0.501
30°	0.518	0.534	0.551	0.568	0.585	0.601	0.618	0.635	0.651	0.667
40°	0.684	0.700	0.717	0.733	0.749	0.765	0.781	0.797	0.813	0.829
50°	0.845	0.861	0.877	0.892	0.908	0.923	0.939	0.954	0.970	0.985
60°	1.000	1.015	0.030	1.045	1.060	1.075	1.089	1.104	1.118	1.133
70°	1.147	1.161	1.175	1.190	1.203	1.218	1.231	1.245	1.259	1.272
80°	1.286	1.299	1.312	1.325	1.338	1.351	1.364	1.377	1.389	1.402

Differences to Be Added for Tenths of a Degree

Angle	0.1°	0.2°	0.3°	0.4°	0.5°	0.6°	0.7°	0.8°	0.9°
0°	0.002	0.003	0.005	0.007	0.009	0.010	0.012	0.014	0.016
10°	0.002	0.003	0.005	0.007	0.009	0.010	0.012	0.014	0.016
20°	0.002	0.003	0.005	0.007	0.009	0.010	0.012	0.014	0.015
30°	0.002	0.003	0.005	0.007	0.008	0.010	0.012	0.013	0.015
40°	0.002	0.003	0.005	0.006	0.008	0.010	0.011	0.013	0.014
50°	0.002	0.003	0.005	0.006	0.008	0.009	0.011	0.012	0.014
60°	0.001	0.003	0.004	0.006	0.007	0.009	0.010	0.012	0.013
70°	0.001	0.003	0.004	0.006	0.007	0.008	0.010	0.011	0.012
80°	0.001	0.003	0.004	0.005	0.006	0.008	0.009	0.010	0.012

To best illustrate this important construction, let us proceed with a typical problem.

PROBLEM

Construct an angle of 44.4°.

Solution: Using some point O as a center, scribe a circle with a radius equal to 1 inch. Draw a line from point O to some point on the circumference (point A). We must now locate some point B on the circumference so that the angle (central angle) *BOA* is equal to 44.4°. Referring to Table 2-1, the chordal constant for 44°, with a radius equal to 1 is:

$$44° = 0.749$$

The constant for 0.4° is found in the 40° column:

$$0.4° = 0.006 \text{ inch}$$

Total chord of 44.4° = 0.749 + 0.006 = 0.755 inch
Setting a compass to 0.755 inch and using point *A* as a pivot, strike an arc on the circumference of the circle. This is point *B*. Scribe a line from point *B* to *O*. This angle now formed by *BOA* is the required angle of 44.4°.

Construction of Triangles

This section considers the construction of a triangle under three separate cases: when three sides are known, when a side and its two angles are known, and finally when two adjacent sides and the angle between them are known.

Here again, as with all geometric construction, the end results as far as accuracy and correctness are concerned depend on the use of proper and careful procedures. Let us now begin our first triangular construction.

Construct a Triangle When Three Sides are Known

The procedure for constructing a triangle when three sides are known is performed with the following set of steps:

Given: Three lengths of a triangle:

length *a* = 6 inches,   length *b* = 7 inches

and length *c* = 8 inches

Find: Graphically construct a triangle of these three sides.

Solution: Select any one of the three given lengths and draw it to scale. For example, let us choose side *a* = 6 inches. Select another length, e.g., 7 inches, and set a compass to this length. Select either end of the 6-inch line as a center and scribe an arc with the 7-inch radius. Set the compass to the last length, 8 inches, and using the opposite end of the 6-inch line, scribe an arc with the 8-inch radius. The two arcs cross one another, forming a point of intersection. By drawing a line from this point to each end of the 6-inch line, a triangle is constructed having the three given side lengths.

Construct a Triangle When a Side and Its Two Angles Are Known

The solution to this problem basically involves the use of angle construction (chord method). The following problem has been chosen to illustrate the point that no matter what system, English or metric, a degree equals a degree and thus construction of angles are the same:

Given: A line AB equal to 6 cm with angles of 32° and 47° (Fig. 2-15).

Find: Construct a triangle and find the lengths of the other two sides graphically.

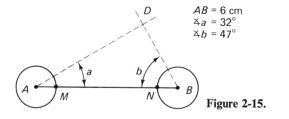

$AB = 6$ cm
$\angle a = 32°$
$\angle b = 47°$

Figure 2-15.

Solution: Construct the line AB equal to 6 cm (Fig. 2-15). Next refer to Table 2-1 on chords of angles to aid in the construction of the angles. Since the table is based on a radius equal to 1 inch, draw a circle with a 1-inch radius at each endpoint of line AB. The circles cross line AB at points M and N. Choose one of the angles, say 32°, and construct it at the A end of the line. Table 2-1 shows a chord length equal to 0.551 inch (radius $= 1$ inch). Using this figure, a compass, and point M as a pivot, scribe an arc on the circle. Draw a line from A through the point and some distance further. The same procedure is used in construction of the 47° angle (its chord constant is equal to 0.797).

Both of these angle sides are then extended until they meet at a point called D. Using a scale, both AD and DB lengths can be measured from our problem.

$$AD = 4.3 \text{ cm}$$

$$BD = 3.2 \text{ cm}$$

Construction of a Triangle When Two Sides and the Angle Between Them Is Known

Given: Two lines AC equal to 5 cm and AB equal to 18 cm that form a vertex angle $CAB = 17°$.

Find: Construct a triangle and graphically find the side CB.

Solution: Choose side AB for the base of the triangle and accurately lay it off. With the vertex angle A as its center, draw a circle of radius equal to 1 inch. It crosses AB at point M. Consulting Table 2-1 for the chord length of a 17° central angle, we find a value of 0.296 inch. Set a compass as close as possible to this value and with M as a center, draw an intersecting arc to the circle. A line drawn from A through the point of intersection is thus a 17° angle. Extend this line for some distance. Using a divider or scale, lay off the 5-cm length that is side AC. Draw a line connecting C and B. Measure its length in centimeters.

$$CB = 13.4 \text{ cm}$$

Construction of a Hexagon (Regular Polygon)

The construction of a hexagon is included in this discussion because of the frequency of its use in the technical world. The hexagon is a six-sided figure with a unique relationship. This relationship can best be illustrated in the following construction: using a radius of any size, draw a circle (Fig. 2-16). With the radius still set on the compass, place the pivot point on some point *A* of the circumference. Scribe an arc; the intersection of the arc and a point on the circumference of the circle is point 1. Using the same compass setting, repeat this process using point 1 as its new center and so on until the circumference is divided into six parts. By drawing lines from point *A* to point 1 and so on, a hexagon is constructed having both radius (*R*) and sides (*A*, 1) (1, 2) etc. equal to each other. This is correct since a hexagon is made of six equilateral triangles.

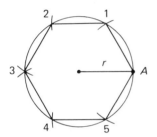

Fig. 2-16. Construction of a hexagon using radius equal to *r*.

This concludes the discussion on geometric construction. It is strongly suggested by the authors that these graphic procedures be performed by the student technician so that the procedures become more familiar to him.

SOLID GEOMETRY

At the beginning of this chapter we introduced a solid figure for the sole purpose of showing that it is composed of a series of component parts such as lines, angles and planes of varied shapes. In our discussion of these shapes, we covered a variety of points such as their description, construction and areas. These figures only had dimensions of width and depth.

We now describe solid figures and will define both the total surface area and the volume of these figures.

2-8. Surface Area of Solid Objects

Refer to Fig. 2-17, which will help to define two important terms: surface area, and volume. The object in Fig. 2-17 is a rectangular solid which, like all solids, has three dimensions: a height equal to three units, a width equal to four units, and a depth equal to five units. When total surface area is mentioned, it means the total area of all of the planes (flat or curved) making up a closed geometric solid.

In the example (Fig. 2-17), there are a total of six planes or three pairs of planes: one pair composed of the top and bottom planes, one pair composed of front and back planes, and the third pair composed of right and left side planes. The surface area of this figure is therefore equal to the sum of the areas of the six planes or the sum of the areas of the three pairs of planes.

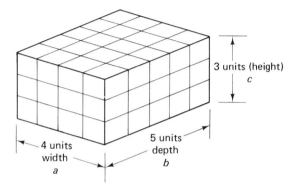

3 units (height)
c

4 units
width
a

5 units
depth
b

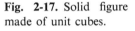

Fig. 2-17. Solid figure made of unit cubes.

The surface area of Fig. 2-17 is therefore

1. Area front plane $= 3 \times 4 = 12$ sq units
2. Area back plane $= 3 \times 4 = 12$ sq units
3. Area right side plane $= 3 \times 5 = 15$ sq units
4. Area left side plane $= 3 \times 5 = 15$ sq units
5. Area top plane $= 5 \times 4 = 20$ sq units
6. Area bottom plane $= 5 \times 4 = 20$ sq units

Total Surface Area $=$ 94 sq units

2-9. Volume of Solid Objects

When we speak of the word *volume*, we are talking about the number of cubic units of measure that are enclosed by the plane surfaces of a solid. These *cubic units* are in the same notation that the figure is measured in. If the dimensions of a solid figure are measured in centimeters, then the volume answer is in cubic centimeters. In inches, the volume is expressed in cubic inches.

Returning to the solid object in Fig. 2-17, notice that it is composed of a number of square blocks or cubes which are one unit by one unit by one unit. The front face is composed of a section of these one-unit blocks stacked three high by four blocks wide. The volume of this section is therefore equal to 3 units times 4 units times 1 unit, which equals 12 cubic units. The whole depth of the object however, is 5 units deep. The total volume of this figure is therefore 3 units times 4 units times 5 units, which equals 60 cubic units. Written mathematically, this is:

$$\text{Volume of a rectangle} = h \times w \times d \qquad (2\text{-}63)$$

where

$$h = \text{height}$$
$$w = \text{width}$$
$$d = \text{depth}$$

This whole idea of finding the volume of a geometric figure projected through a distance d is extended to such problems as finding the volumes of cylinders, squares, prisms, or any plane shape projected through a specific distance.

Prism

A prism is a solid whose polygon-shaped end planes or bases are both equal in size and shape and are also parallel to each other (Fig. 2-18). The faces or sides of a prism are parallelograms. Prisms can have plane shapes, squares, rectangles, triangles, pentagons, hexagons or any polygonal shape for their ends. The altitude of a prism is equal to the depth or distance between the end planes.

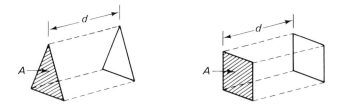

Fig. 2-18. Prisms: parallel and equal. Polygonal planes are separated by some distance d.

The lateral area of a prism is defined as the total area of the sides of the figure and does not include the area of the end planes or bases.

Cylinder

By far, the most common occurring problem relates to a right circular cylinder. This is a solid whose circular ends are mutually perpendicular to the sides of the cylinder (Fig. 2-19). The altitude of this figure is the distance d between the circular endplanes. The total surface area of this figure is simply the sum of the two end circles plus the area of the plane surface connecting the end circles. Mathematically, this is written as:

$$\text{surface area} = 2(\pi r^2) + (2\pi r \times a) \qquad (2\text{-}64)$$

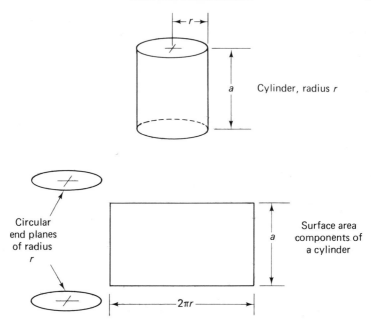

Fig. 2-19. Total surface area of a cylinder.

where

$$r = \text{radius}$$
$$a = \text{altitude}$$

The lateral area, again is the area of the connecting plane:

$$\text{Lateral area of a cylinder} = 2\pi r \times a \qquad (2\text{-}65)$$

where

$$r = \text{radius}$$
$$a = \text{altitude}$$

The volume of a cylinder is, like the volume of the rectangular solid,

volume of cylinder = the area of the face plane times depth:

or

$$\text{volume cylinder} = \pi r^2 a \qquad (2\text{-}66)$$

where

$$r = \text{radius}$$
$$a = \text{altitude}$$

Many answers may be derived from the knowledge of volumes. The weight of an object can be calculated very accurately by merely knowing the weight per cubic unit of that material (density). To illustrate this point, the following problem is presented.

Given: A piece of rolled copper rod 6 inches in diameter and 12 inches long is to be purchased at the cost of $2.50 a pound. What is the cost of the material?

Solution: This is a right circular cylinder.

$$\text{volume} = \pi r^2 \times a \text{ [from Eq. (2-66)]}$$

where

$$r = 3 \text{ inches}$$

$$a = 12 \text{ inches}$$

$$\text{volume} = 3.1416\,(9) \times 12$$

$$= 339.6 \text{ cubic inches}$$

From Table 2-2, we find that copper weighs 0.319 pounds per cubic inch.

TABLE 2-2. WEIGHT PER UNIT VOLUME OF COMMON ENCOUNTERED MATERIALS

Name of Substance	Pounds per cu inch	Pounds per cu foot
Air		0.0795
Aluminum		162
Anthracite coal, broken		52 to 60
Antimony		418
Asphaltum		87.3
Birchwood		41
Brass, cast (copper and zinc)		506
Brass, rolled		525
Brick, common		125
Brick, pressed		150
Bronze, copper 8, tin 1 (gun metal)		531
Chalk		156
Coal, bituminous, broken		47 to 56
Coke, loose		23 to 32
Corundum		
Copper, cast		542
Copper, rolled	0.319	555
Cork		15
Elmwood		35
Glass		186
Gold	0.695	
Granite		170
Ice		57.5
Iron, cast	0.26	450
Iron, wrought	0.28	480
Lead	0.412	712

TABLE 2-2. (CONT'D.)

Name of Substance	Pounds per cu inch	Pounds per cu foot
Maplewood		49
Marble		168.7
Mercury	0.49	
Nickel	0.318	
Oakwood, red		46
Pinewood, white		28
Pinewood, yellow		38
Platinum		
Quartz		165
Silver	0.379	655
Steel	0.29	490
Tin		459
Water, distilled at 32°F		62.417
Water, distilled at 62°F		62.355
Water, distilled at 212°F		59.7
Zinc		438

Therefore, the total weight of the copper rod is:

$$\text{total weight} = \text{vol. (cu. inches)} \times \text{weight (lbs per cu. inch)}$$
$$= 340 \times 0.319$$
$$= 108 \text{ pounds}$$

Since the cost is $2.50 a pound, the copper rod will cost

$$\text{total cost} = 108 \text{ lbs.} \times \$2.50$$
$$\text{total cost} = \$270.00$$

The weight of liquids and also the capacity in quarts, liters, or gallons can also be calculated by the above procedure and Table 2-2.

Consider a horizontal cylindrical tank with an inside diameter of 4 feet and a length of 8 feet. The tank is filled with water to the halfway point. Find both the volume in gallons and the weight of the water.

Solution: The volume of the water in the tank is equal to

$$\text{volume of water} = \tfrac{1}{2}(\pi r^2) \times \text{length}$$

where

$$r = 2 \text{ feet}$$
$$l = 8 \text{ feet}$$

$A = \tfrac{1}{2}\pi r^2$; since the water creates a semi-circular endplane.

$$\text{volume of water} = \tfrac{1}{2}(3.1416 \times 4) \times 8 \text{ feet}$$

$$= \tfrac{1}{2}(12.58) \times 8 \text{ feet}$$

$$= 6.29 \times 8$$

$$\text{vol. water} = 50.32 \text{ cu. feet}$$

From the Appendix it is found that 1 cu. foot of water = 7.48 U. S. gallons. The total gallons is

$$\text{total gallons} = 7.48 \times 50.32$$

$$\text{total gallons} = 376.4 \text{ gallons}$$

The total weight of the water is obtained by using Table 2-2 and the volume in cubic feet.

$$1 \text{ cu. foot water} = 62.4 \text{ pounds}$$

$$\therefore \quad \text{total weight water} = \text{volume} \times \text{weight (cu. feet)}$$

$$= 50.32 \times 62.4$$

$$= 3140 \text{ pounds}$$

NOTE

This problem was selected to encompass the mathematical knowledge described to this point. Solutions to problems where liquid levels are lower or higher will require both the use of the area of a circular segment [Equations (2-18) and (2-19)] and the use of trigonometry (to be described) to find both the central chord angle and the length of arc sustained by the liquid.

---

volumes (V) and areas (A) of solids

Volume (V) of Cube

V = volume

$V = s^3$ (2-67)

where s = side

Volume (V) of Square Prism

V = volume

$V = abc$ (2-68)

where a = width

b = height

c = depth

Volume (V) of Prism

V = volume

A = area of end surface

$$V = d \times A \qquad (2\text{-}69)$$

where d = distance between ends

Volume (V) and Area (A) of Cylinder

V = volume

S = area of cylindrical surface.

$V = 3.1416r^2a = 0.7854d^2\,a$

$S = 6.2832ra = 3.1416da$

Total Surface Area:

$$A = 6.2832r\,(r + a) = 3.1416d\,(1/2d + a) \qquad (2\text{-}70)$$

where r = radius of cylinder

d = diameter of cylinder

a = length of cylinder

Volume (V) of Hollow Cylinder

V = volume

$$\begin{aligned}
V &= 3.1416h(R^2 - r^2) = 0.7854h(D^2 - d^2) \\
&= 3.1416ht\,(2R - t) = 3.1416ht\,(D - t) \\
&= 3.1416ht\,(2r + t) = 3.1416ht\,(d + t) \\
&= 3.1416ht\,(R + r) = 1.5708ht\,(D + d)
\end{aligned} \qquad (2\text{-}71)$$

Volume (V) and Area (A) of Cone

V = volume

A = area of conical surface

$$V = \frac{3.1416r^2h}{3} = 1.0472r^2h = 0.2618d^2h \qquad (2\text{-}72)$$

$$A = 3.1416r\sqrt{r^2 + h^2} = 3.1416rs = 1.5708ds \qquad (2\text{-}73)$$

$$s = \sqrt{r^2 + h^2} = \sqrt{\frac{d^2}{4} + h^2}$$

Volume (V) and Area (A) of Sphere

V = volume

A = area of surface

$$V = \frac{4\pi r^3}{3} = \frac{\pi d^2}{6} = 4.1888r^3 = 0.5236d^2 \qquad (2\text{-}74)$$

$$A = 4\pi r^2 = \pi d^2 = 12.5664r^2 = 3.1416d^2 \qquad (2\text{-}75)$$

$$r = \sqrt[3]{\frac{3V}{4\pi}} = 0.6204\sqrt[3]{V}$$

where r = radius

d = diameter

Volume (V) of Hollow Sphere

V = volume

$$V = \frac{4\pi}{3}(R^3 - r^3) = 4.1888(R^3 - r^3) \qquad (2\text{-}76)$$

where R = outside radius

r = inside radius

TRIGONOMETRY

The pythagorean formula, which states $c = \sqrt{a^2 + b^2}$, is a strong and much-used relationship in right triangles. This formula, however, is only used to find the length of one side of a right triangle when the values of the other two sides are given.

Consider a problem where only one side and one angle A of a right triangle are given. Up to this point, the only other information we could deduce is the value of the third angle B:

$$180° - (90° + \angle A) = \angle B$$

We could not find the value of the other two sides of the triangle. Trigonometry was created to solve this kind of problem.

2-10. Trigonometric Functions (Angle, Side Relationship)

Consider a right triangle ABC (Fig. 2-20) with angles α, and angle β such that angle α plus angle β plus $90° = 180°$. Let us look at angle α and see how many ways this angle can be described by using any two of its sides. There are six methods that can describe any acute angle in a right triangle. The six ways are called trigonometric functions. To separate and define them, the following names have been assigned:

        1. $\sin \alpha$ = (sine alpha)

        2. $\cos \alpha$ = (cosine alpha)

        3. $\tan \alpha$ = (tangent alpha)

        4. $\cot \alpha$ = (cotangent alpha)

        5. $\sec \alpha$ = (secant alpha)

        6. $\csc \alpha$ = (cosecant alpha)

Each of these angle terms defines a specific combination of two sides

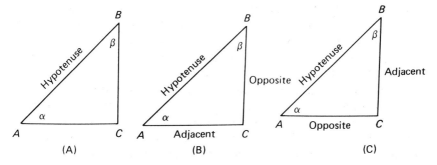

Fig. 2-20. Side to angle relations for trigonometric values.

of the triangle. When dealing with any of the acute angles of a right triangle, a basic set of rules for relating the sides to that angle are formed. Figure 2-20 shows this relationship, which should be committed to memory. The side directly opposite the angle being considered is called the *opposite side*. The hypotenuse is the *hypotenuse*, and finally the remaining side forming the angle with the hypotenuse is called the *adjacent side*.

When considering angle β and applying the above rule, we see the side relationship change to meet this new angle (Fig. 2-20). A great many errors result in the failure to select the proper sides to the angle under consideration.

Returning to the right triangle [Fig. 2-20(B)] and using the side relationship, the six trigonometric functions for some angle α can be defined:

$$1. \ \sin \alpha = \frac{\text{opposite side}}{\text{hypotenuse}} \ \text{or} \ \frac{BC}{AB}$$

$$2. \ \cos \alpha = \frac{\text{adjacent side}}{\text{hypotenuse}} \ \text{or} \ \frac{AC}{AB}$$

$$3. \ \tan \alpha = \frac{\text{opposite side}}{\text{adjacent side}} \ \text{or} \ \frac{BC}{AC}$$

$$4. \ \cot \alpha = \frac{\text{adjacent side}}{\text{opposite side}} \ \text{or} \ \frac{AC}{BC}$$

$$5. \ \sec \alpha = \frac{\text{hypotenuse}}{\text{adjacent}} \ \text{or} \ \frac{AB}{AC}$$

$$6. \ \csc \alpha = \frac{\text{hypotenuse}}{\text{opposite side}} \ \text{or} \ \frac{AB}{BC}$$

These six functions, coupled with the trigonometric values for any angle (procured from a table) allow the solving of basically all of the right triangle problems the technician will come into contact with.

2-11. Determination of Functions of Angles (Natural Functions)

When we speak of a function of an angle, we are speaking of a number value assigned to a specific angle function. It has no dimensions whatsoever. Two types of tables are available for evaluating functions of angles: a logarithmic table and a natural function table. The natural function table is by far the fastest and most error-free method for general use.

Natural trigonometric tables are set up to cover a range of degrees from 0 to 180°. Several forms of these tables will be encountered by the technician. He is cautioned to read the set-up so as to know the correct way of using the tables. Table 2-3 shows a typical set-up for a trigonometric function table. In looking at this table, the first item that is noticed is that there are four different degrees located at each corner of the table—30, 149, 59, and 120°. On the left side is a minute column with the minutes increasing from 0 to 60' from top to bottom. The right side minute column increases from 0 to 60′ from bottom to top. Along the top of the table, from left to right, are listed the six trigonometric functions, sin, cos, tan, cot, sec, csc, while the bottom of these columns reads from left to right cos, sin, cot, tan, csc, and sec. Disregard Vrs. Sin. and Vrs. Cos. columns.

Given any angle from 0 to 180°, how does one enter into the table to obtain the proper value trigonometric function?

Table 2-4 shows how to enter this particular table to find the trigonometric value of an angle.

TABLE 2-3. SAMPLE TRIGONOMETRIC TABLE

30°				Natural Trigonometric Functions					149°
M	Sine	Cosine	Tan.	Cotan.	Secant	Cosec.	Vrs. Sin.	Vrs. Cos.	M
0	0.50000	0.86603	0.57735	1.7320	1.1547	2.0000	0.13397	0.50000	60
1	.50025	.86588	.57774	.7309	.1549	1.9990	.13412	.49975	59
2	.50050	.86573	.57813	.7297	.1551	.9980	.13426	.49950	58
3	.50075	.86559	.57851	.7286	.1553	.9970	.13441	.49924	57
4	.50101	.86544	.57890	.7274	.1555	.9960	.13456	.49899	56
5	0.50126	0.86530	0.57929	1.7262	1.1557	1.9950	0.13470	0.49874	55
6	.50151	.86515	.57968	.7251	.1559	.9940	.13485	.49849	54
7	.50176	.86500	.58007	.7239	.1561	.9930	.13499	.49824	53
8	.50201	.86486	.58046	.7228	.1562	.9920	.13514	.49799	52
9	.50226	.86471	.58085	.7216	.1564	.9910	.13529	.49773	51
10	0.50252	0.86457	0.58123	1.7205	1.1566	1.9900	0.13543	0.49748	50
11	.50277	.86442	.58162	.7193	.1568	.9890	.13558	.49723	49
12	.50302	.86427	.58201	.7182	.1570	.9880	.13572	.49698	48
13	.50327	.86413	.58240	.7170	.1572	.9870	.13587	.49673	47
14	.50352	.86398	.58279	.7159	.1574	.9860	.13602	.49648	46
15	0.50377	0.86383	0.58318	1.7147	1.1576	1.9850	0.13616	0.49623	45
16	.50402	.86369	.58357	.7136	.1578	.9840	.13631	.49597	44
17	.50428	.86354	.58396	.7124	.1580	.9830	.13646	.49572	43
18	.50453	.86339	.58435	.7113	.1582	.9820	.13660	.49547	42
19	.50478	.86325	.58474	.7101	.1584	.9811	.13675	.49522	41
20	0.50503	0.86310	0.58513	1.7090	1.1586	1.9801	0.13690	0.49497	40
21	.50528	.86295	.58552	.7079	.1588	.9791	.13704	.49472	39
22	.50553	.86281	.58591	.7067	.1590	.9781	.13719	.49447	38
23	.50578	.86266	.58630	.7056	.1592	.9771	.13734	.49422	37
24	.50603	.86251	.58670	.7044	.1594	.9761	.13749	.49397	36
25	0.50628	0.86237	0.58709	1.7033	1.1596	1.9752	0.13763	0.49371	35
26	.50653	.86222	.58748	.7022	.1598	.9742	.13778	.49346	34
27	.50679	.86207	.58787	.7010	.1600	.9732	.13793	.49321	33
28	.50704	.86192	.58826	.6999	.1602	.9722	.13807	.49296	32
29	.50729	.86178	.58865	.6988	.1604	.9713	.13822	.49271	31
30	0.50754	0.86163	0.58904	1.6977	1.1606	1.9703	0.13837	0.49246	30
31	.50779	.86148	.58944	.6965	.1608	.9693	.13852	.49221	29

TABLE 2-3. (CONT'D.)

30°			*Natural Trigonometric Functions*					149°	
M	Sine	Cosine	Tan.	Cotan.	Secant	Cosec.	Vrs. Sin.	Vrs. Cos.	*M*
32	.50804	.86133	.58983	.6954	.1610	.9683	.13867	.49196	28
33	.50829	.86118	.59022	.6943	.1612	.9674	.13881	.49171	27
34	.50854	.86104	.59061	.6931	.1614	.9664	.13896	.49146	26
35	0.50879	0.86089	0.59100	1.6920	1.1616	1.9654	0.13911	0.49121	25
36	.50904	.86074	.59140	.6909	.1618	.9645	.13926	.49096	24
37	.50929	.86059	.59179	.6898	.1620	.9635	.13941	.49071	23
38	.50954	.86044	.59218	.6887	.1622	.9625	.13955	.49046	22
39	.50979	.86030	.59258	.6875	.1624	.9616	.13970	.49021	21
40	0.51004	0.86015	0.59297	1.6864	1.1626	1.9606	0.13985	0.48996	20
41	.51029	.86000	.59336	.6853	.1628	.9596	.14000	.48971	19
42	.51054	.85985	.59376	.6842	.1630	.9587	.14015	.48946	18
43	.51079	.85970	.59415	.6831	.1632	.9577	.14030	.48921	17
44	.51104	.85955	.59454	.6820	.1634	.9568	.14044	.48896	16
45	0.51129	0.85941	0.59494	1.6808	1.1636	1.9558	0.14059	0.48871	15
46	.51154	.85926	.59533	.6797	.1638	.9549	.14074	.48846	14
47	.51179	.85911	.59572	.6786	.1640	.9539	.14089	.48821	13
48	.51204	.85896	.59612	.6775	.1642	.9530	.14104	.48796	12
49	.51229	.85881	.59651	.6764	.1644	.9520	.14119	.48771	11
50	0.51254	0.85866	0.59691	1.6753	1.1646	1.9510	0.14134	0.48746	10
51	.51279	.85851	.59730	.6742	.1648	.9501	.14149	.48721	9
52	.51304	.85836	.59770	.6731	.1650	.9491	.14164	.48696	8
53	.51329	.85821	.59809	.6720	.1652	.9482	.14178	.48671	7
54	.51354	.85806	.59849	.6709	.1654	.9473	.14193	.48646	6
55	0.51379	0.85791	0.59888	1.6698	1.1656	1.9463	0.14208	0.48621	5
56	.51404	.85777	.59928	.6687	.1658	.9454	.14223	.48596	4
57	.51429	.85762	.59967	.6676	.1660	.9444	.14238	.48571	3
58	.51454	.85747	.60007	.6665	.1662	.9435	.14253	.48546	2
59	.51479	.85732	.60046	.6654	.1664	.9425	.14268	.48521	1
60	0.51504	0.85717	0.60086	1.6643	1.1666	1.9416	0.14283	0.48496	0
M	Cosine	Sine	Cotan.	Tan.	Cosec.	Secant	Vrs. Cos.	Vrs. Sin.	*M*
120°									59°

TABLE 2-4. ENTRY INTO TRIGONOMETRIC TABLES (2-3)

Angle Range	Enter Table for Degrees and Functions	Minutes
0–45°	At top	Left column
45–90°	At bottom	Right column
90–135°	At bottom	Left column
135–180°	At top	Right column

To illustrate Table 2-4 and the sample page of trigonometric functions,

Table 2-3, the sine values for the following angles are:

$$\sin 30° \qquad = 0.50000$$
$$\sin 59°17' \quad = 0.85970$$
$$\sin 149° \qquad = 0.51504$$
$$\sin 120°45' = 0.85941$$

2-12. Angle Functions with Seconds

In problems where extreme accuracy of measurement is involved, some angle values include seconds. For example, consider the solution for the angle whose sine is 30°17′35″. Since all trigonometric function tables are set up for degrees and minutes of a degree, a method called *interpolation* must be used to solve for seconds. Remember that there are 60 seconds between each minute; this is the basis for solving the function of an angle containing seconds.

Refer to Table 2-3. The values of the sin 30°17′ and of 30°18′ can be immediately located. But, the required value for the sin of 30°17′35″ is not shown; by deduction, it is known that the sin of 30°17′35″ lies somewhere between the sine of 30°17′ and the sine of 30°18′.

The standard set up for a solution of this type is as follows.

1. Remove the seconds from the particular angle and divide them by 60 seconds. This gives a fractional value in terms of minutes. In the case of the sample problem, sin 30°17′35″,

$$\text{fractional value} = \frac{35''}{60''} = \frac{7'}{12}$$

The significance of this fraction of a minute will soon become apparent.

2. The second part of this problem is to find the difference of the function of the angle (less seconds) and the function of the angle (less seconds) with one minute added to it. Written in short form:

(1)   Function of the angle + 1 minute =
(2)   − Function of the angle   =
       _____
(3)   Total difference due to 1 minute =

NOTE

Lines (1) and (2) omit all seconds as this is what is being solved for. Thus,

(1) $\sin 30° (17' + 1') = \sin 30°18' = 0.50453$
(2) $-$ $\sin 30°17' = 0.50428$
       _____
(3) Total difference due to 1 minute = 0.00025

It was originally found that a value of $35''$ was equal to $\frac{7}{12}$ of a minute. Therefore:

$$\text{line (3)} \times \text{fractional part of minute} =$$

$$0.00025 \times \tfrac{7}{12} = 0.00014$$

This value, which is $\frac{7}{12}$ of the value between $30°17'$ and $30°18'$ is then added to the lowest angle. The final solution of the problem is therefore

$$\sin 30°17'35'' = 0.50442$$

2-13. The Law of Sines

The law of sines to be presented shortly, is used to compute angles or sides of a triangle when several quantities are known about the triangle. The law in written form states the following: *any side is to the sine of the angle opposite that side as any other side is to the sine of the angle opposite it* (Fig. 2-21).

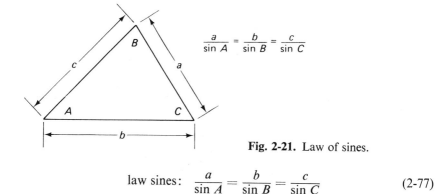

$$\frac{a}{\sin A} = \frac{b}{\sin B} = \frac{c}{\sin C}$$

Fig. 2-21. Law of sines.

$$\text{law sines:} \quad \frac{a}{\sin A} = \frac{b}{\sin B} = \frac{c}{\sin C} \qquad (2\text{-}77)$$

To properly use Eq. (2-77) and obtain a result, three of the four values in any combination of two parts of the equation must be known. To illustrate this equation consider the following problem: Two observers report seeing a meteor. Observer A records an angle of $52°$ while observer B, 6000 yds. away, records an angle of $58°$. Calculate the distance from A to the meteor C in yards at the instant it was observed.

$$\angle C = 180° - (52° + 58°)$$

$$C = 70°$$

Using the law of sines [Eq. (2-77)]

$$\frac{c}{\sin C} = \frac{b}{\sin B}$$

$$c = 6000 \text{ yds}, \qquad b = AC = \text{unknown}$$
$$\angle C = 70°, \qquad \angle B = 58°$$
$$\therefore \quad \frac{6000}{0.9396} = \frac{b}{0.8480}$$
$$0.9396b = 5088$$
$$AC = b = 5041 \text{ yd}$$

2-14. Law of Cosines

The law of cosines is used for the solution of any triangle when two sides of a triangle and the angle between them is known. In written form, the law of cosines states that *in any triangle, the square of any side is equal to the sum of the squares of the other two sides minus twice the product of the two sides times the cosine of their included angle:*

$$a^2 = b^2 + c^2 - 2bc \times \cos A \tag{2-78}$$

Since the approach to a solution of the law of cosines is very close to that as shown in the law of sines, no sample problem is shown.

2-15. Trigonometric Identities

The following partial list of trigonometric identities are offered to help the technician in his solutions of trigonometric problems. Their real use is in applying them to find the value of particular angles from these identities:

$$\sin^2 A + \cos^2 A = 1 \tag{2-79}$$

$$\sin A = \sqrt{1 - \cos^2 A} \tag{2-80}$$

$$\cos A = \sqrt{1 - \sin^2 A} \tag{2-81}$$

$$\tan A = \frac{\sin A}{\cos A} = \frac{1}{\cot A} \tag{2-82}$$

$$\cot A = \frac{\cos A}{\sin A} = \frac{1}{\tan A} \tag{2-83}$$

$$\sin (A + B) = \sin A \cos B + \cos A \sin B \tag{2-84}$$

$$\sin (A - B) = \sin A \cos B - \cos A \sin B \tag{2-85}$$

$$\cos (A + B) = \cos A \cos B - \sin A \sin B \tag{2-86}$$

$$\cos (A - B) = \cos A \cos B + \sin A \sin B \tag{2-87}$$

There are other identities other than the above list, but for simplicity, only the most frequently used identities are presented.

2-16. Trigonometric Solutions to Right Triangles and Oblique Triangles

Table 2-5 is offered both as a summary and as a ready reference for problem solving using trigonometry as a tool.

Throughout this section, we have made use of the natural functions of angles. This use allows the use of side values of the triangle directly in the formulas with only an angle function value to be determined from the tables.

The second method in these type of solutions involves the use of logarithms. In this case, all information such as sides and angles must be first reduced to logarithmic notation. A table in logarithmic notation for the

TABLE 2-5. RIGHT TRIANGLE SOLUTIONS

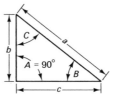

Sides and Angles Known	Formula for Sides and Angles to be Found		
Sides a and b	$c = \sqrt{a^2 - b^2}$	$\sin B = b/a$	$C = 90° - B$
Sides a and c	$b = \sqrt{a^2 - c^2}$	$\sin C = c/a$	$B = 90° - C$
Sides b and c	$a = \sqrt{b^2 - c^2}$	$\tan B = b/c$	$C = 90° - B$
Side a; angle B	$b = a \text{-} \times \sin B$	$c = a \times \cos B$	$C = 90° - B$
Side a; angle C	$b = a \times \cos C$	$c = a \times \sin C$	$B = 90° - C$
Side b; angle B	$a = b/\sin B$	$c = b \times \cot B$	$C = 90° - B$
Side b; angle C	$a = b/\cos C$	$c = b \times \tan C$	$B = 90° - C$
Side c; angle B	$a = c/\cos B$	$b = c \times \tan B$	$C = 90° - B$
Side c; angle C	$a = c/\sin C$	$b = c \times \cot C$	$B = 90° - C$

Solution of Oblique-Angled Triangles

One side and two angles are known.
The known side is a, the angle opposite it A, and the other known angle B. Then
$$C = 180° - (A + B)$$
$$b = \frac{a \times \sin B}{\sin A} \qquad c = \frac{a \times \sin C}{\sin A}$$
$$\text{area} = \frac{a \times b \times \sin C}{2}$$
If angles B and C are given, but not A, then
$$A = 180° - (B + C).$$

TABLE 2-5. (CONT'D.)

Solution of Oblique-Angled Triangles

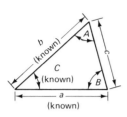

Two sides and the angle between them are known.
The known sides are a and b, and the known angle between them C. Then

$$\tan A = \frac{a \times \sin C}{b - (a \times \cos C)}$$

$$B = 180° - (A + C) \quad c = \frac{a \times \sin C}{\sin A}$$

Side c may also be found as below:

$$c = \sqrt{a^2 + b^2 - (2ab \times \cos C)}$$

$$\text{area} = \frac{a \times b \times \sin C}{2}$$

Two sides and the angle opposite one of the sides are known.
The known angle is A, the side opposite it a, and the other known side b. Then

$$\sin B = \frac{b \times \sin A}{a} \quad C = 180° - (A + B)$$

$$c = \frac{a \times \sin C}{\sin A}$$

$$\text{area} = \frac{a \times b \times \sin C}{2}$$

All three sides are known.
The sides a, b, and c, and the angles opposite them, A, B, and C, are known. Then

$$\cos A = \frac{b^2 + c^2 - a^2}{2bc}$$

$$\sin B = \frac{b \times \sin A}{a} \quad C = 180° - (A + B)$$

$$\text{area} = \frac{a \times b \times \sin C}{2}$$

trigonometric angle functions must be used. The technician is cautioned in the following statement. *When using true values of the sides of a triangle for the solution to a problem, always use natural functions of angles; conversely, when solving problems by the use of logarithms, use the logarithmic tables for functions of the angles.*

SUMMARY

This chapter has presented some basic principles of mathematics involving plane geometry, solid geometry, and trigonometry. The basic geometric designs of the triangle, the circle, the rectangle, and other polygons are basic to the mechanical designs of many workpieces. Thus, with an understanding of the basic designs, the technician can apply his knowledge to the more complex geometric designs, whether plane or solid, with which he may come into contact. The technician should practice the various practical construction procedures provided in this chapter so that he will be able to rapidly solve problems from a practical geometric standpoint.

The fundamentals of trigonometry are likewise necessary because often the technician will encounter a problem where he has nearly all of the information that he needs, but he must make calculations to find some unknowns.

REVIEW QUESTIONS

1. Construct a right triangle with a base of 2 inches and an altitude of $1\frac{1}{2}$ inches. Determine by direct measurement the length of the hypotenuse.

2. Lay out a triangle that has one side (AB) equal to 6 cm, an angle at point A of 24.4°, and an angle at point B of 45°. Determine the third angle of the triangle and also the length of the other two sides. Use Table 2-1 to construct the angles.

3. Determine the area of a sector of a circle with a radius of 5 inches. The central angle of the sector is 12°.

4. Solve for the area of the plane geometric shape shown in Fig. 2-22.

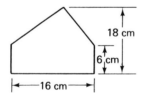

18 cm

6 cm

|←——16 cm——→|

Figure 2-22.

5. A right triangle has a hypotenuse equal to 12 feet and a base equal to 6 feet. Find the value for the altitude.

6. Given the trapezoidal plate shown in Fig. 2-23, the plate has a square hole and a round hole machined into it. The diagonal of the square is

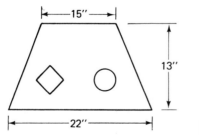

|←——15"——→|

13"

|←————22"————→|

Figure 2-23.

known to be 3 inches, and the radius of the circular cutout is 2.75 inches. Calculate the area of the plate less the cutouts.

7. A solid steel shaft 12.5 inches in diameter is machined with hemispherical ends to an overall length of 3.5 feet. Calculate the weight of the shaft and the cost of the material (assume the cost of the steel to be $1.20 per pound).

8. A drum that has a diameter of 2.5 feet and an overall height of 4.6 feet is $\frac{7}{16}$ full of water. Calculate the volume and weight of the water in the drum.

9. Calculate the surface area of a sphere with a diameter equal to 2.5 meters.

10. Find one side of a cube, in inches, that will contain 1825 pounds of copper.

11. Using Fig. 2-24, list the trigonometric functions of angle θ.

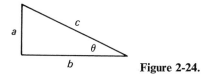

Figure 2-24.

12. In natural functions of angles, are the table values pure numbers?

13. Fig. 2-25 shows a right triangle with an exterior angle of 158°. In terms of the sides a, b, and c, list the cos, cot, and csc of angle θ. What is the value of angle θ?

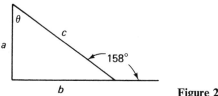

Figure 2-25.

14. Find the value of angle α in Fig. 2-26.

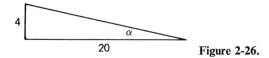

Figure 2-26.

15. Enter Table 2-3 and find the values of the following angles: tan 30°46′; cos 59°10′15″; sin 120°30′20″; tan 149°18′.

3

ENGINEERING DRAWINGS AND SKETCHES

Engineering drawings are used daily by the mechanical technician because drawings are his "roadmap" to completion of his job assignment. Upon a new assignment, the mechanical technician's supervisor hands him a set of engineering drawings that details the part(s) (hereafter called the workpiece) that he is to make. The engineering drawing is the specification that accurately describes the workpiece's shape, dimensions, tolerances, construction materials, and finishes. Thus, the engineering drawing is a record of design information and is used to transfer design information from the engineer's mind to the mechanical technician's mind, with the end result being that the mechanical technician builds the proposed/designed workpiece.

A sketch is a "freehand" pencil drawing used by the mechanical technician for the purpose of aiding in mentally visualizing the workpiece described by the engineering drawing. While engineering drawings are formal drawings made exclusively in the engineering department by engineers and draftsman, the sketch is an informal drawing made by the mechanical technician. It is interesting to note, however, that the engineer's *original* design ideas were more than likely in a sketched form before being made into engineering drawings by the draftsman. In addition, the draftsman may have made sketches to help him in the planning and layout of the engineering drawings prior to his actual drawing of the engineering drawings.

Often, on "one of a kind" prototype (the original or model after which other workpieces are made) workpieces, only sketches will be given to the technician to work from. A field technician finds that sketching is an important tool because he often must make a sketch of a needed replacement part, or a sketch noting critical dimensions needed for a workpiece. The sketch may be retained for future use by the technician, or it may be sent to his manufacturing plant for the manufacture of a replacement part.

Thus the mechanical technician is required to make sketches, but he is not required to make engineering drawings. Hence, this chapter deals first with sketching methods that can then be applied by the reader in a later portion of this chapter, which describes types of engineering drawings. Since it is essential for the technician to be able to completely and accurately read engineering drawings, this chapter expounds upon the information that is detailed on the drawings and presents problems to help the reader put that information into practice.

The major areas discussed in this chapter are: sketching; engineering drawings; projections; drawing formats; line characteristics, dimensions, and tolerances; and symbols, abbreviations, and notes.

SKETCHING

As previously mentioned, a sketch is an informal pencil drawing that is made without the use of elaborate drawing instruments. The sketch is used to help visualize the finished product in your mind. For example, the engineer or your supervisor may draw a sketch of a workpiece which he wants you to make. It normally is marked with dimensions, finish marks, workpiece material, and some notes; but you must be aware of which dimensions are true, because lines which are not parallel to the drawing axes (width, height, depth), are not true lengths. Or perhaps you are given a three-view detailed drawing of a complex workpiece and you can't seem to mentally visualize the finished shape of the workpiece. Then, a "three-dimensional" sketch is your answer. Thus, the sketch, like the engineering drawing, is used to pictorially express information or ideas by use of lines, points, arcs, and angles.

3-1. Sketching Tools and Materials

All of us have sketched (or doodled) to some extent and the only items needed were something to sketch with and something to sketch on. However, engineering sketching is more sophisticated, as we are concerned that our sketch shows similarity and proportion to the workpiece to be eventually built. Therefore, it is suggested that the following tools and materials be procured for sketching: transparent graph paper, a six-inch steel pocket scale, a medium lead pencil, and a soft eraser.

For more accurate freehand sketches and where it is expected that a great deal of sketching will be done, transparent coordinate graph paper of $\frac{1}{4}$, $\frac{1}{8}$, or $\frac{1}{10}$ inch grid divisions and *isometric* paper should be obtained. The coordinate paper will be used to sketch front (plan), top, and end views of workpieces where only two dimensions are shown in each view. These views are developed through *orthographic* projection, which is discussed in Sec. 3-7. The isometric paper is used to sketch an *isometric* view of a workpiece. An isometric view is a three-dimensional or solid view in which measurements on the three axes are of true dimension. Isometric views are discussed in Secs. 3-8 to 3-13. The six-inch steel pocket scale is used to measure the workpiece, dimensions on engineering drawings, and as a straight edge, if desired, in drawing axes or lines in the sketches.

Very often a technician must go into the "field" to measure and sketch a particular item to be made in the shop. Always discuss the size and shape of the job first so that proper measuring tools can be gathered and carried to the job.

3-2. Sketching Procedures

There are no hard and fast sketching procedures. There are, however, some hints which will enable you to sketch more "accurately."

1. Make initial lines, especially curved lines, very light. Then, when the correct path of the line has been lightly made, darken the line.

2. In making a line or arc between two points, keep your eye on the points to which the line is being drawn (i.e., don't watch the pencil point).

3. Do not fasten the paper down. Rotate or move the paper, as required, to aid you in drawing straight or curved lines.

4. In making short lines, move the pencil by pivoting your wrist. For making longer lines, pivot your forearm at the elbow. Make short connecting strokes. Never attempt to sketch a long line in one stroke.

5. Practice drawing vertical lines from top-to-bottom and bottom-to-top. Choose the method which is best for you and stick to it. You may rotate the paper to allow an easier method of sketching vertical lines. Practice drawing horizontal lines left-to-right and right-to-left; choose the method best for you and stick to it.

6. Practice dividing lines (without use of a scale) into equal segments of two, four, and eight units. If using coordinate paper, the squares may be used as an aid in dividing horizontal and vertical lines. Practice dividing lines which are at angles to the coordinates of the graph paper. Also practice dividing (without the use of a protractor) a 90° angle into three 30° angles. Next practice dividing 90° angles into two 45° angles.

Sketches are important in technical work. The sketches must contain the information required to completely describe the workpiece that the technician wants to describe; the information must be conveyed! However, it should also be emphasized that the sketch should be as simple as possible. Think of the workpiece only; don't become confused with extraneous parts.

ENGINEERING DRAWINGS

Engineering drawings, also called working drawings, are the formal (pictorial or textual) drawings which totally and accurately describe the physical and functional requirements of a workpiece. The engineering drawings for a complete unit may consist of *assembly* drawings, *detail* drawings and *parts lists* (bill of materials). The assembly and detail drawings are generally carefully made scaled drawings of the workpiece and are made using drafting instruments which include a drawing board, T-square, triangles, curves, and compasses.

Some of the other types of engineering drawings that are drawn are: electrical schematic diagrams, hydraulic schematic diagrams, pneumatic schematic diagrams, mechanical schematic diagrams, exploded view diagrams, and top assembly drawings.

3-3. Assembly Drawings

An assembly drawing (Fig. 3-1) depicts the assembled relationship of two or more parts, a combination of parts and subassemblies, or a group of subassemblies that make up an assembly. The assembly drawing contains sufficient views to show the relationship between each subordinate assembly and part shown. Parts or assemblies are often indentified by an item number enclosed in a circle, called a "bubble," which is related by the same item number to a parts list. Details of parts or subassemblies including dimensions are not shown on the assembly drawings but are shown in subassembly or detail drawings. In summary, the assembly drawing pictorially shows the technician how the parts and/or subassemblies are put together into a complete assembly. The parts or subassemblies may be manufactured by the technician's company, or commercial subassemblies and parts such as bolts, nuts, "0" rings, etc., may be purchased.

3-4. Detail Drawings

A detail drawing (Fig. 3-2) depicts the complete requirements for the manufacture of the part(s) shown on the drawing. Sufficient selected views and notes are shown to fully and clearly describe and dimension the part so

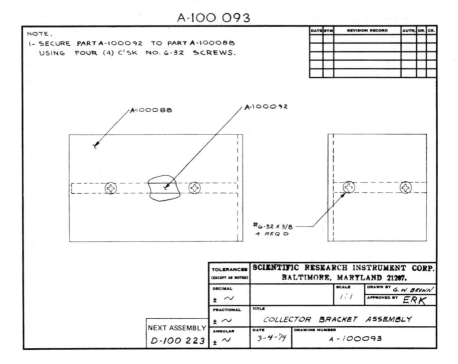

Fig. 3-1. An example of an assembly drawing. (Courtesy of Scientific Research Instrument Corp.)

that the technician can produce the part without further discussions with the engineer or draftsman.

3-5. Parts Lists (Bill of Materials)

All of the parts shown on an assembly drawing are called out on a parts list (or bill or material), which may be an integral part of the assembly drawing (located directly above the title block, which is described in Sec. 3-17) or may be a separate listing. In either case, each contains the same information. Each part on the drawing is identified by an item number which refers to a corresponding item number in the list. Once the item number is located in the list, the following information is shown describing that part: the number of times that the part called out by the item number is used in the subassembly or assembly; the manufacturer's name (or a code identification number); part number or identification number; and nomenclature or description of the part.

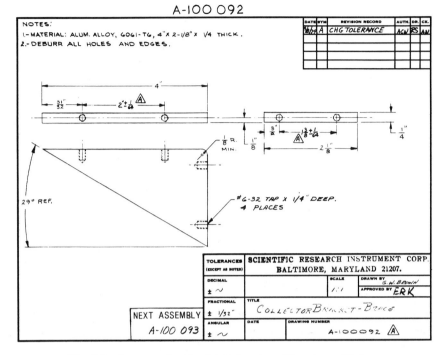

Fig. 3-2. An example of a detail drawing. (Courtesy of Scientific Research Instrument Corp.)

3-6. Other Types of Engineering Drawings

An electrical schematic diagram shows, by the use of graphical symbols, the electrical connections and functions of a specific circuit arrangement. The schematic diagram facilitates tracing the circuit and its function without regard to the actual physical size, shape, or location of the parts. Hydraulic and pneumatic schematic diagrams graphically depict the interconnection of piping, tubing, or hose between motors, pumps, gauges, and valves. Sequential flow, direction, and pressures are often shown.

A mechanical schematic diagram graphically illustrates the operational sequence or arrangement of a mechanical device. An exploded view diagram is a pictorial "three-dimensional" drawing and has the parts "blown" apart as if disassembled. A thin single dashed line between the centers of the "blown" away parts show where the parts go back together. A top assembly drawing lists all of the subassemblies and parts that make up the final completed article.

PROJECTIONS

By definition, a projection is a view of an object (workpiece) obtained upon a plane by projecting lines perpendicular to the plane. The views are drawn on paper so that persons using the drawing may know the shape and size of the object to be made.

The first part of this section describes orthographic projections and pictorial projections. The latter part of this section describes procedures for sketching, reading, and drawing those types of projections that are most often used by mechanical technicians. Finally, sectional views and auxiliary views are described.

3-7. Orthographic Projections

Engineering drawings and sketches provide the complete information required for production of the workpiece that is to be built. The major portion of the drawing or sketch consists of a view or views known as *orthographic projections*. The number of views selected is the least number of views required to completely describe the workpiece. Thus, an object such as a ball only requires one view to completely describe it; a piece of flat metal requires only one view; a symmetrical workpiece such as a cylinder or a cone normally requires but two views; and an unsymmetrical workpiece usually requires three views. Complex workpieces may need as many as six orthographic projection views and perhaps even a sectional or auxiliary view to completely describe it. In the vast majority of drawings, three views are used. Orthographic projections, used entirely in the United States, are often called third-angle projection.

Orthogonal is defined as pertaining to or involving right angles or perpendicular lines. *Projection* is defined as the act of visualizing and regarding an outline or the like as objective reality. An *orthographic projection* is a projection in which the projecting lines are perpendicular to a viewing plane. For example, when you hold a box and look directly at the edge of it [Fig. 3-3(A)] such that the lines of view from your eye to the box are perpendicular to the box, the view that you see is an orthographically projected view. In order for the lines of view to be theoretically perpendicular to the box from the eye, the distance between the eye and the box has to be assumed to be infinite. If the box is rotated in all directions so that one by one, each surface of the box comes into view in the perpendicular line of sight of the viewer to the box, there are six views of the box, which are called the front, rear, top, bottom, left side, and right side views [Fig. 3-3(B)]. These views are the orthographically projected views of the box; in common engineering drawing and sketching practice however, the drawing need show only the number of views necessary to completely describe the workpiece. Thus, in this example, since the top and

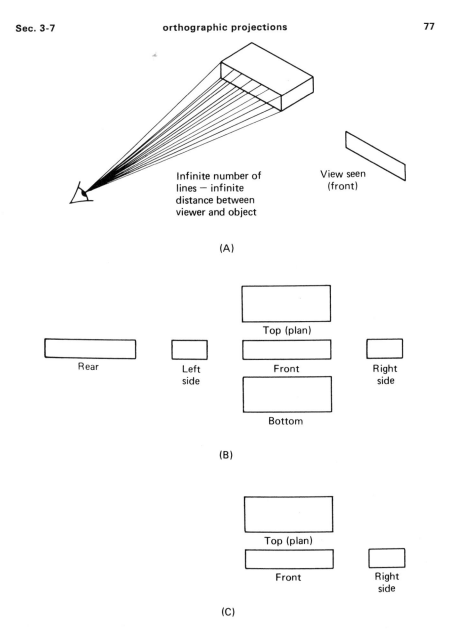

Infinite number of
lines — infinite
distance between
viewer and object

View seen
(front)

(A)

Top (plan)

Rear   Left   Front   Right
        side   side

Bottom

(B)

Top (plan)

Front   Right
         side

(C)

Fig. 3-3. Orthographically projected views.

bottom views are the same, the left and right side views are the same, and the
front and rear views are the same, it is only necessary to show three views.
Through long usage, it has become common practice to usually show the
front, top (also called the *plan* view), and the right side views. The views,
by convention, are laid out on the drawing board in the positions shown in

Fig. 3-3(C). The "front" of a workpiece is determined by the designer and is usually considered to be the shape which is narrow and long. The front view is with the workpiece sitting in a manner in which you would normally look at it. Once the front is established, the other views are easily established (Fig. 3-4).

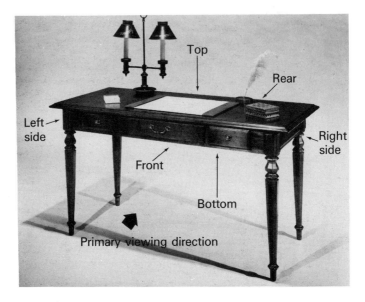

Fig. 3-4. The front of a workpiece is determined by the designer and is usually considered to be the shape which is narrow and long. (Courtesy of Ethan Allen, Inc.)

Any workpiece having three dimensions must have these three dimensions shown on the orthographically projected views. A sheet of metal is usually not considered as having a height (thickness) dimension and quite often, the thickness is simply noted on the drawing. Figure 3-5 shows the standard dimensioning of the views. The front view shows true width and height (thickness); the top view shows true width and depth; and the side view shows true depth and height (thickness). Note that each view shows two dimensions of the workpiece; thus, the view shown is the view seen when you are looking at the workpiece in only two planes—it is not three dimensional.

Assuming that the observer's primary viewing direction is toward the front of the workpiece, the width is considered the left-to-right distance between two points. The height is considered as the elevation between two points, and the depth is the front-to-rear distance between two points.

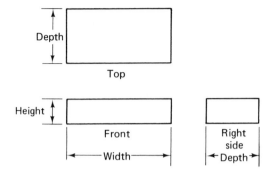

Fig. 3-5. Each view is shown in the two dimensions illustrated. Which dimensions are shown in the left side view? In the bottom view? In the rear view?

3-8. Pictorial Projections

Workpieces are often shown by engineering drawings or sketches in a three-dimensional pictorial view. A pictorial view is created to provide a mental image of the finished workpiece. The pictorial projection may be one of five types of representations: *dimetric, trimetric, isometric, oblique,* and *perspective.* A *dimetric* projection is a projection in which the workpiece is rotated into any position in which two edges (axes) are *equally* foreshortened (the length of the axes are reduced in length to give perspective). A *trimetric* projection is a projection of a workpiece that is placed in any position in which all three dimensional axes are *unequally* foreshortened. Pictorial projections in which all three axes are true dimensions are *isometric* projections. *Oblique* projections are projections in which the projectors make an angle of other than 90° with the viewing plane. A *perspective* drawing represents a workpiece as it appears to an observer stationed at a particular position relative to the workpiece.

Of the five methods of pictorially representing a workpiece, the isometric is the most commonly used by the mechanical technician because it is the easiest to draw and the dimensions are true on all three axes: width, height, and depth.

To accomplish the laying out of true dimensions on the three axes, the three axes must be laid out at three equal angles to each other; this results in axes (isometric axes) which are 120° apart from one another as shown in Fig. 3-6(A). The three axes are established by use of the 30–60–90° triangle and T-square. Since the drawing of isometric drawings is usually started from the bottom and drawn to the top of the workpiece, the height axis is drawn above the intersection of the width and depth axes as shown in Fig. 3-6(B).

Isometric lines are those lines which are parallel to one of the three isometric axes; nonisometric lines are not parallel. Hidden lines, which define surfaces or edges behind the outside views, are usually left out on isometric drawings except in cases where they are needed to clarify a drawing.

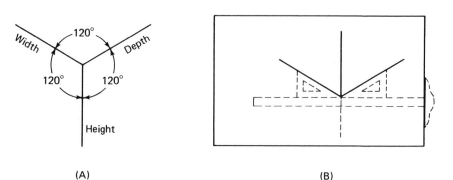

Fig. 3-6. Isometric drawings—layout of axes.

Pictorial illustrations (in this case, isometrics) are often shaded to show detail better. In shading, the illustrator picks some point from which a light source is supposedly located; the upper left front is most often used. Shading is then added accordingly. Thus, the area in the front would have the least shade—areas further away have more shading. The side away from the light source is drawn with heavier lines.

NOTE

The next part of this section covers sketching isometric projections of an object, reading orthographic projections, drawing and sketching orthographic projections, reading isometric projections, and sketching isometric projections from orthographic views. Since these procedures all interact with one another, the instructor may choose to cover the material in an order other than is presented here.

Remember that the technician is not a draftsman, but he must be able to read and interpret engineering drawings prepared by an engineer or draftsman. The technician must also be able to sketch both orthographic and isometric projections.

3-9. Sketching Isometric Projections of an Object

Refer to Fig. 3-7, which represents the shape of a worn part that must be replaced in the field by a technician. The technician must sketch an isometric view of the part and send it to the factory for the manufacture of a new part.

The first step in making the isometric sketch is to gather the materials and tools described in Sec. 3-1. On either plain transparent paper or isometric

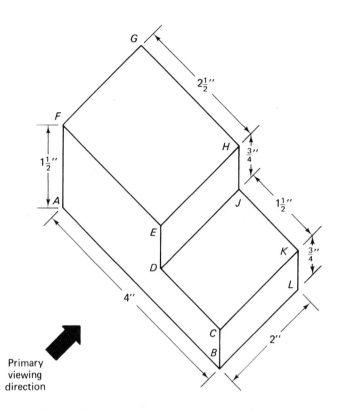

Fig. 3-7. The diagram represents a worn part. The technician is to sketch the part in an isometric view.

transparent paper (preferred), lay out the three isometric axes by making light pencil lines [Fig. 3-8(A)]. Pick the view that is the easiest for you to visualize and begin the drawing. This is usually the front or top view. You may choose to measure and note dimensions on a line-for-line basis or fill them in after completion of the sketch.

NOTE

In preparing the isometric sketch, the technician would obtain the dimensions of the workpiece by direct measurement with a scale, micrometer, or calipers, depending upon the accuracy required. The dimensions would then be roughly measured off on the isometric axes with the scale and the actual dimensions labeled. For discussion purposes, the dimensions of the workpiece have been indicated on the drawing of Fig. 3-7.

The next step after drawing the three axes is to draw the baselines of the workpiece that are parallel to the isometric axes. The starting point for these lines would of course be point B of Fig. 3-7, the intersection of the three axes. Figure 3-8 illustrates the logical steps in the completion of the lines of the isometric sketch. Note that long light lines are first drawn; then, using the grid line paper, lengths are judged and dark lines are drawn.

Look at familiar subjects and practice drawing isometric sketches of them. Remember, *dimensions are true only when measured on lines parallel to one of the three isometric axes.* Some familiar subjects you might try sketching are a textbook, radio, loudspeaker enclosure, table, and file cabinet. Note that only "rectangular" shapes have been suggested; curved and round shapes are more difficult and are covered in more depth in Sec. 3-13. You should learn the basics of drawing isometric sketches of rectangular shapes, because it will help you in the understanding of subsequent topics in this section.

3-10. Reading Orthographic Projections

As previously discussed, orthographic projections are shown in the least number of views that completely describe a workpiece. There are usually two views for symmetrical workpieces and three views for unsymmetrical workpieces. The views normally used are front, top or plan, and the right side. Dimensions, tolerances, and notes together with the orthographically projected views provide all of the information necessary for the technician to make the workpiece. It is of utmost importance that the technician be able to read orthographic projections. He must develop the ability to visualize the finished three-dimensional workpiece from the views presented on the engineering drawings or sketch. Reading an orthographic projection is the process of interpreting (visualizing) the shape of an object from the views presented.

The first step in reading an orthographic projection is to determine which views are present. Front? Top? Right or left sides? Rear? Bottom? A line or surface cannot be determined from only one view—you must look at two views to determine what the line or surface represents. Likewise, a shape can't be assumed from one or two views. All views must be carefully read.

For example, Fig. 3-9(A) and (B) have exactly the same front and top views. It is absolutely necessary to have a side view provided that further defines the lines and surfaces shown in the front and top views.

The second step in reading an orthographic projection is to look at each view one at a time. "Look" at the front view as if you were standing in *front* of the workpiece; then the top as if you were standing on top; then do likewise for the other views provided. Look at the dominant lines—look for center

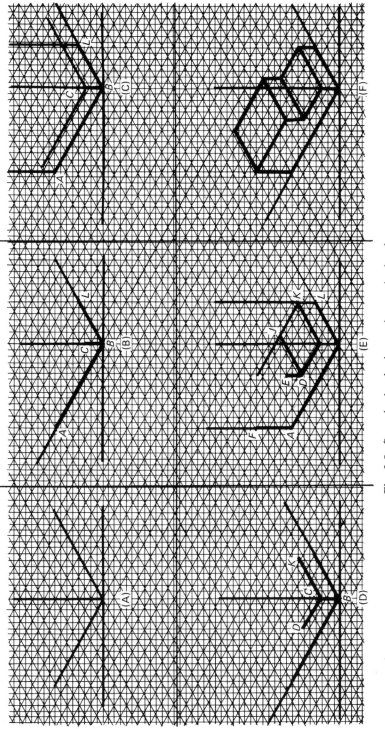

Fig. 3-8. Steps in developing an isometric sketch.

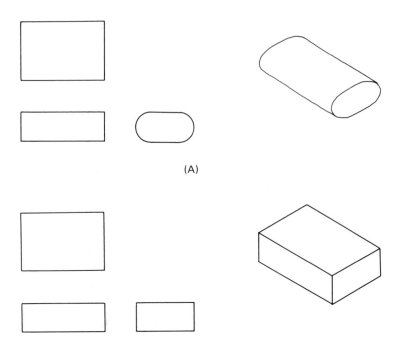

(A)

(B)

Fig. 3-9. The top and front orthographic views of the objects in (A) and (B) are the same. The third view (right side) is required to completely describe the shape of the object.

lines—look for dashed lines (hidden lines). Note the numbers of holes, depth of holes, thicknesses, etc.

Let your eyes scan from view to view. Piece together a mental picture. If necessary, make an isometric sketch using the knowledge that you learned from Sec. 3-9. Start the sketch with a corner or with an area of the workpiece that you recognize. Once the sketch is done, look hard at the sketch to see if the lines of the sketch are shown in the views. Isometric sketching helps you to learn orthographic projection rapidly, gives you practice in sketching, and trains you to visualize three dimensionally.

To aid the technician in learning to read orthographic projections, he may want to label common points (and hence lines) on different views of the drawing and on any isometric sketch he may make. This is illustrated in Fig. 3-10, which should be studied in detail at this time. Which points show up in all three orthographic views? In two views? In only one view? Could you sketch the left side view?

It is important for the technician to learn to study and read lines and

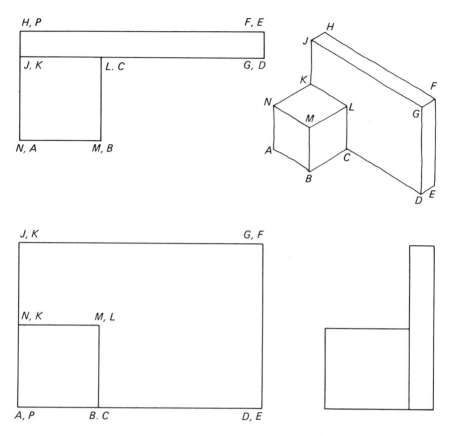

NOTE: The first letter given is the point actually "seen"; the second letter is the point on the other end of the line which is coincident with the first point.

Fig. 3-10. Compare the points and lines shown on the orthographic views with those shown on the isometric view.

surfaces on orthographic projections. For example, some views of curved surfaces appear as flat surfaces (Fig. 3-11). Surfaces inclined to the planes of projection appear as flat surfaces in some views. On inclined surfaces, such as surface *BCDE*, note that the *true* dimensioned *surface* does not appear on the top, front, or side views. To show the true surface shape in proper dimension, an auxiliary view is required (described in Sec. 3-15).

Note that any time two surfaces (different planes) intersect, there is a line common to both surfaces (Fig. 3-12 and Figs. 3-7 through 3-11) shown. Since there are no planes intersecting along the surface of *CBED* in Fig. 3-11, no lines show in the top view except for the intersections of the curved surfaces

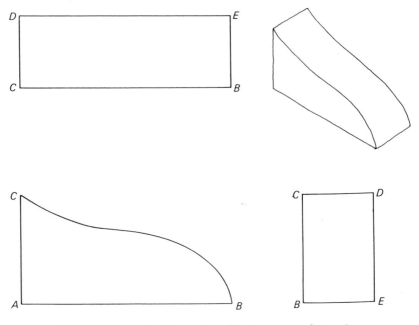

Fig. 3-11. Curved surfaces of an object appear as flat surfaces in some views.

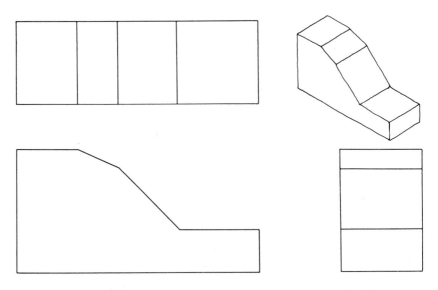

Fig. 3-12. When two planes (surfaces) intersect, a line is shown.

with the left side and bottom surfaces. Thus, a line indicates the edge of a surface, an intersection of two surfaces, or a surface limit.

Skew (oblique—not perpendicular to the axes) surfaces shown as an area in more than one view appear in a similar shape in the other views. Figure 3-13 illustrates such an area.

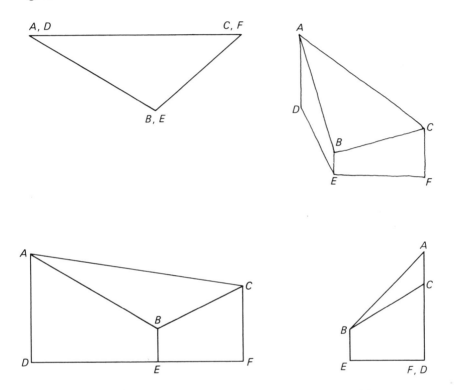

Fig. 3-13. Skew (oblique) surfaces appear as an area with a similar shape in more than one view. Here, skew surface *ABC* appears as a triangular shape in each view.

3-11. Drawing and Sketching Orthographic Projections

The most difficult parts of this section on projections and in fact this chapter, have been covered. At this point, the technician has learned the fundamentals of sketching, the types of engineering drawings, orthographic and pictorial projections, sketching isometric projections of an object, and reading orthographic projections. We can now use this knowledge, but reverse some of the procedures to learn the information required to conclude this section on projections.

To draw accurate orthographic projections, several drawing tools are required. The drawing tools include a small drawing board (approximately 18×24 inches), a T-square, a 30–60–90° triangle, a mechanical engineer's scale, drawing paper, 2H or 3H pencils or lead, and an eraser. A mechanical engineer's scale is usually one foot long and is divided and numbered so that fractions of an inch represent inches. The most common scales are $\frac{1}{8}$, $\frac{1}{4}$, $\frac{1}{2}$, and 1 inch to the inch.

To use the scale, select the scale that allows the views to most fully fit the drawing paper. Use the 1 scale for 1-to-1 drawings; use the $\frac{1}{2}$ scale for a reduction of $\frac{1}{2}$ times—thus, $\frac{1}{2}$ inch on the drawing equals 1 inch on the workpiece. When a $\frac{1}{8}$ scale is used, $\frac{1}{8}$ of an inch on the drawing represents 1 inch on the workpiece, and so on for the other scales. If a scaled-down drawing is made, be sure to indicate the scale on the drawing.

There are two decisions to be made in drawing or sketching orthographic projections. First, what is the primary viewing direction? It is the direction in which the observer is viewing the front of the workpiece and is usually considered to be the shape which is narrow and long. Thus, in Fig. 3-14, the front view is shown by the arrow. The second decision to be made is how many views are necessary? Enough views, complete with dimensions, tolerances, and notes added, must be provided to completely describe the workpiece. In Figure 3-14, three views are required. Finally, the drawing or sketch

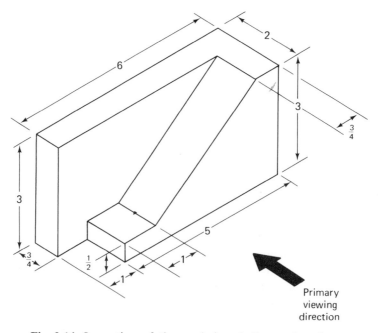

Primary
viewing
direction

Fig. 3-14. Inspection of the workpiece indicates that three views are necessary to describe it: front, top, and left side.

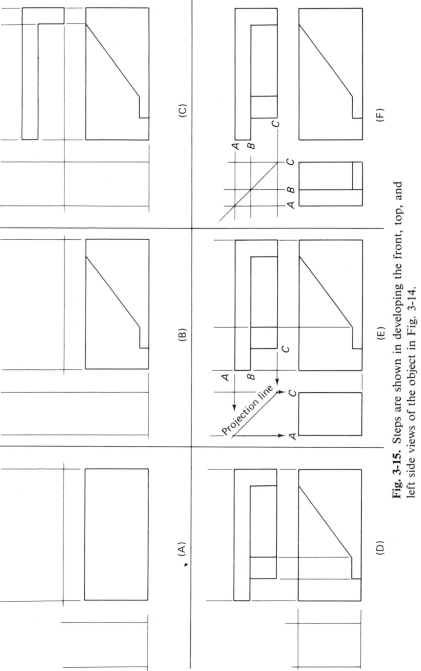

Fig. 3-15. Steps are shown in developing the front, top, and left side views of the object in Fig. 3-14.

must be made. The technician should make several "accurate" orthographic drawings so that he learns the basics in drawing—the actual making of finished drawings is the draftsman's job.

Referring to Fig. 3-14, the primary viewing direction has been chosen (looking into the front view). Inspection of the workpiece indicates that in addition to a front view in orthographic projection, the best selection of views to completely describe the workpiece is a top and a left-side view.

Using the drawing tools previously described, draw light guide lines to roughly outline the three views of the workpiece (Fig. 3-14) onto the drawing paper [Fig. 3-15(A)]. In one of the views that you feel confident that you can understand and draw, darken the lines. Areas or lines that you are uncertain about may be more clearly developed in another view and then transferred to the view of uncertainty. If known at this time, draw in the other lines [Fig. 3-15(B)].

Now, imagine you are looking from another direction, as the top. Project dimensions from the front view toward the top view, measure as required, and draw the outline lines for the top view [Fig. 3-15(C)]. Continue to complete the top view [Fig. 3-15(D)]. As you have probably gathered by now, the orthographic views are the views seen looking at one surface—thus, only two planes (dimensions) are seen. If you look at the front view, you see width and height, but not depth. In the top (plan) view, you see width and depth, but not height. Finally, in the end view, you see height and depth, but not width.

By projecting light guide lines from the front and top views and by drawing a 45° projection transfer line as shown in Fig. 3-15(E), the end view outline may be drawn. Finally, by projection of lines and/or measuring, complete the end view and hence the complete set of orthographic views necessary to describe the workpiece [Fig. 3-15(F)]. Now, and this is extremely important, visually and mentally check each view against the workpiece (or isometric view) to ensure that each orthographic view is accurate. A very effective check is to use the created orthographic views, and without referring to the workpiece, draw an isometric view of the workpiece. Any error in the isometric reflects an error in one (or more) of the orthographic views.

3-12. Reading Isometric Projections

Figure 3-16 illustrates a workpiece that, for example, was requested by the field technician as a replacable part for a machine. The technician forwarded the isometric sketch to the factory, complete with dimensions properly shown. To produce the workpiece, the factory technician felt that he should first make an orthographic drawing of the part to ensure that all dimensions shown were true.

In Fig. 3-16, it is noted that the field technician did the following steps correctly in making the isometric sketch: the isometric axes are properly laid out 120° apart; the "front" of the workpiece was chosen as being the narrow and long shape; and the dimensions were correctly shown by dimension lines along the isometric axes.

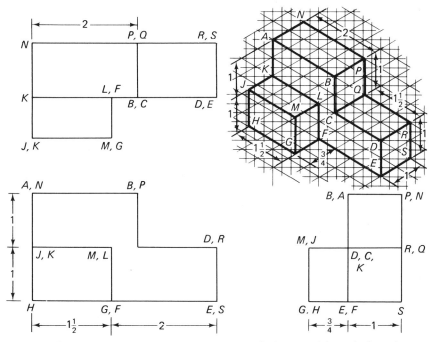

Note: the first letter given is the point actually "seen"; the second letter is the point on the other end of the line which is coincident with the first point.

Fig. 3-16. Given an isometric view, read and draw the orthographic views.

NOTE

To aid in discussing the reading of the isometric in Fig. 3-16, points have been labeled on the isometric view with corresponding points shown on the orthographic views.

The procedures for the drawing of the orthographic views are identical to those described for Figs. 3-14 and 3-15 and thus are not expanded here. Study the points (Fig. 3-16) on the orthographic views in relationship to those on the isometric views. Are all the areas shown? Are all the lines shown? Are enough dimensions shown on the isometric view? On the orthographic views?

Figure 3-16 illustrated a simple isometric view that was very easy to translate into orthographic views. But supposing that the workpiece had curved surfaces and circles shown in the isometric view, as shown in Fig. 3-17. To accurately draw the orthographic views from the isometric views, you would need to know the distances of selected points on the curved surfaces from at least two of the three axes X, Y, and Z; those distances would also have to be measured along lines parallel to the three axes because, as previously

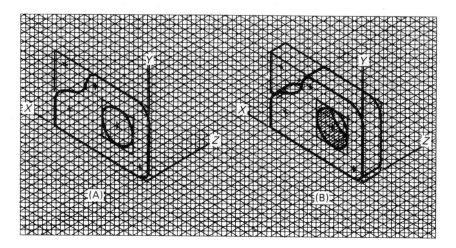

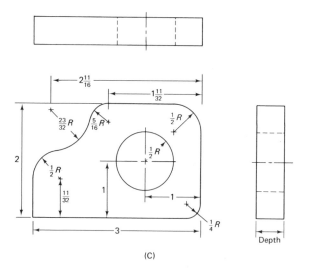

Fig. 3-17. A more complex isometric view. Note that it would be impossible for a field technician to accurately describe the part by a sketch. In this case, the field technician would have to make isometric views correctly dimensioned.

described, isometric dimensions are only true along lines parallel to the three reference axes *X*, *Y*, and *Z*. By plotting several points, the curved surface can then be drawn.

3-13. Sketching Isometric Projections from Orthographic Views

To sketch an isometric drawing from orthographic views, first study the orthographic views and try to picture the workpiece in your mind [Fig. 3-18(A)]. Pick the one of the three orthographic views that seems easiest to you to interpret or is most important to display. Let's assume that in Fig. 3-18, you understand the front view the best, or decide it's the most important to present. Using isometric paper for sketching, sketch the front view [Fig. 3-18(B)]. Then change your "visual picture" of the workpiece to the end and top views and draw additional lines [Fig. 3-18(C)]. Draw in the rest of the lines [Fig. 3-18(D)]. Finally, "view" the completed isometric drawing from the front, top, and end, and compare your "visual" orthographic view with the original orthographic view. Make any required changes in the drawings.

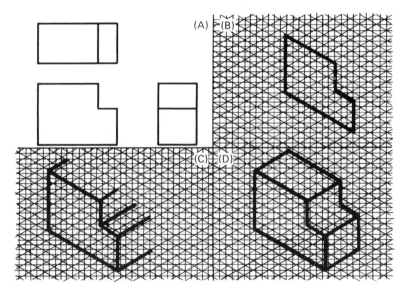

Fig. 3-18. Given the orthographic views of a workpiece, an isometric sketch is made so that the workpiece is "visualized" by the technician.

In sketching more complex isometric projections, especially of curved areas [Fig. 3-17(A)], you may draw one of the views with light lines in isometric. You can then "box" in the rest of the drawing with light lines so that you know the outermost dimensions. By going back and forth with your eyes between the orthographic and the isometric, you can fill in the box. Add lines or erase lines as required until the sketch is complete. Then darken the required lines.

Angles are not true in isometrics because the sides of the angles are not parallel to the isometric axes. Therefore, the vertices of angles and the sides of angles must be plotted from points on the orthographic views and must be transferred to the isometric, always measuring along lines which are parallel to one of the isometric axes. Center lines must be located in the same manner. Thus, circles take the shape of ellipses in isometric work. Curves are not true. This is shown in Fig. 3-17(B) and (C).

3-14. Sectional Views

The orthographic views (top, bottom, front, rear, right side, left side) discussed thus far are used to describe the exterior shape of a workpiece. Dashed lines are often used to show hidden internal areas. But quite often there is a need to know more about the interior of a workpiece. To describe this additional information about the workpiece interior, a *sectional* view is used.

A *sectional* view exposes the interior features of the workpiece as if the portion of the object between the observer and the previously hidden area had been cut away; it appears as if a cutting plane has sliced off the front of the workpiece (for example, as the end of an unsharpened wooden pencil appears).

There are three types of sectional views: cross section, longitudinal section, and an oblique section. Each type of sectional view is an orthographic projection used to show shapes and materials with more detail and definition than is possible in the front, top, and side views. A cross section (half-full section) is a section shown by a plane cutting at right angles to the longitudinal axis of the workpiece. A longitudinal section (full section) is a section shown by a plane passing through the longitudinal axis of the workpiece. An oblique section is a section showing a part of a workpiece that is cut by a plane at an oblique angle (the section is at an angle to all three planes).

Figure 3-19 illustrates sectional views of different sections of the workpiece. Cutting planes are defined by a line with arrows and letters; thus, AA and BB define the sections and the arrows indicate the direction in which the observer is looking. The sectional view may be drawn adjacent to the complete view or at a distance away and is always denoted by section letters.

Often times, only a partial sectional view is shown (Section BB of Fig. 3-19 shows a half section). This technique is quite often used on drawings of symmetrical workpieces so that the exterior and interior views may both be shown in one drawing. A sectional view is normally drawn to the same scale as the external views.

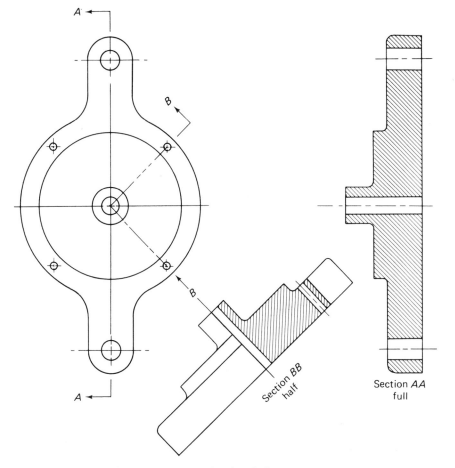

Fig. 3-19. Sectional views.

In addition to showing the shape and construction of a workpiece, sectional views are also used to distinguish the individual components of an assembly. The metal symbol, which is a set of parallel lines drawn at an angle of 45° and spaced $\frac{1}{16}$ inch apart, is used to identify the metallic portions of the sectioned view. Where two metals are used, such as in an assembly, the lines of the second piece of metal are drawn at 45° in the opposite direction to the first metal piece sectioning lines. Thin materials such as sheet metals and gaskets that are too thin for section lines are shown in solid black. Where two or more thicknesses of thin sections are shown, a very narrow space is left between them. Exact material descriptions are given in notes, specifications, and parts lists.

3-15. Auxiliary Views

There are often inclined surfaces on workpieces which do not appear in true shape and dimension in any of the six orthographic views because the inclined lines and surfaces are not parallel to any plane of projection. When a surface is inclined to two of the projection planes but is perpendicular to the third, it can be projected onto an auxiliary plane placed parallel to it. The auxiliary plane is then revolved into the plane of the paper. Thus, the auxiliary view is a projection of an inclined surface of a workpiece which shows the true size and shape of the inclined surface.

The auxiliary view is drawn from a reference line (Fig. 3-20) that is parallel to the inclined surface. All true dimensions on the auxiliary view are referenced to the reference line *RL*. Thus, the width of the auxiliary view of surface *BCHG* is the width of the edge *BC* of the front view. The depth of surface *BCHG* is the true depth shown in the end view by *BG* and *CH*. The dimensions must be measured from the reference line. Note that the circle is not in the center of *BCHG*; its center is measured along a line which is perpendicular to the reference line *RL*.

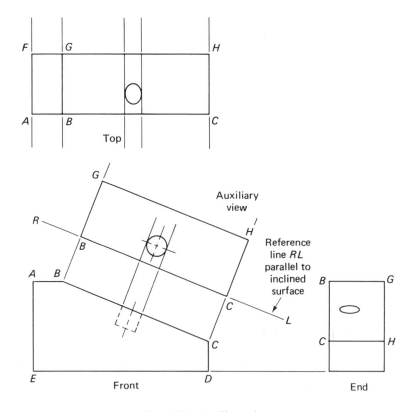

Fig. 3-20. Auxiliary view.

3-16. First-Angle Projection System

With increasing trade and manufacture becoming more prevalent between the different countries of the world, it is necessary to discuss the first-angle projection system. Many countries use this system. First-angle views are directly opposite the system of third-angle views used in the United States and some other countries. In third-angle projections, the views of the sides are shown as if hinged about the top surface edge. First-angle projection views, however, are formed as if the pivot were located at the bottom surface edge.

Figure 3-21 illustrates a comparison between third and first angle projections. The basic object is a solid cylinder with a flat down along one side. Located at 90° to this flat and halfway down the cylinder is a keyway.

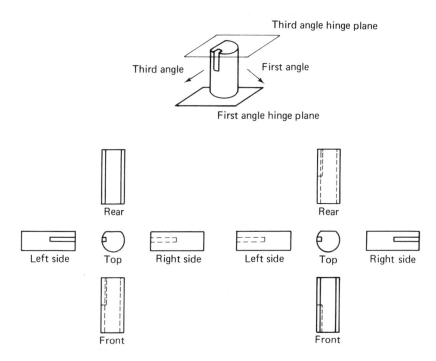

Fig. 3-21. Comparison of third angle and first angle projections.

This concludes the discussion on projections, but a few points must still be made. First, there is no set manner in the "order" of reading or drawing isometric or orthographic projections. The order is up to the individual because all individuals think differently and compare what they are viewing to other similar views they have seen. Second, the technician should begin to "see" orthographic views in isometric; likewise he should "see" the two-dimensional orthographic views of the three-dimensional objects with which he comes into contact daily. The technician should practice sketching and visualizing isometric and orthographic views throughout the rest of his training.

DRAWING FORMATS

Engineering drawings are drawn on special sizes (Table 3-1) of formatted paper. The format consists of a *title* block, a *revision* block, and *zones*, although many drafting sheets do not use zoned layout.

3-17. Title Blocks

The information contained in the title block generally includes: name and address of company making the unit, assembly, or part; drawing title; drawing number; drawing size (letter designation—refer to Table 3-1); sheet number; number of items required for each assembly; drawing scale; next assembly number; calculated weight; material (that the part is to be made from); general tolerance note; mechanical properties; a drafting room record that includes the names and drawing approval dates of the draftsman, checker, engineer, project manager, etc.; and, in the case of a drawing prepared under a military contract, special government information such as contract number, agency, and security classification. Of all the information in the title block, the most important blocks to the mechanical technician are the drawing scale, material, mechanical properties, general tolerance blocks and number of pieces to be made.

TABLE 3-1. SIZES OF DRAWINGS

Size Designation	Width (*inches*)	Length (*inches*)
A	8-1/2	11
B	11	17
C	17	22
D	22	34
E	34	44
F	28	40
G	11	42 to 144 (roll size)
H	28	48 to 144 (roll size)
J	34	48 to 144 (roll size)
K	40	48 to 144 (roll size)

The drawing scale block shows the scale to which the drawing was prepared. Four methods of indicating scales are shown in Table 3-2. In the event the drawing is not drawn to any scale, the word "NONE" or the letters "NTS" (not to scale) appear in the scale block.

TABLE 3-2. METHODS OF INDICATING SCALE, ENGLISH SYSTEM

Spelled Out	Decimal	Equivalent	Fractional
Full size	1.00 = 1.00	1 = 1	1/1
Half size	0.50 = 1.00	1/2 = 1	1/2
Quarter size	0.25 = 1.00	1/4 = 1	1/4
Twice size	2.00 = 1.00	2 = 1	2/1
Twenty times size	20.00 = 1.00	20 = 1	20/1

NOTE

Some drawings have scales such as 1" = 10', etc. It is important to always inspect the scale block of the drawing to properly interpret the scale.

Scale drawings in the metric system are full size (1 to 1), 1 to 2, 1 to $2\frac{1}{2}$, 1 to 5, 1 to 10, 1 to 20, 1 to 50, and 1 to 100. The unit of measure is the millimeter and dimension numbers on the drawings are understood to be in millimeters.

The material block provides a material description or refers to a note on the drawing that provides material information. The mechanical properties block shows the physical or mechanical properties of the material. The general tolerance block provides manufacturing tolerances of the part. These are the tolerances or deviations acceptable when measuring decimals, fractions, and angles. This block may also provide a dimension note such as "unless otherwise specified, all dimensions are in inches."

3-18. Revision Block

The revision block is used to record revisions made to the drawing. For example, if it is necessary to alter the dimension of a part, the change is noted on the field of the drawing by a supervisor and is then sent to the engineering department for review. It is assigned a revision letter (A, B, C, etc. for succeeding changes); in the description block, the draftsman notes the change, as "changed length from 5.125 ± 0.002 to 5.225 ± 0.002." After approval, a "revised" blueprint is reissued to the shops. Date blocks and approval blocks are available for engineering approval of the revision. If a zone block is provided, it indicates the zone (Sec. 3-19) of the field of the drawing where the change was made. If, for example, this was revision A, the revision is also noted in the field of the drawing by placing an A in a circle

Ⓐ next to the area revised (as in the example, next to the dimension 5.225 ± 0.002″).

3-19. Zones

Zones are used to locate a certain item within a "block area" in the field of a drawing (the same as zones on a road map). The zones are useful if a report is written about the drawing and you want to help the reader find the area you are describing. Alphabetical letters are along the horizontal axis and arabic numerals (1, 2, 3, etc.) are on the vertical axis.

3-20. Engineering Drawing Paper—Originals and Prints

Engineering drawings are drawn on a translucent or transparent material which becomes known as the "original" or "master" drawing. Translucent and transparent materials are chosen because they readily allow for any type of reproduction of the original drawing. Reproduction methods include sepia (brown line), ozalid (blue line), photostat, Xerox*, camera, or microfilm.

The original drawing may be on tracing paper, pencil cloth, tracing cloth, or a plastic material. The tracing paper is a thin translucent material, called vellum, and may be drawn on with either pencil or ink. Pencil cloth is a transparentized fabric which "takes" pencil. Tracing cloth is a fine threaded fabric that is sized and transparentized with a starch. It has a smooth side, which is the working side; however draftsman use the dull side which takes pencil as well as ink. The plastic material, called Mylar†, is a translucent material with a matte (working side) finish; it is for both pencil and ink.

The draftsman draws the original drawing; prints (copies) are then made and are transmitted to persons requiring the print. The original is placed in a file. Thus, the technician receives a print of the original which may be slightly smaller (due to shrinkage) or may be a reduced or enlarged print of the original. Thus, the technician must *never* measure dimensions on the print and use them as dimensions for his finished product. He *must* obtain the correct dimension of the part he is making from the dimensions provided along the dimension lines of the drawings and shown in one or more views.

If one of the dimensions is missing on a print, the technician must notify the draftsman or engineer and receive the correct dimension from him. In this case, the draftsman would issue a revision to his drawing to document the new (previously unknown) dimension.

A word of caution: the engineering department normally doesn't give

---

*Xerox is the registered trademark of the Xerox Corporation.
†Mylar is the registered trademark of the DuPont Corporation.

out originals, but, if the technician does use the original drawing for making his workpiece, he must take care not to damage or soil it. In addition, he must not make *any* changes to the original. If the technician feels there is an error, he should have a copy made of the original. He should then mark up the copy and return both the copy and the original to the engineer for evaluation. The technician often retains a copy of the workpiece drawing in his file because he may have put notes on it that have helped him in constructing the work-pieces. Whenever a "repeat" workpiece is to be made, the technician should always check that his file print is the same revision letter and date as the latest revised engineering drawing.

LINE CHARACTERISTICS, DIMENSIONS, AND TOLERANCES

Line shapes have special meanings on engineering drawings and sketches; heavy, medium, thin, dashes, long and short dashes—all have special meaning. Dimension lines and dimension formats with tolerances are also shown on drawings and sketches. This section provides the description and use of these items to further aid the technician in his evaluation of a drawing or sketch.

3-21. Line Characteristics

Thick solid lines are used to outline the visible shape of parts shown on engineering drawings and sketches. Thick solid lines outline the parts in plan, top, side, rear, bottom, auxiliary, and other views in orthographic, isometric, and other projections. But, in addition to the lines to show the visible shape of the parts, there are many other types of lines shown where applicable on a drawing or sketch. These other types of lines include: *hidden, center, dimension, extension, phantom, break, cutting plane* or *viewing plane* lines and *section* lines.

Line characteristics are determined by both weight (also called thick-ness) and by shape; shape may be solid, dashed, broken, etc. Line char-acteristics and uses of the various types of lines found on engineering drawings and sketches are described and shown in Fig. 3-22. The technician should be familiar with line characteristics so that he is able to readily read engineering drawings and sketches.

Dashed lines, sometimes called hidden lines, are used on drawings to show the outline of features within the workpiece or on the back face of a particular view. Dashed lines are used to show slots, countersunk and counter-bored holes, etc. or to show any feature that is hidden from view but is impor-tant in understanding the drawing.

NAME	CONVENTION	DESCRIPTION AND APPLICATION	EXAMPLE
VISIBLE LINES		HEAVY UNBROKEN LINES USED TO INDICATE VISIBLE EDGES OF AN OBJECT	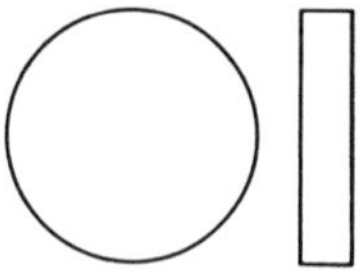
HIDDEN LINES		MEDIUM LINES WITH SHORT EVENLY SPACED DASHES USED TO INDICATE CONCEALED EDGES	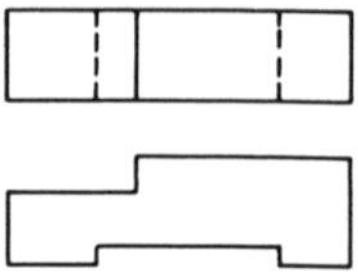
CENTER LINES		THIN LINES MADE UP OF LONG AND SHORT DASHES ALTERNATELY SPACED AND CONSISTENT IN LENGTH USED TO INDICATE SYMMETRY ABOUT AN AXIS AND LOCATION OF CENTERS	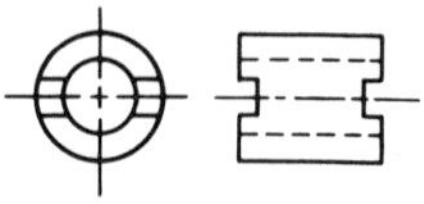
DIMENSION LINES		THIN LINES TERMINATED WITH ARROWHEADS AT EACH END USED TO INDICATE DISTANCE MEASURED	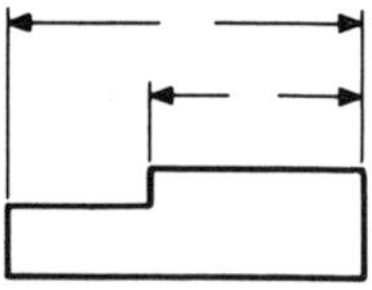
EXTENSION LINES		THIN UNBROKEN LINES USED TO INDICATE EXTENT OF DIMENSIONS	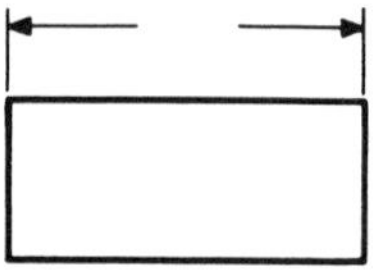

Fig. 3-22. Line characteristics.

3-22. Dimensions

A *dimension* is a numerical value expressed in appropriate units (inches, centimeters, degrees, etc.) of measure and indicated on an engineering drawing or sketch along with other lines, symbols, and notes to define the geometrical shape and dimensions of an object. Some other terms that should also be mentioned are *basic dimension, reference dimension, nominal* and *basic sizes, allowance,* and *tolerance.*

A *basic dimension* is a theoretical value used to describe the exact size,

NAME	CONVENTION	DESCRIPTION AND APPLICATION	EXAMPLE
LEADER		THIN LINE TERMINATED WITH ARROW-HEAD OR DOT AT ONE END USED TO INDICATE A PART, DIMENSION OR OTHER REFERENCE	
PHANTOM		MEDIUM SERIES OF ONE LONG DASH AND TWO SHORT DASHES EVENLY SPACED ENDING WITH LONG DASH USED TO INDICATE: ALTERNATE POSITION OF PARTS, REPEATED DETAIL OR TO INDICATE A DATUM PLANE	
BREAK (LONG)	(WOOD)	THIN SOLID RULED LINE WITH FREE-HAND ZIG-ZAGS USED TO REDUCE SIZE OF DRAWING REQUIRED TO DELINEATE OBJECT AND REDUCE DETAIL	
BREAK (SHORT)		THICK SOLID FREE-HAND LINES USED TO INDICATE A SHORT BREAK	
CUTTING OR VIEWING PLANE VIEWING PLANE OPTIONAL		THICK SOLID LINES WITH ARROWHEAD TO INDICATE DIRECTION IN WHICH SECTION OR PLANE IS VIEWED OR TAKEN	
CUTTING PLANE FOR COMPLEX OR OFFSET VIEWS		THICK SHORT DASHES USED TO SHOW OFFSET WITH ARROW-HEADS TO SHOW DIRECTION VIEWED	

shape, or location of a feature (hole, face, edge, etc.). A *reference dimension* is a dimension used only for information (reference) purposes. *Nominal size* is the designation used for the purpose of general identification. The *basic size* is the size from which the maximum and minimum sizes are determined by the application of allowances and tolerances. For example, a tube may be called out as a $\frac{1}{2}$ inch diameter with a $\pm$ 0.010 inch diameter variation. Thus, an *actual* size is the measured size; the maximum and minimum sizes become the *limits.*

An *allowance* is an intentional measured difference between the maximum material limits of a finished part or parts. A *tolerance* is the total difference that a specific dimension may vary between its limits.

Dimensions are used on drawings to indicate heights, widths, depths, radii, angles, and diameters of parts. Dimensions (and tolerances) to describe a part must be shown on at least one view. The common practice in the United States and the United Kingdom is to use inches and common fractions, but where more accuracy is required, decimal fractions are used. Use of decimal fractions in dimensioning allows for tolerances to thousandths of an inch (0.001 inch), or to ten-thousandths of an inch (0.0001). When dimensions and tolerances are specified and then met in manufacturing, parts mate properly and identical parts are interchangeable with one another.

Dimensions of parts on drawings up to 72 inches are normally expressed in inches (as $58\frac{3}{4}''$); over 72 inches, feet and inches are used (as 8'-3''). Fractional parts of inches are expressed in common fractions (as $\frac{3}{4}''$) or decimal fractions (as 0.50'').

Dimensions are located outside of the main outline of a particular view of the drawing of the part. Figure 3-23 illustrates the location of dimensions to identify the sizes of surfaces (lines), circles, angles, and radii.

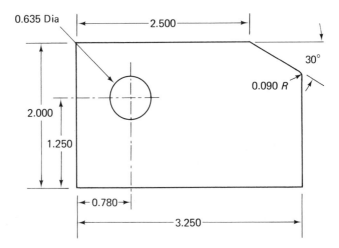

Fig. 3-23. Location of dimensions.

3-23. Tolerances

As previously defined, a *tolerance* is the total difference that a specific dimension may vary between its limits. Thus, when a dimension is given as 3.175 ± 0.002 inches, the tolerance is 0.004 inch (the difference from −0.002 to +0.002 inch is 0.004 inch).

All detail drawings of parts indicate dimensions and tolerances. The tolerance may be general or specific. A general tolerance is a tolerance that is applicable to all dimensions on the drawing; the general tolerance is used whenever possible and is written on the drawing in the title block or in a note. Specific tolerances are shown as required and are given with the dimension. If no tolerance is given in the title block, in a note, or with the dimension, then the tolerance is accepted to be $\pm\frac{1}{64}$ inch and $\pm\frac{1}{2}°$ in the fractional notation; and $\pm$ the nearest significant figure for decimal notation. Thus, in decimal, the tolerance would be ± 0.01 for a two-place decimal and ± 0.001 for a three-place decimal. Some companies, however, have a ± 0.005 tolerance for three-place decimals.

Example

With no tolerance specified on the drawing, what are the upper and lower limits for the dimension:
(a) 2.54 inches?   (b) 12.765 inches?   (c) 3-1/32 inches?
    (a) 2.54 $\pm$ 0.01 = 2.53 to 2.55 inches
    (b) 12.765 $\pm$ 0.001 = 12.764 to 12.766 inches
    (c) 3-1/32 $\pm$ 1/64 = 3-1/64 to 3-3/64 inches

Dimensions with tolerances are presented on drawings on dimension lines between extension lines as: $\boxed{\dfrac{3.175}{3.177}}$. When the dimension is an internal dimension (a dimension on the interior of the workpiece, such as a hole diameter) the minimum size is shown above the line; the maximum size is shown below the line. In the example then, the dimension is an internal dimension with a maximum dimension of 3.177″ and a minimum dimension of 3.175″. The tolerance is 0.002 inch (3.177″ minus 3.175″). If the dimension is an external dimension (an outside dimension, such as the outside diameter of a pin), the maximum size is shown above the line. This method of dimension marking is the preferred method and is the method usually used in mass production.

Another method of dimensioning with tolerances is: $\boxed{4.250^{+.002}_{-.001}}$.

Here the total range is 4.249–4.252 inches. If one limit is shown, the other is assumed to be zero: $\boxed{\dfrac{1.375}{+.002}}$. In this example, the tolerance range is from 1.375 to 1.377 inches. This is called a *unilateral* tolerance because variation of dimension is permitted only in one direction from the specified dimension. Other methods of indicating decimal unilateral tolerances are shown by example as: $1.150 + 0.005$; $3.125^{+0.005}_{-0.000}$. Fractional unilateral tolerances are shown by examples as: $1\frac{1}{2} - \frac{1}{32}$; $\frac{3}{4}{}^{+0}_{-1/32}$. Angular unilateral tolerances are shown by examples as: $35°18'45'' + 0°45'0''$; $35°18'45''{}^{+0°45'0''}_{-0°\ 0'0''}$.

A *bilateral* tolerance is a tolerance in which variation in both directions from the specified dimension is permitted. Thus, a bilateral tolerance in decimal is expressed by examples as: $3.500^{+0.002}_{-0.002}$; 3.500 ± 0.002. Fractional bilateral tolerances are shown as: $4\frac{1}{2}{}^{+1/32}_{-1/64}$; $5\frac{1}{4} \pm \frac{1}{64}$. Angular bilateral tolerances are shown as: $45°0'{}^{+0°15'}_{-0°10'}$; $55°0' \pm 0°15'$.

3-24. True Position Tolerancing

Fast becoming a standard in many organizations is *true position* tolerancing. The true position tolerancing setup produces a zone within which a hole, slot, etc. may exist and still make the part acceptable. In coordinate dimensions, the tolerance of a point location varies at 90° to the point. True position, however, denotes a theoretical position of a feature. This theoretical position is in reality a circular zone or area in which the feature is located. In the case of slots, they may vary on either side of the true position line, while a hole can vary within the limits of a circular zone. When a feature is allowed to vary in any direction within a zone, it should be noted in the following way: 4 holes located at true position within 0.008 dia. or 4 holes located within 0.004 *R* of true position. Slots may be noted in the following form: 4 slots located at true position within a 0.015 wide zone.

TABLE 3-3. SOME COMMON DRAWING SYMBOLS

Symbol	Interpretation	Examples
T.I.R.	Total Indicator Reading	T.I.R. ± .002''
	Diameter	2.500 ± .005''
	Roundness	within 0.002''
	Concentricity	within 0.002'' T.I.R.
X	Times or places	Drill 1/4'' dia. holes thru 10X
R	Radius	12' R.
	Perpendicular	Side to base within 1/64''
	Straightness	Plate within 1/16'' all over
	Parallelism	Plate within 1/8'' all over
	Square	4'' bar-2' long
	Center line	
	Surface finish rating	
	Satin finish	
	Polished finish	
	Flatness	
	Cylindricity	
	Angularity	
	Runout	
	True position	
	Symmetry	
'	Minute or foot	
''	Second or inch	
# or No.	Number	

SYMBOLS, ABBREVIATIONS, AND NOTES

Standard symbols, standard abbreviations, and notes are used on engineering drawings to lessen the draftsman's work, to reduce repetition, to save space, and to aid in preventing clutter. Table 3-3 shows the most frequently used symbols placed on drawings of parts to be shop made. Figure 3-24 shows the representations of threads. Table 3-4 lists the most frequently used abbreviations. Notes are used to supplement the drawing in a condensed and systematic manner. The notes may be located in the immediate area of the point of application or may be grouped and assigned consecutive numbers.

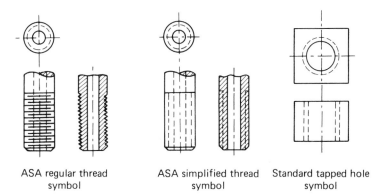

ASA regular thread   ASA simplified thread   Standard tapped hole
symbol   symbol   symbol

Fig. 3-24. Thread representations.

TABLE 3-4. ABBREVIATIONS USED ON DRAWINGS

AWG	American wire gauge
BRS CAST	Brass casting
BRZ CAST	Bronze casting
C	Course (screw threads)
C to C	Center-to-center
CBORE	Counterbore
CHFR	Chamfer
CI	Cast iron
CP	Circular pitch
CRS	Cold rolled steel
CSK	Countersink
CYL	Cylinder
DIA	Diameter
DIAG	Diagonal
DR	Drill
F	Fine (screw threads)
FAO	Finish all over

TABLE 3-4. (CONT'D.)

FI or FIL	Fillet
FIN or FNSH	Finish
FORG	Forge
GA	Gage
GI	Galvanized iron
GR	Grind
HDN	Harden
HT TR	Heat treat
ID	Inside diameter
LAT	Lateral
LH	Left hand
LONG	Longitudinal
LUB	Lubricate
M	Meter
MACH	Machine
MAX	Maximum
MC or MAL CAST	Malleable casting
MI or MAL I	Malleable iron
MIN	Minimum
MM	Millimeter
MS	Machine steel
N	National form (screw threads)
NC	National course (screw threads)
NF	National fine (screw threads)
OD	Outside diameter
P	Pitch
PD	Pitch diameter
PRESS ST or PRSD STL	Pressed steel
R or RAD	Radius
RD	Round
RH	Right hand
RM	Ream
RPM	Revolutions per minute
RPS	Revolutions per second
SC or STL CAST	Steel casting
SEC	Section
SQ FT	Square foot
SQ IN	Square inch
ST STAMP or STL STP	Steel stamping
STD	Standard
STL	Steel
STL C	Steel casting
T	Teeth
THD or THR	Thread
TS	Tool steel

TABLE 3-4. (CONT'D.)

U	Unified (screw threads)
UNC	Unified national fine
WI	Wrought iron
YDS	Yards

For example, the 60° ✓ finishing mark touching against a line indicates that the surface described by that line is to be machined and hence allowance must be made for finishing. A letter designation such as *S* or *P* designates the finish as satin or polished, respectively. The surface quality (roughness) is indicated by a numeral in microinches, e.g. ³²✓.

3-25. Surface Finish Notation

On all engineering drawings, a surface finish symbol is found. This symbol (Fig. 3-25) is used to describe pertinent facts about the final finish of the workpiece. This text only discusses the importance of the value known as "roughness height." Quite often, the number for the roughness height will determine the machining procedure to be used. As an example, suppose that a particular hole is to be machined into a plate with a roughness height equal to 63 (⁶³✓).

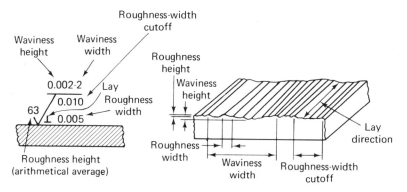

Fig. 3-25. American standard symbols indicating surface finish characteristics.

There are three ways to produce this hole: drill, drill and ream, or bore. Refer to Table 3-5. Under the 63 microinch average, we find that drilling produces a ¹²⁵✓ finish, while reaming or boring produces a finish in the range we require. The real key to remember in judging surface finish is that the range runs from 1000 to 0 microinches. The larger the number, the greater the roughness; the smaller the number, the smoother. Generally, a surface finish of 125 microinches is most often used. A 63 microinch finish is used where smoothness and appearance are needed. A 32 microinch finish is used for such things as finished grooves for rubber "0" rings, bearings, or any use where extreme smoothness of finish is required for a specific function.

TABLE 3-5. SURFACE ROUGHNESS

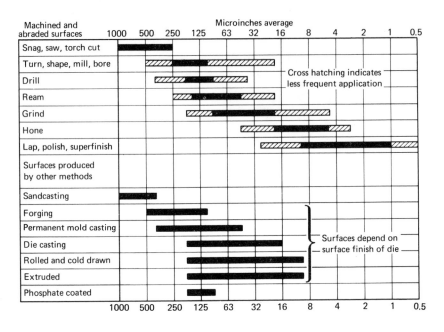

Machined and abraded surfaces	1000	500	250	125	63	32	16	8	4	2	1	0.5
Snag, saw, torch cut												
Turn, shape, mill, bore												
Drill							Cross hatching indicates less frequent application					
Ream												
Grind												
Hone												
Lap, polish, superfinish												
Surfaces produced by other methods												
Sandcasting												
Forging												
Permanent mold casting												
Die casting							Surfaces depend on surface finish of die					
Rolled and cold drawn												
Extruded												
Phosphate coated												

Microinches average

1000 500 250 125 63 32 16 8 4 2 1 0.5

SUMMARY

Figure 3-26 illustrates a complete parts drawing in English units with dimensions and tolerances correctly added. Review this drawing.

A·100 088

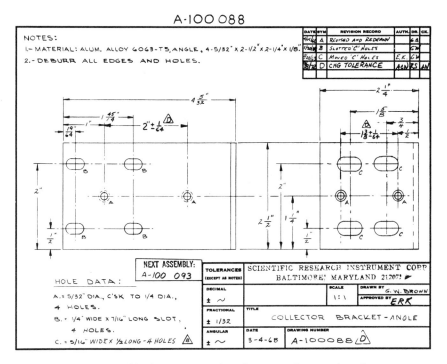

Fig. 3-26. This is an example of a complete engineering drawing—shown in English units. (Courtesy of Scientific Research Instrument Corp.)

REVIEW QUESTIONS

1. Figure 3-27 shows two of three orthographic views. Sketch the two views given and then sketch the third view.

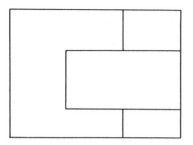 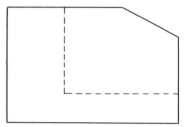

Fig. 3-27. Sketch the two views given and add (sketch) the third.

2. Figure 3-28 shows three orthographic views of each workpiece. Sketch isometric views of the workpieces.

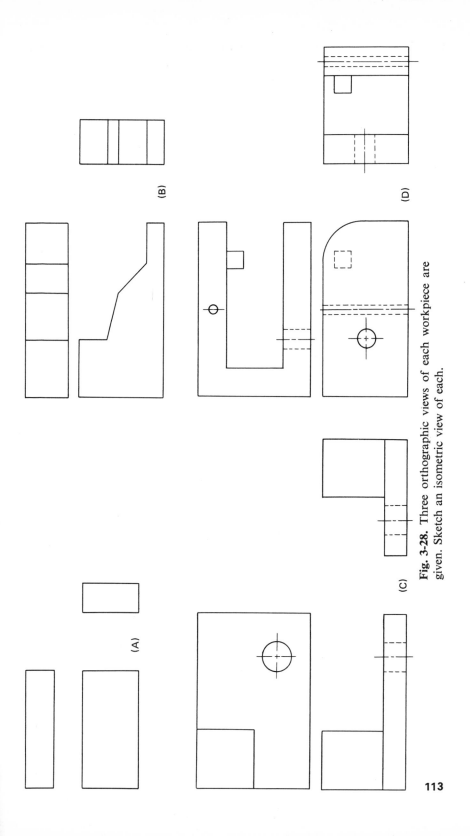

Fig. 3-28. Three orthographic views of each workpiece are given. Sketch an isometric view of each.

(A)

(B)

(C)

(D)

113

3. Figure 3-29 shows isometric views of workpieces. Draw the orthographic views.

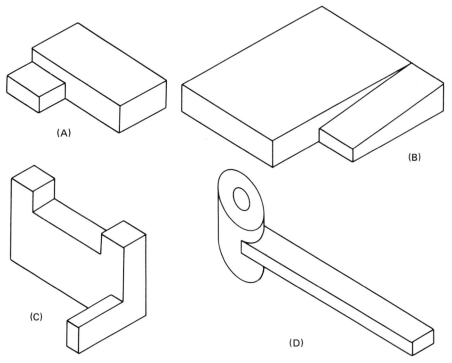

Fig. 3-29. An isometric view of each workpiece is given. Draw the orthographic views.

4. Make a sectional view of the surfaces indicated by the arrows in Fig.
3-30.

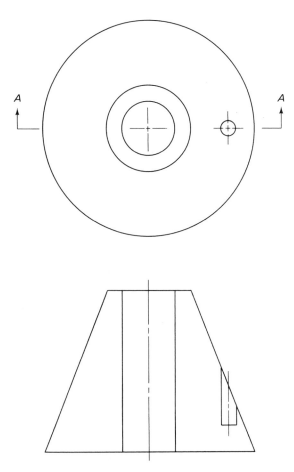

Fig. 3-30. Draw a section view of the cone.

5. Make auxiliary views of the inclined surfaces shown in Fig. 3-31.

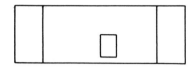

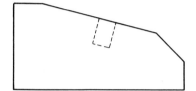

Fig. 3-31. Draw auxiliary views of the inclined surfaces.

6. List the types of information described by an engineering drawing.

4
GENERAL HAND TOOLS

Tools are the instruments used by the mechanical technician to perform a job. Hand tools are those tools which are powered only by a man's hand, whereas power tools (Chapter 9) use electrical, pneumatic, or hydraulic power to aid man in performing the task.

In this discussion of the application, use, and care of general hand tools, the hand tools have been categorized into tool groups according to their general functions: *torquing* (screwdrivers, wrenches); *striking* (hammers, punches); and *holding/slicing/shearing* (pliers, locking plier wrenches, chisels, files, file card/brush, emery cloth, hacksaws, nibbler, tin snips).

The hand tools included in this chapter are the tools that are used most often by the mechanical technician. This chapter does not cover every tool nor every variation of a tool because of the large number of various designs by the many manufacturers. Through study of this chapter, the technician will gain an understanding of the basic hand tools specifically required in his job. He should then consult various tool catalogs and visit hardware stores to become further acquainted with the variety of designs. In regard to the materials of the hand tools discussed herein, only quality characteristics are discussed.

This chapter also describes general hand tool care and procedures for hardening and tempering tools that have lost these characteristics because of overheating during grinding. Finally, the chapter describes a recommended set of hand tools for the beginning mechanical technician.

Each technician generally owns his own set of hand tools. These are purchased on his own, but his employer is often able to order the tools for him at a discounted price. Tools are chosen by need and by personal preference of size, weight, material, accuracy required from the tool, and price. It is more economical, over a time period, to purchase quality tools. If available in sets, it is also more economical to buy tools in sets rather than one tool at a time.

HAND TOOLS THAT APPLY TORQUE

General hand tools that are used by the mechanical technician to provide torque include screwdrivers, screw extractors, and wrenches. Screwdrivers and wrenches are normally sold in sets of different sizes and types. Replacements for broken, worn, or lost tools may be purchased as required.

4-1. Screwdrivers

Screwdrivers are tools for driving or withdrawing screws or slot-headed bolts. The screwdriver blade head is inserted into the head slot and is rotated to drive (usually clockwise) or withdraw (usually counterclockwise) the screw or bolt. Screwdrivers are named by the type of tip as either conventional (for slotted screws), Phillips, Reed and Prince, or offset (Fig. 4-1).

Screwdriver blades are made of tempered alloy steel and are available with chrome plating to retard rusting. The tip end is cross ground and the sides are hollow ground. The blade is anchored deep in a shockproof plastic or wood handle. Wide flutes on the handle provide a firm gripping surface.

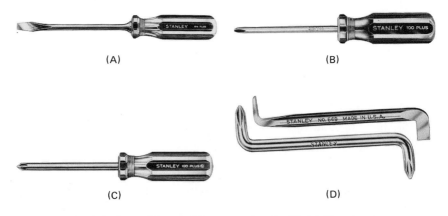

(A)   (B)

(C)   (D)

Fig. 4-1. Screwdrivers: (A) conventional, (B) Phillips head, (C) Reed and Prince, and (D) offset. (Courtesy of Stanley Tools)

Conventional Screwdriver

The conventional screwdriver [Fig. 4-1(A)] has a blade from 4 to 12 inches long. The blade may be round or square; a wrench may be used on the square blade to enable the technician to apply more torque to the screw or bolt. Five conventional screwdrivers are sold in a set with blade tip widths, blade lengths, and handles proportioned for the most often used screw and bolt sizes.

Phillips Head Screwdriver

A Phillips head screwdriver [Fig. 4-1(B)] is distinguished by the tip, which is designed to fit into Phillips head screws or bolts. The tip has about 30° flukes and a blunt end. The screwdriver tips are made in four sizes to match Phillips head screws as follows: size 1 for no. 4 and smaller screws; size 2 for no. 5–9; size 3 for no. 10–16; and size 4 for no. 18 and larger. Blade lengths vary in length from 1 to 18 inches.

Reed and Prince Screwdriver

The Reed and Prince screwdriver [Fig. 4-1(C)] is similar to the Phillips screwdriver except that the tip is defined as a cross-point. The tip has 45° flukes and a sharp-pointed end. The outline of the end of the Reed and Prince screwdriver is close to a right angle. This type screwdriver is used on Frearson screws. Blade lengths vary from 3 to 12 inches.

Offset Screwdriver

The offset screwdriver [Fig. 4-1(D)] is used to drive or withdraw screws or bolts that cannot be aligned with the axis of the conventional or Phillips head screwdrivers because of physical space limitations. The offset screwdriver is also used where high torque is required to tighten or loosen a screw or bolt. Offset screwdrivers are made in a variety of sizes with different size conventional and Phillips heads. On models with four tips, two tips are offset at 90°, and two tips at 45°. This offset allows for quarter or eighth revolution torquing.

4-2. Use of Screwdrivers

Use the longest blade screwdriver that is convenient and has the proper size tip for the job. The width of the tip should equal the length of the screw or bolt slot, and the tip thickness should equal the width of the screw or bolt slot (Fig. 4-2). The screwdriver is gripped firmly by the handle; the blade is steadied and guided in the screw or bolt slot with the other hand.

It is often difficult to determine the difference between the Phillips head and Reed and Prince screwdrivers. Review and remember the tip descriptions previously given. Also, the Phillips screw has beveled walls between the slots, whereas the Reed and Prince has straight, pointed walls. The Phillips screw

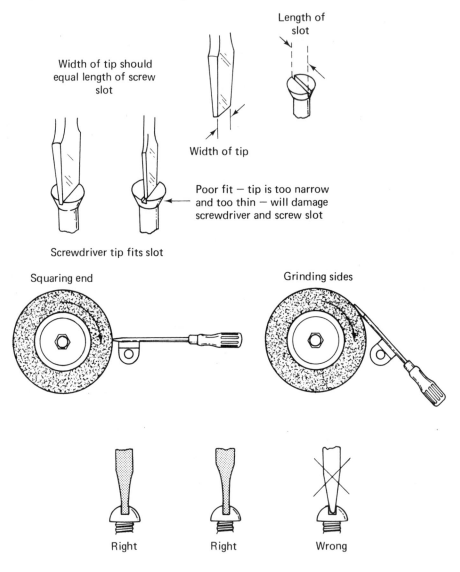

Width of tip should equal length of screw slot

Length of slot

Width of tip

Screwdriver tip fits slot

Poor fit — tip is too narrow and too thin — will damage screwdriver and screw slot

Squaring end

Grinding sides

Right   Right   Wrong

Fig. 4-2. Grinding conventional screwdriver blade tip.

slots are not as deep as the Reed and Prince. You may check that you are using the proper screwdriver as follows: hold the screw to be inserted in a vertical position with the head up. Place the screwdriver tip in the head; if the screwdriver tends to stand up unassisted, it is probably the proper screwdriver.

Never use a screwdriver for prying or chiseling. Never use pliers on the blade of the screwdriver; if additional torque is required, use a square blade screwdriver turning it with the proper size wrench.

4-3. Care of Screwdrivers

If the tip of a conventional screwdriver becomes worn or nicked so that the tip no longer fits the screw or bolt slot, the tip must be reground. When using a grinder, adjust the tool rest (refer to Chap. 9) to hold the screwdriver against the wheel to produce the desired shape. Square the tip (Fig. 4-2); then grind both sides of the tip until the tip is the required thickness. During grinding, frequently dip the screwdriver into water to prevent overheating, which can cause a loss in the steel temper. If a blue color appears at the tip during grinding, the metal has lost its temper and grinding beyond this area is required for maximum strength.

If the tip of a Phillips head or a Reed and Prince screwdriver becomes burred, place the blade in a vise and carefully remove the burr with an oil stone. If the tip becomes sufficiently worn so that it does not properly fit the screw or bolt head, then the screwdriver must be replaced.

4-4. Screw Extractor

A screw extractor (Fig. 4-3) is used to quickly remove broken screws, bolts, studs, pipe fittings, etc. without damaging the material or the threaded hole. Screw extractors are left-hand spiral shaped and are available in sets. Extractor sizes 1 to 6 are used for machine shop and garage uses.

Fig. 4-3. Screw extractor (and accompanying drill). (Courtesy of J.H. Williams & Co.)

Use of Screw Extractor

To remove a broken screw, bolt, stud, pipe fitting, or other fastener from a material, first remove any visible chips with a small brush or a sharp-pointed tool. Select an extractor that is slightly smaller than the small diameter of the broken item being removed. Then drill the recommended size hole, which is larger than the smaller dimension of the extractor, in the center of the broken fastener to a depth such that the extractor will not touch the bottom. Insert the extractor into the hole and turn it counterclockwise with a tap wrench. This forces the extractor spirals into the side of the hole, causing the screw to be turned and withdrawn. Table 4-1 lists the screw extractor size (number) and drill size for the removal of the indicated fastener.

TABLE 4-1. SCREW EXTRACTOR AND DRILL SIZES

For Screws & Bolts Sizes (*inches*)	For Pipe Sizes (*inches*)	Extractor No.	Size Drill to Use (*inches*)
$\frac{3}{16}$ to $\frac{1}{4}$		1	$\frac{5}{64}$
$\frac{1}{4}$ to $\frac{5}{16}$		2	$\frac{7}{64}$
$\frac{5}{16}$ to $\frac{7}{16}$		3	$\frac{5}{32}$
$\frac{7}{16}$ to $\frac{9}{16}$		4	$\frac{1}{4}$
$\frac{9}{16}$ to $\frac{3}{4}$	$\frac{1}{8}$, $\frac{1}{4}$	5	$\frac{17}{64}$
$\frac{3}{4}$ to 1	$\frac{3}{8}$	6	$\frac{13}{32}$

Care of Screw Extractor

Keep the screw extractors clean at all times. After use, coat the extractors with a light film of oil and place in the storage case or store separately so that the extractors do not contact each other or other metal, as this may cause the gripping edges on the flutes to become dull.

4-5. Wrenches

Wrenches are tools that are used to tighten or loosen bolts, nuts, headed screws, and pipe plugs. Special wrenches are made to grip round stock such as pipe, studs, and rods. Special *torque* wrenches indicate how tight the bolt, nut, headed screw or pipe plug is tightened. Spanner wrenches are used to turn cover plates, rings, and couplings.

There are many types of wrenches. This chapter describes the wrenches normally used by the mechanical technician: open end; box; combination; adjustable open end; pipe; strap; socket; T-handle socket; offset L-handle socket; striking socket; torque; spanner; set screw; and nut drivers.

Wrenches are made of hardened and tempered steel. Many are chrome plated to prevent rusting. On fixed nominal opening wrenches (such as open end, box, combination, and sockets), the size of the opening is stamped near the opening; sizes are available in English units and metric units. The openings are from 0.005 to 0.015 inch larger than the size marked on the wrench so that the wrench fits easily over the nut or bolt.

Open End Wrench

The open end wrench [Fig. 4-4(A)] is usually a double ended wrench with different size openings at each end. Wrenches are available with typical openings from $\frac{1}{4}$ to $1\frac{3}{4}$ inches and 6 to 28 mm. Midget wrenches have openings ranging from $\frac{13}{64}$ to $\frac{3}{8}$ inch. The common open end wrench has ends that are angled at 15° to permit complete rotation of 30° hex nuts. There are other open end wrenches having ends at 45, 60, 75, or 90°. The length of the wrench is proportioned to the size of the opening; the greater the length, the greater

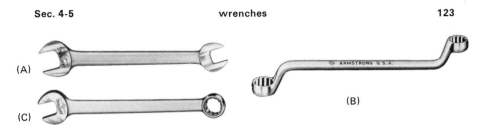

Fig. 4-4. Wrenches: (A) open end, (B) box end, and (C) combination (open end and box end). (Courtesy of Armstrong Brothers Tool Co.)

the leverage. Other types of open end wrenches are single open end and S-shape handled.

It is important to be sure that the open end wrench fits the nut or bolt so that the wrench does not slip and round off corners. Offset open end wrenches make it possible to turn a bolt or nut that is recessed or is in a limited space where there is little room to turn the wrench. The wrench should be turned over, often after each partial turn, to aid in turning the nut or bolt in limited areas. If possible, always pull the wrench toward you to move the nut or bolt in the desired direction. If you must push on the wrench, push with your palm (knuckles out of the way) so that if the wrench slips, your knuckles won't be skinned. To avoid slipping on the final tightening of the nut or bolt, technicians should make the final turns with a box wrench.

Box Wrench

Box wrenches [Fig. 4-4(B)] completely "box" the nut or bolt to be tightened or loosened and hence is a safer tool than the open end wrench. The box opening contains 6 or 12 points; the 12-point end allows the wrench to be used for $\frac{1}{12}$ turns in close quarters. The box wrench is usually a double ended wrench with different size openings in each end. The openings range from $\frac{1}{4}$ to $2\frac{3}{4}$ inches and 8 to 32 mm. Midgets are available from $\frac{3}{16}$ to $\frac{11}{32}$ inch. The height of the box is correctly proportioned for all nut sizes and series; handle lengths vary from $4\frac{1}{2}$ to 18 inches. Longer lengths provide greater leverage. The handle may be straight, but it is often offset 15° from the box opening to allow for hand clearance. Sometimes the box end is offset.

Always select the proper size box wrench for the nut or bolt to be turned. The box wrench does not slip off, and hence is preferred over the open end wrench. A swing through an arc of 15° (a 12-point wrench) is sufficient to completely tighten or remove a nut or bolt. Remove the wrench completely from the nut or bolt and then replace it when repositioning the wrench for another partial turn. This method is slower than with an open end wrench, but the box wrench enables you to make tighter fittings or to remove already tight fittings. Thus it is desirable to rapidly spin down a nut or bolt with an open end wrench and then set it tight with a box wrench. For this reason, many technicians prefer the combination wrench.

Combination Wrench

The combination wrench [Fig. 4-4(C)] has an open end at one end of the handle and a 12-point box end at the other; nominal openings at both ends are the same. The open end is often angled at 15° from the handle and the box end is often offset 15° from the plane of the handle to give hand clearance. Combination sets are made in various combinations of sizes of openings, lengths, offsets, and angles as discussed for the open end and box wrenches. The combination wrench is often the choice of technicians because it combines the open end for quick spin down of nuts or bolts and the box end for final tightening.

Adjustable Open End Wrench

Adjustable open end wrenches [Fig. 4-5(A)] are used to tighten an infinite number of English and metric size nuts, bolts, headed screws, and pipe plugs. This wrench is normally used when the proper size open end, box, or combination wrench is not available. The wrench has one fixed jaw and one movable jaw and is shaped such that the jaws form four sides for firm gripping of hex nuts or bolts. The wrench head is thin and angled from the handle for use in tight areas.

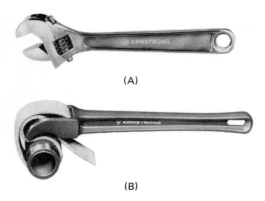

(A)

(B)

Fig. 4-5. Wrenches: (A) adjustable and (B) strap. (Courtesy of Armstrong Brothers Tool Co.)

The movable jaw is adjusted by rotating a knurled nut. Some models include a locking device in the form of a push rod to hold the adjustment. Typical wrench handles are 4 to 16 inches long with jaw capacities of approximately $\frac{1}{2}$ to $1\frac{3}{4}$ inches. Longer handles provide the best leverage.

Always place the adjustable open end wrench on the nut or bolt to be tightened or loosened so that the force used is applied to the fixed jaw. After placing the wrench into position, tighten the knurled adjustment nut until the jaws tightly fit the nut or bolt. If the jaws do not fit tightly, the wrench will slip and the corners of the nut or bolt will be rounded; the knuckles may also be skinned.

Strap Wrench

The strap wrench [Fig. 4-5(B)] is used for rotating a round object such as chrome pipes without allowing any damage to the surface. The wrench handle (approximately 12 inches) is made of cast iron. The strap, generally made of canvas, has a capacity of $\frac{1}{8}$ to 2 inches and can be replaced when worn.

Pipe Wrench

A pipe wrench (Fig. 4-6) is used to rotate a round workpiece such as a pipe or the head of a worn nut or bolt that has had its corners rounded; thus, the pipe wrench is not to be used on good nuts and bolts. The pipe wrench has two hardened jaws (with milled teeth) that are not parallel. The outer jaw, which is adjustable by means of a knurled nut, is called the hook jaw. This jaw is made with a small amount of play which provides a tight grip on a pipe when the wrench is turned in the direction of the movable jaw. The jaws always leave marks on the workpiece. The heal jaw, which is the fixed jaw, is replaceable. It is also spring loaded to give it full floating action to increase the wrench grip. Pipe wrenches used by mechanical technicians are available in lengths from 6 to 14 inches with capacities of $\frac{3}{4}$ to 2 inches.

Fig. 4-6. Pipe wrench. (Courtesy of Snap-On Tools of Canada, Ltd.)

The pipe wrench will work in only one direction; the wrench is turned in the direction of the opening of the jaws. Force is applied to the back of the handle. Since the top jaw has slight angular movement, the grip on the workpiece is increased by the pressure on the handle.

Socket Wrench

A socket wrench consists of a drive handle and one of a various number of English or metric sized sockets. The socket wrench is used to tighten or loosen nuts, bolts, and headed screws. The socket wrench is made from parts within a socket wrench set.

When a technician requests a $\frac{3}{4}$ inch socket wrench, he is normally requesting a $\frac{3}{4}$ inch detachable socket plus a drive handle which is usually either a ratchet drive handle or a flex drive handle. Prior to his request, he has determined the size of and number of nuts or bolts to be tightened or loosened. He has also determined his working space.

A basic socket wrench set usually consists of various sized 12-point sockets, a ratchet drive, a flex handle, a slide bar handle, and an extension. The socket wrench drives are available in $\frac{1}{4}$, $\frac{3}{8}$, $\frac{1}{2}$, $\frac{3}{4}$, and 1 inch sizes and should be selected according to their required use. A $\frac{1}{4}$ inch drive is for light-duty work; $\frac{3}{8}$ and $\frac{1}{2}$ inch for automotive, industrial, and aircraft work; and $\frac{3}{4}$ and 1 inch drives are for heavy-duty use, including machinery and heavy equipment installations.

There are four basic types of socket wrench drives: ratchet, flex handle, speeder, and slide bar handle. These drives attach to a variety of sizes of sockets which may be 4- or 8-point sockets for square head nuts and bolts and 6- or 12-point sockets for hexagonal head nuts and bolts. Accessories such as socket adapters, universal joints, and extensions extend the capability of the socket wrenches.

The ratchet drive [Fig. 4-7(A)] is used with various sized English and metric sockets to tighten or loosen nuts and bolts. The ratchet drive saves time in performing its functions because the socket does not have to be removed from the nut or bolt. It is particularly useful in close quarters; a fine tooth action ratchet drive permits handle movement in increments of 4°.

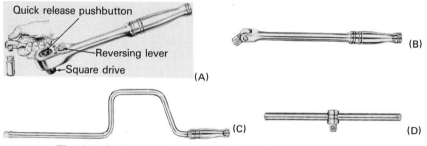

Fig. 4-7. Socket wrench drives: (A) ratchet, (B) flex handle, (C) speeder, and (D) slide bar handle. (Courtesy of Snap-On Tools of Canada, Ltd.)

Ratchet drives contain two and sometimes three "controls." The quick-release pushbutton releases spring pressure on the spring tensioned ball in the square drive, allowing the socket to slide easily off of the drive; a different socket readily slips on and locks over the spring tensioned ball. The reversing lever allows the ratchet to slip (ratchet) in one direction for tightening and the opposite direction for loosening. A knurled speeder (on the ratchet) allows nuts to be quickly spun onto a bolt. The speeder accessory may be added to a ratchet drive if it is not on the original model (not shown in Fig. 4-7).

The flex handle drive [Fig. 4-7(B)] is used with various sized English and metric sockets to tighten or loosen nuts and bolts. With the flex handle drive in a straight (perpendicular to the work) configuration, nuts or bolts are spun snugly. The flex handle is then turned 90° (parallel to the work). This provides additional leverage for final tightening of the fastener. In confined areas where the flex handle cannot be turned 90°, a sliding bar may be placed through a cross hole in the handle; this provides additional leverage for final tightening of the fastener.

The speeder drive [Fig. 4-7(C)] is used with various sized English and metric sockets as a rapid means of spinning (running) a number of nuts and bolts on or off. Final tightening or breaking of tight fasteners must be done with the ratchet or flex handle drives, which provide for additional leverage.

The slide bar handle, often called a T or L handle drive [Fig. 4-7(D)], is used with various sized English and metric sockets to tighten or loosen nuts and bolts. For additional leverage and for working in confined areas, the drive may be positioned anywhere along the handle; thus, a T or an L shape is formed.

Sockets [Fig. 4-8(A–C)] have two openings: a square hole that fits the socket wrench drive, and a circular hole with notched sides to fit the nut or bolt head to be turned. The square hole is $\frac{1}{4}$, $\frac{3}{8}$, $\frac{1}{2}$, $\frac{3}{4}$, or 1 inch across to mate with the respective socket wrench drive. The circular hole may be notched with 4 or 8 notches for square shaped nuts and bolt heads, or with 6 or 12 notches for hexagon shaped nuts and bolt heads. Socket sizes usually range from $\frac{1}{8}$ to $\frac{9}{16}$ inch for $\frac{1}{4}$ inch drives; $\frac{3}{8}$ to $\frac{3}{4}$ inch for $\frac{3}{8}$ inch drives; $\frac{3}{8}$ to $1\frac{1}{4}$ inches for $\frac{1}{2}$ inch drives; $\frac{3}{4}$ to $2\frac{1}{4}$ inches for $\frac{3}{4}$ inch drives; and $1\frac{1}{16}$ to $3\frac{1}{8}$ inches for 1 inch drives. Metric sizes are also available.

Regular sockets are shown in Fig. 4-8(A). Deep sockets, which are use-

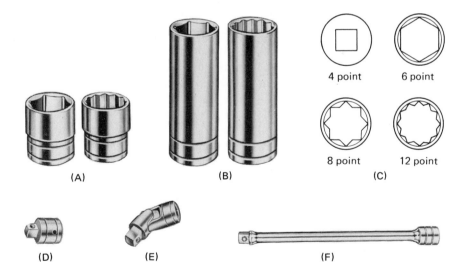

4 point   6 point

8 point   12 point

(A)   (B)   (C)

(D)   (E)   (F)

Fig. 4-8. Sockets and accessories: (A), (B), and (C) sockets, (D) socket adapter, (E) universal joint, and (F) extension. (Courtesy of Snap-On Tools of Canada, Ltd.)

ful for spinning nuts onto long bolts because they allow a length of the bolt into the socket, are shown in Fig. 4-8(B). Socket adapters [Fig. 4-8(D)] are attached to a socket wrench drive when it is desired to change from one square drive size to another square drive size. Socket adapters are available to change socket wrench drives of $\frac{3}{8}$ to $\frac{1}{4}$; $\frac{1}{4}$ to $\frac{3}{8}$; $\frac{3}{8}$ to $\frac{1}{2}$; $\frac{1}{2}$ to $\frac{3}{8}$; $\frac{1}{2}$ to $\frac{3}{4}$, $\frac{3}{4}$ to $\frac{1}{2}$, 1 to $\frac{3}{4}$, and $\frac{3}{4}$ to 1 inch.

A universal joint [Fig. 4-8(C)] makes it possible to turn nuts or bolts where a straight wrench could not be used because of lack of clearance. One end of the joint is attached to a socket wrench drive; the other end attaches to the square hole end of a socket. The joint swings up to 90°. Spring tension holds the joint at any desired angle. Universal joints are also available having a $\frac{1}{2}$ inch square drive with 12-point sockets of $\frac{1}{2}$, $\frac{9}{16}$, $\frac{5}{8}$, $\frac{11}{16}$, and $\frac{3}{4}$ inch. These fixed-size universal joint sockets are useful for applications where the same size universal joint is required for extended work.

Extensions [Fig. 4-8(F)] extend the length of socket wrench drives. The extension is connected between the drive and the socket.

To use a socket wrench, first select the shape (for square or hex nut or bolt) and the size of the socket that is required. Hand try the socket for proper fit. Next, inspect the clearance area to determine which type of drive is the most satisfactory for the job. If the ratchet drive is selected, snap the socket onto the square drive. Flip the reversing lever to the applicable position for either ratcheting in the clockwise or counterclockwise direction. Use the ratchet wrench speeder to quickly spin the nut or bolt finger tight. Finally, apply force to the handle and torque the nut taut. It may be necessary to swing the handle back and forth, but the socket does not have to be removed from the nut. In loosening, reverse the process.

If the flex handle is selected, snap the proper size socket onto the square drive. With the drive in a straight line with the handle, spin the handle with the fingers until the nut or bolt is snug. Turn the handle 90° (parallel to the workpiece) and apply pressure to the handle to tighten the fastener.

The T or L handle drive is used because of its sliding variable handle length. The speeder is used in the same manner as a brace. One hand holds the rear handle and applies a force toward the drive; the other hand rotates the speeder causing the nut or bolt to be tightened or loosened.

T Handle Socket Wrench

The T handle socket wrench [Fig. 4-9(A)] is a special purpose socket wrench. It is available with a square opening ($\frac{3}{16}$ to $1\frac{1}{2}$ inches) or hexagonal openings ($\frac{5}{16}$ to $2\frac{3}{8}$ inches). A removable pin handle that is perpendicular to the wrench enables additional leverage to be put on the nut or bolt to be tightened or removed. The top of the wrench has a hexagonal shape for attaching another wrench to give additional leverage, as required.

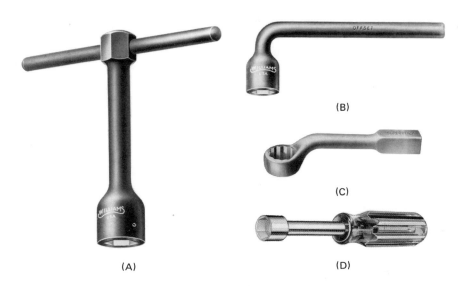

Fig. 4-9. Wrenches: (A) T handle socket, (B) offset L handle socket, (C) striking socket, and (D) nut driver. (Courtesy of J.H. Williams & Co.)

Offset L Handle Socket Wrench

The offset L handle socket wrench [Fig. 4-9(B)] is a special-purpose socket wrench of one piece construction. It is available with a square opening (typically, $\frac{1}{4}$ to $1\frac{1}{4}$ inches) or hexagonal openings (typically, $\frac{5}{16}$ to $2\frac{3}{8}$ inches). The wrench handle is offset 90° for applying leverage in tightening and loosening nuts and bolts.

Striking Socket Wrench

The striking socket wrench [Fig. 4-9(C)] is a special-purpose one-piece wrench used where large nuts and bolts must be set up tight or frozen nuts and bolts loosened. The wrench is available in a hexagonal shape angled at 15° from the handle and with typical nominal openings of $3\frac{3}{8}$ to $4\frac{5}{8}$ inches; the handle length is from 18 to 21 inches. The far end of the handle contains a flat striking surface which is hit with a hammer or sledge.

Nut Driver

The nut driver [Fig. 4-9(D)] is similar to the design of a screwdriver except that the tip is a six-point socket for hexagonal shaped nuts and bolts. The nut driver speeds up the turning of nuts. Portions of the shanks are often hollow to accept the end of a bolt as the nut is turned down onto the bolt. There are nut drivers to fit sizes from approximately $\frac{3}{16}$ to $\frac{5}{8}$ inch. The handles are shockproof butyrate plastic or wood and are fluted for a more positive grip; the handles are often color coded according to size. Nut drivers are available in midget pocket clip styles, stubby, regular, and extra long.

The nut driver is used a great deal by the electronic technician to rapidly spin nuts and hex headed sheet metal screws. It is used like a screwdriver. When the nut is tightened, care must be taken so that the nut driver does not round the corners of the nut and likewise, that the corners of the nut driver do not become stripped.

Torque Wrench

A torque wrench (Figs. 4-10 to 4-12) is used to tighten nuts, bolts, cap screws, etc. to specified pressures. The torque wrench is used on machines, automobile engines, and airplane engines in applications where one or more bolts or nuts are to be tightened to the same specified pressure to assure alignment, equal pressure for sealing, etc.

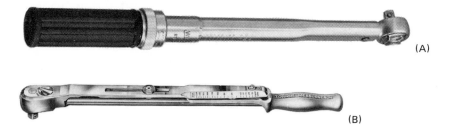

(A)

(B)

Fig. 4-10. Torque wrench: (A) Torque limiting type and (B) deflecting beam–converging scale type. (Courtesy of J.H. Williams & Co.)

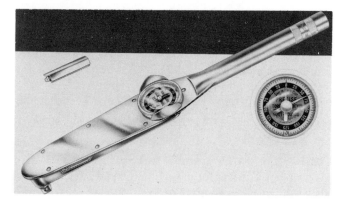

Fig. 4-11. Torque wrench: dial type. (Courtesy of Snap-On Tools of Canada, Ltd.)

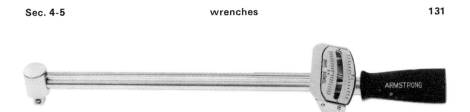

Fig. 4-12. Torque wrench: round beam deflecting type. (Courtesy of Armstrong Brothers Tool Co.)

Several types of torque wrenches are available: torque limiting type; deflecting beam-converging scale type; the dial type, and the round beam deflecting type [Fig. 4-10(A–D)]. Each is basically the same with the exception of the readout of the torque value. Each may also incorporate one or more of the following features: square drives of $\frac{1}{4}$, $\frac{3}{8}$, $\frac{1}{2}$, $\frac{3}{4}$, or 1 inch; quick-release pushbutton for releasing the various English and metric size sockets available (refer to the description of *sockets*); ratchet and nonratchet drives; reversing levers to change the ratchet direction for tightening or loosening; torque measurement in one or both directions; and a scale which is calibrated to a National Bureau of Standards standard. Torque wrenches are available in calibrated units of inch-ounces, inch-pounds, foot-pounds, centimeter-kilograms, and meter kilograms. Handle lengths vary from 9 to 40 inches.

The torque limiting type torque wrench [Fig. 4-10(A)] is set to the predetermined specified torque by rotating the micrometer style handle to the value specified, as indicated on the scale. A snap lock holds the setting. The nut or bolt to be tightened is then turned. When the set torque pressure is reached, you feel and hear an audible click in the handle. There is no scale to read, which makes this wrench ideal for use in the dark, in blind spots, and even under water or oil. The wrench automatically resets to the set value for the next operation.

On deflecting beam-converging scale torque wrenches [Fig. 4-10(B)], the torque pressure value is read out on the scale as the nut or bolt is torqued. If the simple sound device is set for a predetermined specified torque value, there is also an audible sound when the predetermined value is reached. The accuracy of this type of torque wrench is not dependent upon delicate gears, levers, or dials, but is dependent on rugged sections of high tensile strength. Only right-hand torque is measured; however, the action is reversible for left-hand turning.

Dial-type torque wrenches (Fig. 4-11) provide a "meter" readout for torque values in either direction. Round-beam deflecting type torque wrenches (Fig. 4-12) also read out the torque values in either direction, but the readout is on a moving metal scale. The pointer actually remains stationary; the handle (and hence scale) bend as the torquing force is applied.

This scale type is also available with an audible sound and a snapping feeling to the hand.

One other type of torque wrench that is available is the preset value torque wrench. This wrench is very useful on a production line where many nuts or bolts are to be torqued to the same value. It is also valuable in quality control verification of torque pressure value. The torque value is preset by means of an adjustment screw in the handle. A special plug then closes off the handle opening so that the preset value cannot be changed without damage to the plug. In operation, when the preset torque value is reached, an audible snap signal is produced.

Torque wrenches are operated the same as the socket wrench composed of a ratchet drive and socket. First, if a torque wrench requiring a setting is used, the specified torque value is set into the wrench. Second, the proper size socket is selected and snapped onto the torque wrench drive. The ratchet reversing lever, if available, is placed to the proper position for tightening. The socket is then placed fully over the head of the nut or bolt to be tightened. The wrench is operated back and forth, tightening the fastener until the proper torque pressure is indicated by either an audible click or by a visual indication on a gauge.

Spanner Wrench

Spanner wrenches (Fig. 4-13) are of two types: fixed and adjustable. Fixed spanner wrenches are used to turn flush and recessed retaining rings, cover plates, plugs, hose couplings, and parts which have holes or slots for insertion of the wrench pins or lugs. Adjustable spanner wrenches perform the same functions as the fixed spanner wrenches; the adjustable legs or hook may be adjusted to fit a variety of cover plates, plugs, couplings, and other parts. Types of fixed spanner wrenches are pin, face, and hook; there are also adjustable hook and face spanner wrenches.

Various sized fixed pin spanner wrenches [Fig. 4-13(A)] are used to adjust cylindrical nuts with diameters of 1 to 6 inches. The pin of the wrench fits into a drilled hole in the cylindrical nut. The pin face of the wrench is kept flush with the part surface when the nut is turned. Enough pressure must be exerted so that the pin does not come out of the hole.

The fixed face spanner wrench [Fig. 4-13(B)] has two forged pins that are inserted into two holes in nuts such as those used on surface and tool grinders. Different size wrenches are available with pin centers between 1 and 4 inches apart. The adjustable face spanner wrench [Fig. 4-13(E)] performs the same function but the legs open and close such that the pins can fit the various distances found in many types of cover plates and plugs. Adjustable face spanner wrenches are available with extreme capacities of 2, 3, and 4 inches.

The fixed hook spanner wrench [Fig. 4-13(C)] is used to service face plate draw nuts used on taper-nose spindles for lathes and general work. Hook

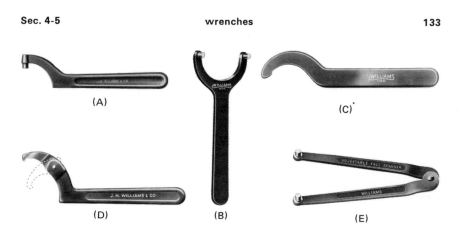

Fig. 4-13. Spanner wrench: (A) pin, (B) face, (C) hook, (D) adjustable hook, and (E) adjustable face. (Courtesy of J.H. Williams & Co.)

spanner wrenches are available for circle diameters of $4\frac{1}{2}$ to $12\frac{1}{2}$ inches. The overall length varies from 13 to 31 inches. The adjustable hook spanner wrench [Fig. 4-13(D)] is used for adjusting collars, lock nuts, rings, spindle bearings, etc. used in various types of machine tool equipment. The wrench is also for spindle bearings of lathes, milling machines, and grinding machines. Various models are available with typical capacities for circle diameters as follows: $\frac{3}{4}$ to 2 inches; $1\frac{1}{4}$ to 3 inches; 2 to $4\frac{3}{4}$ inches; and $4\frac{1}{2}$ to $6\frac{1}{4}$ inches. Lengths are $6\frac{1}{2}$ to 12 inches.

Key Setscrew Wrench

Key setscrew wrenches [Fig. 4-14(A)], also called Allen wrenches and hex keys, are L shaped, made of tool steel and have a hexagonal section. The key setscrew wrench is used to tighten or loosen socket setscrews, socket-head capscrews, button head capscrews, socket-head shoulder screws, flat head capscrews, and tru-round pressure plugs having hexagonal sockets. Dimensions across the flats are available from approximately 0.028 to 2 inches in

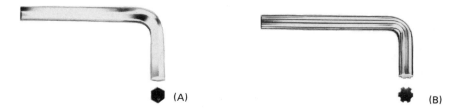

Fig. 4-14. (A) Key setscrew and (B) splined setscrew wrench. (Courtesy of Snap-On Tools of Canada, Ltd.)

English and metric units. A typical set includes key setscrew wrenches of: 0.028, 0.035, 0.050, $\frac{1}{16}$, $\frac{5}{64}$, $\frac{3}{32}$, $\frac{7}{64}$, $\frac{1}{8}$, $\frac{9}{64}$, $\frac{5}{32}$, $\frac{3}{16}$, $\frac{7}{32}$, $\frac{1}{4}$, and $\frac{5}{16}$ inch. In addition to L shape, key setscrew wrenches are available with loop handles, as a flexible driver, and in a holder similar to a pocket knife.

When using the key setscrew wrench, be sure that the proper size wrench is selected and is inserted as far as possible into the fastener to be tightened or loosened. The short end of the wrench is used to give a final tightening or to break a tight fastener loose. The long end is used to turn the screw rapidly.

Splined Setscrew Wrench

Splined setscrew wrenches [Fig. 4-14(B)] are made of round stock with ends to fit little flutes or splines in headless setscrews. When using the wrench, be sure the correct size is selected and is inserted completely into the setscrew. The short end of the wrench is inserted for final tightening or to break a tight setscrew loose.

4-6. Use of Wrenches

Be sure to use the proper size wrench for the nut, bolt, or other fastener to be loosened or tightened. Use of the wrong size wrench will cause the points of the nut or bolt to be rounded, making it very difficult to further tighten or to remove. Always use adjustable wrenches so that the pulling force is applied to the fixed jaw; never to the adjustable jaw. Do not use a pipe or other device to extend the length of the handle of a wrench; instead, obtain a wrench with a longer handle. Do not strike wrench handles with a hammer to tighten or loosen a bolt or nut.

If a rusted nut, bolt, or other type fastener is to be removed, apply penetrating oil or a mixture of kerosene and lubricating oil and allow time for the oil to penetrate before attempting to turn with a wrench. Pull on the wrench (instead of pushing), whenever possible, to prevent skinning the knuckles if the wrench slips. Do not exert a hard pull on a pipe wrench until it has a firm grip on the workpiece. Finally, always keep your wrenches clean and free of oil or grease.

4-7. Care of the Wrenches

The knurled nuts of adjustable open end, pipe and monkey wrenches should occasionally be lubricated with graphite. The healjaw of the pipe wrench may be replaced when damaged by cutting the rivet with a cold chisel, replacing the jaw, and riveting the healjaw into place.

To protect wrenches from becoming damaged, *always* use the proper size wrench. Ensure that adjustable open end, pipe, and monkey wrenches are adjusted for proper fit; this protects both the wrench and the workpiece.

HAND TOOLS THAT FUNCTION THROUGH STRIKING

General hand tools that are classified as striking tools are the hammer and the punch. This section describes the various types of each that are used by the mechanical technician.

4-8. Hammers

Hammers are used by the mechanical technician to drive drift pins and wedges, and to strike chisels, punches, and to shape metal. The types of hammers used by the technician are: ball-peen; rawhide; plastic tip; and brass.

Ball-Peen Hammers

Ball-peen hammers [Fig. 4-15(A)] are used by machine shop and auto mechanic personnel for general purpose work such as striking punches and chisels (but never for driving nails), for shaping metal (peening), and for riveting. The term peen means to indent or compress in order to expand or stretch a portion of metal next to the indention. Peen hammer heads have two ends: a face for the general purpose work and a ball end for peening. Head weights vary from 2 to 40 oz.; 20 oz. for general use. The wooden handle lengths are between 11 and 15 inches. The peen is uniformly beveled and the peen and face are hardened to reduce damage.

Fig. 4-15. Hammers: (A) ball-peen, (B) plastic tip, (C) soft face, and (D) brass. (Courtesy of Snap-On Tools of Canada, Ltd.)

Plastic Tip Hammer

The plastic tip hammer [Fig. 4-15(B)] is a soft faced hammer used where metal parts must be driven together. The soft face prevents marring. The two faces of the hammer are replaceable and are usually different. For example, one face may be plastic for use on iron or steel; the other face may be vinyl for use on aluminum, wood, and polished surfaces. The heads weigh between 2 and 32 oz.; the replaceable tips are resistant to common industrial solvents and acids. Handles are 7 to 13 inches long.

Soft Face Hammer

The soft faced hammer [Fig. 4-15(C)] is used where a light nonmarring blow is required to form or shape sheet metal. The replaceable tips prevent chipping, flaking, and mushrooming. Tip materials include rubber, wood, lead, rawhide, and copper.

Brass Hammer

The brass hammer [Fig. 4-15(D)] is used for adjusting and setting tools on screw machines, turret lathes, and all types of turning machines. It is also recommended when a tapping force is needed in fitting a bearing, as it does not chip or crack hardened bearing rings. Like the plastic tipped hammer, the heads of the brass hammer are replaceable. The heads vary in hardness from super soft to extra hard. The heads are $1\frac{1}{2}$ to 2 inches in diameter with 5- or 7-inch lengths; handles are 13 inches. The hammer head is loaded with shot to give about 30% more driving power. In addition to brass, heads of aluminum alloy and die-cast zinc alloy are also available.

Rawhide Hammer

The rawhide hammer is a soft faced hammer used where a light blow is required to form or shape sheet metal. The hammer head is a split head design which permits easy replacement of the water buffalo rawhide. Other soft materials available include molded rubber, plastic, copper, babbitt, and nylon. The face diameters range between 1 and $3\frac{1}{2}$ inches; handles are approximately 14 inches long.

4-9. Use of Hammers

The hammer is grasped with the hand at the far end of the handle. Light blows are struck almost entirely with wrist motion; heavy blows require the use of the wrist, forearm and shoulder.

Never use any of the technician's hammers to drive nails. Never use soft face hammers on sharp corners because the corners could damage the hammer.

4-10. Care of Hammers

The faces of hammers should be dressed (filed or ground) regularly to remove battered edges. Small burrs should be filed off; worn face surfaces should be ground. Dip the head in water often to prevent the loss of metal temper by overheating during grinding. On double-faced hammers, remove the same amount of material from both sides to maintain the hammer balance.

Split or broken hammer handles should be replaced. Cut off the handle at the head and drive out the remaining portion of the handle. Firmly insert

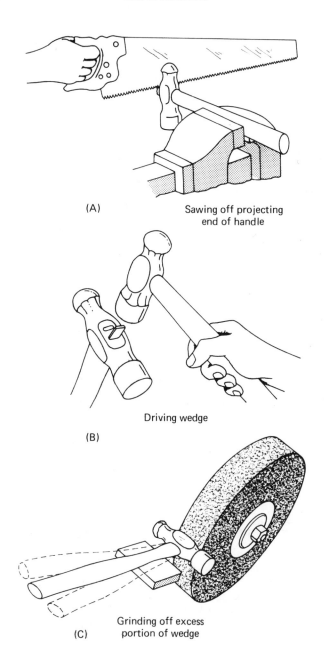

(A)   Sawing off projecting
                 end of handle

Driving wedge

(B)

Grinding off excess
(C)   portion of wedge

Fig. 4-16. Replacing hammer handle: (A) saw off end, (B) drive wedges, and (C) grind off excess wedge.

a new, proper sized handle and cut off excess wood extending from the head. Drive wedges into the handle to expand the wood against the head. Grind off excess wedge material. (See Fig. 4-16.)

If a handle is loose, place the hammer on a bench so that the head is against the bench and the handle is vertical. Tap the end of the handle. If necessary, drive in another wedge as previously described. Never use nails or screws in place of wedges.

4-11. Punches

Punches are used by the mechanical technician to punch indentions in metal and other materials for later drilling of holes; to mark metal for layout; to drive straight or tapered pins and rivets; and to align holes in different sections of metal. The types of punches most often used by the mechanical technician are the prick punch, the center punch, the aligning or drift punch, and the pin punch.

Prick Punch

The prick punch [Fig. 4-17(A)] is used as a starting tool in layout. It is used to punch small conical shaped indentions in the workpiece surface to permanently mark points of a line, the intersections of lines and circles or arcs, and the centers of circles. A typical prick punch is made of approximately $\frac{3}{8}$ inch hardened tool steel and is $4\frac{1}{2}$ to 6 inches long; it has a long tapered cone-shaped point terminating in a conical point with an included angle of 60°.

Fig. 4-17. Punches: (A) prick, (B) center, (C) aligning (or drift), and (D) pin. (Courtesy of J.H. Williams & Co.)

Center Punch

After layout is completed, (layout is discussed in Chap. 5) the center punch [Fig. 4-17(B)] is used to repunch the indentions made by the prick punch where holes are to be drilled in the workpiece. The center punch has a pointed shape with an included angle of 118 to 120°; this angle corresponds to the included angle of the most generally used drill bits. The formation of the 118° indentation in the workpiece surface allows for the direct drilling of a hole without the fear that the drill bit will "walk" over the surface causing an inaccurately drilled hole. A typical center punch is made of $\frac{1}{4}$ to $\frac{3}{4}$ inch hardened tool steel and is $4\frac{1}{2}$ to 7 inches long.

Aligning (or Drift) Punch

Aligning punches, also called drift punches, are used to line up holes in two pieces of metal work [Fig. 4-17(C)]. Aligning punches are made of stock from $\frac{3}{8}$ to 1 inch with tips ranging from $\frac{1}{8}$ to $\frac{3}{8}$ inch; lengths are from 9 to 15 inches.

Pin Punch

Pin punches [Fig. 4-17(D)] are used for driving out straight or tapered pins such as shear pins on a lathe or outboard motor. Pin punch point sizes range in diameter from $\frac{1}{16}$ to $\frac{1}{2}$ inch; stocks are from $\frac{1}{4}$ to $\frac{1}{2}$ inch and lengths are 4 to 9 inches. Pin punches may often be purchased in sets.

4-12. Use of Punches

Use a punch only for its intended use as previously described. In using the prick or center punches, hold the selected punch lightly in the fingers of one hand and slant the punch with the striking end (anvil) away from yourself. Hold the punch point to the layout point and bring the punch to a vertical position with respect to the work surface. Firmly strike the punch anvil with a peen hammer to make an indention in the workpiece surface.

The aligning punch is placed through a drilled hole from the bottom surface to the top surface in one piece of metal; the second piece of metal with a drilled hole is placed on top of the first piece with its drilled hole over the aligning punch. With the holes now aligned, the alignment punch is removed and a fastener may be installed to join the metal pieces.

To remove a straight or tapered pin, select a pin punch having a slightly smaller diameter than the diameter of the pin (or of the hole in which the pin is inserted). In the case of tapered pins, carefully measure both ends to determine which is the smaller diameter; apply the pin punch tip to the smaller end directly against the pin. Tap the pin punch anvil with light blows from a light hammer until the pin is driven from the hole.

4-13. Care of Punches

Aligning and pin punch points are ground so that the end is flat [Fig. 4-18(A)] and at right angles to the center line of the punch. Adjust the grinder rest so that the end of the punch is opposite the center of the wheel [Fig.

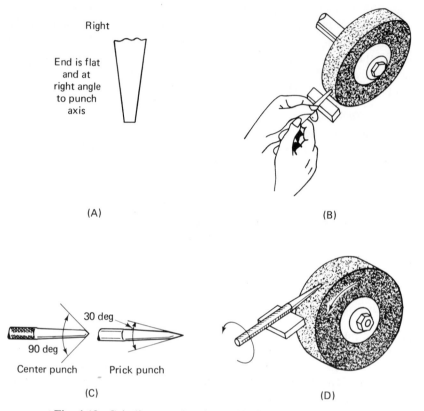

Right

End is flat
and at
right angle
to punch
axis

(A)   (B)

30 deg

90 deg

Center punch   Prick punch

(C)   (D)

Fig. 4-18. Grinding punch points: (A) aligning and pin punch point, (B) grinding an aligning or pin punch point, (C) center punch and prick punch tip angles, and (D) grinding a center or prick punch point.

4-18(B)]. Place the punch on the rest and rotate it as it is fed against the wheel. Keep the point cool by dipping it in water often.

Prick and center punches are ground with conical points [Fig. 4-18(C)]. Adjust the grinder rest so that the end of the punch is at the desired angle [Fig. 4-18(D)]. Place the punch on the rest and rotate it as it is ground. Keep the point cool by dipping it in water during grinding.

If the anvil of a punch becomes "mushroomed" with burrs, file or grind the burrs off. Keep the anvil cool by dipping it in water often.

HAND TOOLS THAT HOLD, SLICE, OR SHEAR

General hand tools that are used by the mechanical technician to hold, slice, or shear metals or other materials include pliers, locking plier wrenches, chisels, files, emery cloth, hacksaws, hand nibblers, and tin snips. Most of these tools are sold on an individual basis and the selection as to type, size, material, etc. for each application is up to the buyer.

4-14. Pliers

Pliers are hand tools used for gripping, cutting, bending, forming, or holding metals and other materials. There are numerous types of pliers available, many with a very specific function as for work in metal or electronic assembly. The general types of pliers most frequently used by the mechanical technician are combination (slipjoint), diagonal, duckbill, needlenose, and side-cutters.

Pliers consist basically of a pair of milled jaws, a pivot point, and a pair of handles. Pliers are made of tool steel and are often chrome plated to prevent rust. Jaws are made narrow to fit tight places. Handles are often knurled for a more positive hand grip. The size of the pliers is determined by the overall length, which is usually between 5 and 10 inches.

Combination (Slip-Joint) Pliers

The combination pliers, also referred to as slip-joint pliers [Fig. 4-19(A)], are used to hold or bend small pieces of flat or round metal stock, tubing, bars, wires, and a variety of other items. The slip-joint allows for adjustment of the serrated jaws to a wider opening, which gives more closing force leverage to the gripping hand. Some combination pliers have a short set of wire cutting jaws near the pivot pin. The capacity of the jaws (the opening for

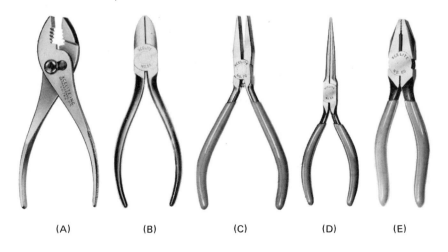

(A)   (B)   (C)   (D)   (E)

Fig. 4-19. Pliers: (A) combination–slip joint, (B) diagonal, (C) duckbill, (D) needlenose, and (E) side cutters. (Courtesy of Xcelite Inc.)

insertion of the object to be held) is approximately $\frac{3}{4}$ inch for a 5-inch (length) plier, and $1\frac{1}{2}$ inches for a 10-inch plier.

Diagonal Pliers

Diagonal pliers [Fig. 4-19(B)] are used to cut soft wire and small stock. Diagonal pliers are also used to cut cotter pins to proper length and to spread the pins once the cotter pin has been inserted through a hole. The diagonal pliers have short jaws with the cutting edges at a slight angle; some are notched for wire stripping. Lengths are approximately 6 inches.

Duckbill Pliers

Duckbill pliers [Fig. 4-19(C)] are used for bending or forming small metal pieces into various shapes and for electronic assembly work on circuit boards and chassis. The flat surfaced jaws are checker milled for positive gripping. The cutting edges are used for cutting soft wires and small stock. Duckbill pliers are approximately 8 inches long.

Needlenose Pliers

Needlenose pliers [Fig. 4-19(D)] are used for mechanical assembly, miniature electronic parts assembly, wire forming, and for reaching into tight spaces. Some models have curved or extra long noses. The needlenose plier jaws may also have a set of cutting edges near the pivot point for cutting soft wire or stock. Plier lengths are from 6 to 8 inches.

Side-Cutting Pliers

Side cutters [Fig. 4-19(E)] are used to grip flat materials and to bend, cut, and strip insulation from wires. The flat serrated jaws are also used to twist wires together. Different size side-cutting pliers are often referred to as electrician's or lineman's pliers; lengths vary from 4 to 8 inches.

4-15. Use of Pliers

When holding or cutting a material with pliers, place the object as close to the pivot point as possible. This allows the gripping hand to squeeze the handles with the greatest force. When using for electrical work, remove electrical power. As a precaution, be sure to insulate plier handles. Special rubber grips are sold commercially and may be slipped over the metal handles, or the handles may be wrapped with several layers of electrical insulating tape.

When stripping insulation from wires, place the wire in the cutter edges or notches provided. If more than one notch is available, choose the proper size notch for the gauge of the wire being stripped. Lightly close the handles so that the cutting edges break into the insulation; rotate the wire in the cutting edges to "score" the insulation. Pull the wire slowly, stripping the insulation off. Inspect the wire to be sure that the cutting edges did not damage it.

Use pliers only for their intended use. Do not try to increase the leverage of plier handles by lengthening the handles with sections of pipe or other extensions.

4-16. Care of Pliers

Damaged or worn serrated jaws may be repaired by filing. If possible, separate the jaws by removing the pivot nut and screw. Place the plier jaw in the protected jaws of a vise. Restore or repair damaged serrations by filing with a three square file.

The cutters on some side cutting pliers can be reground. Before attempting to grind however, inspect the cutter design to determine if the cutting edges will still close after grinding material from the edges. Do not attempt to sharpen pliers (such as diagonal pliers) which are not designed for regrinding.

To regrind plier cutting edges, separate the pliers, if possible, and place in a vise. Grind the cutting edges so that the ground bevel is at right angles with the inside machined bevel. Grind the same amount of stock from both jaws. Cool the jaws frequently in water to prevent loss of temper (which makes the cutters practically useless).

4-17. Locking Plier Wrench

Locking plier wrenches (Fig. 4-20) provide a vise-like grip around any nut, bolt, flat or round metal object placed between the serrated jaws. Advertisers claim that the locking plier wrench is seven tools in one: clamp, gripping tool, pipe or locking wrench, pliers, adjustable wrench, portable vise, and a wire and bolt cutter. The upper jaw is fixed; the lower jaw is adjustable by turning the knurled adjustment screw. A locking handle locks the jaws onto an object with up to a ton of pressure. The lock release lever unlocks the jaws to allow the removal of the wrench. Typical wrench jaws open up to 2 inches; the locking capacity is $1\frac{1}{4}$ to $1\frac{3}{4}$ inches for wrench lengths of $7\frac{1}{2}$ to 10 inches.

Fig. 4-20. Locking plier wrench. (Courtesy of Peterson Manufacturing Co.)

To grip an object with the locking plier wrench, close the jaws part way by squeezing the handles together. Rotate the adjustment screw until the jaw opening just slips over the workpiece. Squeeze the handles together until locked. After the job is completed, remove the wrench from the object by operating the lock release. Practice using the locking plier wrench on several metal workpieces of various sizes until you are fully acquainted with the procedures for locking and unlocking the wrench from the workpieces.

Damaged, unhardened serrated jaws or a damaged adjustment screw can be renewed by filing. Place the jaw or screw into a vise and carefully file with a three square file. If the spring that is housed in the fixed handle becomes stretched or broken, it is easily replaced by following the manufacturer's recommendations. Occasionally place a drop of oil on all points of wear.

4-18. Chisels

Chisels are the simplest of the cutting tools; they are used to chip and cut cold metal. Chisels are made of heat-treated steel and have various shaped cutting edges. The edges are hardened but the head remains soft and chamfered to prevent it from spreading or splintering. The edge is placed against the metal to be chipped or cut and the head is struck with a hammer. The types of chisels used by mechanical technicians are known as cold, cape, round nose, and diamond point.

Cold Chisel

The cold chisel, also called a flat chisel [Fig. 4-21(A)] is used to cut metals and other materials and to shear old bolt heads or nuts. The cold chisel is identified by its flat-shaped cutting edge, which ranges from $\frac{3}{16}$ to 1 inch. Lengths vary from 5 to 12 inches.

Cape Chisel

The cape chisel [Fig. 4-21(B)] is used to cut keyways, channels, and slots in metal and also for "dividing" work so that a flat chisel can be used for finish-

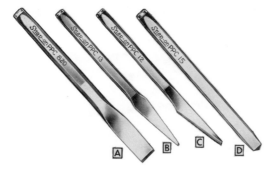

Fig. 4-21. Chisels: (A) cold, (B) cape, (C) round nose, and (D) diamond point. (Courtesy of Snap-On Tools of Canada, Ltd.)

ing. The cape chisel is specified by the cutting edge which ranges from $\frac{1}{8}$ to $\frac{7}{8}$ inch wide. Lengths are 5 to 9 inches.

Round Nose Chisel

The round nose chisel [Fig. 4-21(C)] is used for cutting channels (half-circle grooves), such as oil grooves or similar tasks. The round nose chisel is similar in shape to the cape chisel with one edge ground flat; the other edge makes a round cutting edge. The chisel has its cutting edge ground at an angle of 60° with the axis of the chisel. It is available in sizes of $\frac{1}{8}$ to $\frac{7}{8}$ inch and lengths of 5 to 9 inches.

Diamond Point Chisel

The diamond point chisel [Fig. 4-21(D)] is used to cut V grooves, to draw holes, to turn out broken studs, and to cut holes in flat stock. Diemakers use the diamond point chisel for corner chipping and for correcting drilling errors. The chisel has a solid diamond-shaped point. Cutting edge sizes vary from $\frac{1}{8}$ to $\frac{7}{8}$ inch; the chisels are 5 to 9 inches long.

4-19. Use of Chisels

Chisels are used for three types of cutting operations: chipping, a term which is applied to the removal of metal with a cold chisel and a hammer; cutting wire or round stock; and for cutting sheet or metal plate and rivet heads. Chipping is seldom done in production, but it is still used where machining is difficult or inconvenient.

To chip metal away from a workpiece, first fasten the workpiece in a vise in a manner such that chipping will be done toward the stationary vise jaw. To prevent accidents from flying chips, place a canvass barrier a few feet beyond the vise and wear goggles. Obtain a sharp chisel and ball-peen hammer (about one pound).

Grasp the chisel with one hand and hold the cutting edge of the chisel to the workpiece at an angle of about 35° from the horizontal. All four fingers should encircle the chisel; the third and fourth fingers grasp tightly and the little finger is used to guide the chisel. Strike the chisel anvil with the ball-peen hammer. Use the bevel of the cutting edge as a guide. Reset the chisel after each blow. Cuts of $\frac{1}{32}$ to $\frac{1}{16}$ inch should be made. The depth of cut is determined by the angle of the chisel; raise or lower the chisel to obtain the proper depth of cut. Don't use the chisel to chip to the finished level desired; leave enough material so that it may be finished with a file.

Work should progress from the outer edge of the work toward the center. When a large surface is to be chipped, first cut parallel grooves into the workpiece with a cape chisel. Space the grooves closely enough together so that the material between the grooves is slightly narrower than the width of the cold chisel that will be used to remove the metal left between the grooves.

The edge of the chisel should be wiped with an oil saturated cloth when chipping wrought iron or steel. This will lubricate the contacting surfaces and hence will protect the edge of the chisel. When chipping cast metal, begin at the ends and work toward the center; this prevents the breaking of edges and corners. When chipping keyways with a cape chisel, leave metal for filing on both sides and the bottom of the keyway.

The cold chisel may be used for cutting wire or round stock when a hacksaw is not available. Place the workpiece on an anvil or other working surface. Place the chisel cutting edge on the cutting mark. Lightly tap the chisel with a ball-peen hammer; examine the chisel mark to be sure it is in the correct place. Place the chisel on the mark and strike the chisel with successive hammer blows. The last few blows should be lighter weight. On thick round stock, cut halfway through in one direction. Then turn the piece over and finish the cut.

Sheet or plate metal is cut with a chisel when metal cutters or machines are not available. If the workpiece is small enough, secure the workpiece in a vise so that the guideline is even with the vise jaws. If the piece is large, place it on a suitable work surface. Sheets should be backed up with a wood or metal plate to prevent bending. Start at one edge of the workpiece and work across. Strike the end of the chisel sharply, keeping the chisel cutting edge firmly against the workpiece.

To cut a rivet head off, place the edge of a cold chisel against the rivet head (as near the shaft as possible). Strike the chisel sharply with a ball-peen hammer.

4-20. Care of Chisels

To work effectively, chisels must be kept sharp. Grinding is necessary when the cutting edge has been badly nicked or the bevel has been rounded. The included angle formed by the two bevels should be about 65°. Set the grinder tool rest for the angle and hold the chisel against the grinding wheel as shown in Fig. 4-22(A). Move the chisel back and forth across the wheel with moderate pressure. Grind the cutting edge slightly convex as shown in Fig. 4-22(B). Dip the chisel often in cold water to prevent loss of tool temper with overheating. Grind both bevels, taking equal amounts of material from each side. If a chisel head mushrooms, file or grind the excess metal off.

Always wear goggles when using or grinding chisels. Protect fellow technicians by placing a screen behind your work when chipping. Do not leave chisels unprotected on the bench.

4-21. Files

Files are very important hand tools for use by the mechanical technician. They are used to cut, smooth, and to remove small amounts of material.

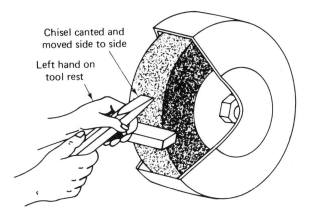

Chisel canted and
moved side to side

Left hand on
tool rest

(A)

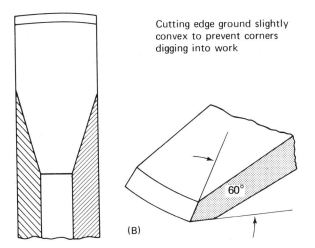

Cutting edge ground slightly
convex to prevent corners
digging into work

60°

(B)

Fig. 4-22. (A) Grinding a cold chisel; (B) the cutting edge of
the cold chisel is ground slightly to convex to prevent corners
from digging into the workpiece.

Often the file is the finishing tool used to make machined parts fit accurately
together in prototype or small production quantity runs. Because files are so
important, it is imperative that the reader learn the parts of a file, file *cuts*,
length, *teeth spacing*, *shape*, and *procedures for using* before he attempts to
begin filing operations.

Figure 4-23 illustrates the various parts of a file. On a file such as a half-
round file, the flat side is termed the face and the opposite side is termed the
back. The edge is said to be *safe* when there are no teeth cut into it. Edges

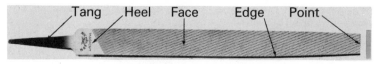

Fig. 4-23. Parts of a file. (Courtesy of Nicholson File Co.) .

may be flat or rounded. The length of a file is the distance from the point to the heel (length excludes the tang). Handles are readily removed and are not normally sold as part of the file.

The *cut* of a file is the layout of file teeth that the file manufacturer has cut into the file. Four cuts are available (Fig. 4-24): single, double, rasp, or curved tooth.

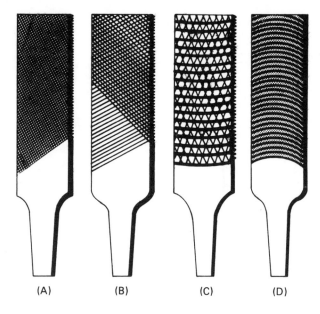

Fig. 4-24. File cuts: (A) single, (B) double, (C) rasp, and (D) curved tooth.

The *single* cut file has a single set of parallel rows of teeth at an angle of 60–80° to the length of the file extending the length of the file. The *double* cut file has a double set of parallel rows of teeth; the first parallel set is cut at about 45° and the second set is at 60–80°. The double cut file is usually used by machinists. The *rasp* cut has many short separate teeth each separately formed during manufacturing by a single pointed tool or punch. The *curved teeth* cut has rows of curved teeth along the length of the file.

File teeth row spacing determines the coarseness of the file. The files are categorized into coarse (greatest distance between rows of teeth), bastard (medium coarse), second cut, and smooth (Fig. 4-25) grade teeth. Two other grades, rough and dead smooth, are also available. Spacing of the teeth changes with file length, increasing proportionately to the increased length of the file; the longer the file, the coarser the file (Fig. 4-26). However, the same relative difference always exists between the cuts for any file in any particular length.

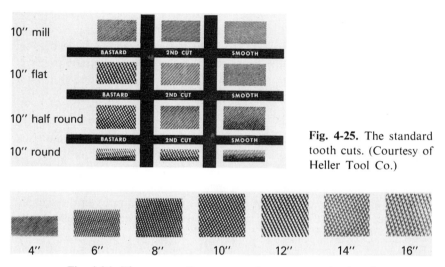

Fig. 4-25. The standard tooth cuts. (Courtesy of Heller Tool Co.)

Fig. 4-26. The range of coarseness increases as the length increases. (Courtesy of Nicholson File Co.)

The *shape* of the file is determined by its cross-sectioned view. Many shapes are available; those most often used by the mechanical technician are the flat, half-round, round, three square, square, pillar, mill, and Swiss pattern. Most files are tapered toward the point, both in width and thickness, and are made of high-carbon tool steel.

Flat File

The flat file [Fig. 4-27(A)] is used by the mechanical technician for rapid removal of metal. It is rectangular in shape and tapers toward the point in both width and thickness. It is double cut on both sides and single cut on the edges. It is available in bastard, second cut, and smooth teeth, from 4 to 18 inches in length.

Mill File

The mill file [Fig. 4-27(B)] is used for draw filing (Sec. 4-22) and for finishing metals. The mill file is also used for sharpening mill and circular saws, machine knives, lawn mower blades, axes, and shears. The two faces are

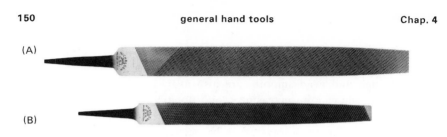

Fig. 4-27. Files: (A) 10″ flat and (B) 8″ mill. (Courtesy of Nicholson File Co.)

parallel and flat; the edges may be flat or rounded. The faces are usually single cut and available in bastard, second cut, or smooth cut. The edges are single cut. A round edged mill file is used where rounded gullets are preferred as compared to sharp corners or squared gullets. Mill files are available in widths from $\frac{7}{16}$ to $1\frac{1}{2}$ inches and lengths from 4 to 16 inches.

Half-Round File

The half-round file [Fig. 4-28(A)] has a flat face and a half round back, which makes it useful for a number of different types of jobs. The flat face is used like the flat file to rapidly remove metal on flat surfaces. The half-round back is used to rapidly remove metal from concave or convex surfaces. The face and the back are single or double cut depending upon the size and type. The width and thickness are tapered toward the point; lengths are from 4 to 16 inches.

Round File

The round file [Fig. 4-28(B)] is used to enlarge or smooth circular openings or concave surfaces. It is tapered toward the point. Files are available in lengths of from 4 to 16 inches and $\frac{5}{32}$ to $\frac{3}{4}$ inch diameters. Longer files are double cut.

Fig. 4-28. Files: (A) half round, (B) round, (C) three square, (D) square, (E) pillar, and (F) handle. (Courtesy of Heller Tool Co.)

Three Square File

The three square file [Fig. 4-28(C)], also called a triangular file, is used to file internal angles that are more acute than a right angle. It is used for cleaning out square corners and for filing taps and cutters. The three square file faces are at 60° angles. The faces taper to a point and have three sharp corners. The faces are double cut; the edges are single cut. Lengths are available up to 12 inches.

Square File

The square file [Fig. 4-28(D)] is used to file square and rectangular slots, keyways, and splines. Some square files have four double cut sides, while others have three double cut sides and a safe side. Square files are available with side dimensions of from $\frac{5}{32}$ to $\frac{3}{4}$ inch and lengths of from 4 to 16 inches; the smaller the dimension of the side, the shorter the handle.

Pillar File

The pillar file [Fig. 4-28(E)] is used for filing narrow slots and keyways. It has a rectangular cross section with one or sometimes two safe edges. Pillar files are of the same coarseness as square files of the same length. The pillar file is available in lengths to 14 inches.

File Handles

File handles [Fig. 4-28(F)] are made of soft wood, *without* a finish (such as varnish, etc.), so that the wood absorbs hand moisture. A metal ferrule (on the end of the handle that slips onto the tang) protects the wood from splitting. Handles range in approximate sizes as follows: 4-inch handle for files 3–6 inches; $4\frac{1}{2}$-inch for files 4–8 inches; 5-inch for files 6–10 inches; $5\frac{1}{2}$-inch for files 8 to 12 inches; and 6-inch handles for files 12–18 inches.

Swiss Pattern Files

Swiss pattern files are used by precision craftsmen for filing delicate instruments and mechanisms for final fitting and for use in tool and die work. The files are narrower in both width and thickness, and the tapered files have fine points. The file tang is shaped into a handle. Cuts are identified differently for the Swiss files. They are made in seven cuts known as 00 (coarsest—34 teeth per inch), 0, 1, 2, 3, 4, and 6 (finest—184 teeth per inch). There are 12 shapes available in lengths from 3 to 12 inches.

Swiss files are used as finishing files for removing burrs, rounding out slots, and cleaning out square corners. The Swiss pattern files are also useful for enlarging small holes, and for shaping and finishing very narrow grooves, notches, and keyways.

4-22. Use of Files

Prior to filing, the technician must first consider his job. What material is to be filed? How tough is the material? Is there a lot of material to be removed or just a little? Is the surface flat? Rounded? Use your knowledge gained from the introduction to this discussion of files and Table 4-2 to determine the answers to these questions. Then select the proper file for your application. Never file without placing the proper handle on the tang. This protects your hand. If a hard surface on a casting is to be removed, use a grinder first. This removes sand grits in the outer cast that would ruin the file.

TABLE 4-2. FILE CUT USE

Use	*Cut Type*
Heavy, rough cutting	Large coarse, double cut
Finishing cuts	Second or smooth cut, single cut
Cast iron	Start with bastard; finish with second cut
Soft metal	Start with second cut; finish with smooth cut
Hard steel	Start with smooth cut; finish with dead smooth cut
Brass, bronze	Start with bastard cut; finish with second or smooth cut
Aluminum, lead, babbitt	Bastard cut curved tooth
Small work	Short file
Medium work	8-inch file
Large work	Use most convenient size

It takes practice to be able to properly file a surface flat, but it is a skill that the mechanical technician must acquire. Practice slowly and carefully.

Clamp the material to be filed into the protected jaws of a vise that is approximately at elbow height (Fig. 4-29). The material should be parallel to the jaws and should protrude slightly above the jaws. Hold the file in one hand, thumb on top, and hold the point of the file with the fingers of the other hand; again the thumb should be on the top. Keep the file strokes in a straight line.

Fig. 4-29. Position of the hands for filing. (Courtesy of Nicholson File Co.)

Apply even pressure with both hands on the forward (away from the body) stroke only. The return stroke does not cut; therefore, if you are filing a hard material, lift the file from the material for the return stroke. This prevents the file from becoming dull faster than it should. If you are filing soft metals, light pressure on the return stroke helps keep the cuts in the file clear of waste metal. Use a rocking motion when filing rounded surfaces. File slowly, lightly, and steadily at a rate of about one stroke per second. The proper speed and pressure come with experience. Do not force the file. Squareness of work must be checked repeatedly with a steel square. When the file cuts become clogged, clean them with a file card or brush. New files should be used carefully. Do not apply too much pressure, because it may break off the teeth.

Draw filing produces a very smooth and true surface. It is used on narrow surfaces and edges. A single cut file with a second cut or smooth teeth spacing is used: successively finer teeth should be used. To draw file, place the material in a vise as previously described. Grasp the file with both hands (Fig. 4-30) close together (to prevent bending and breaking the file) and draw the file across (at right angle to) the workpiece. Work in both directions. Use moderate pressure on both the draw and back strokes. For an extra smooth surface, wrap a piece of emery cloth around the file and stroke in the same manner.

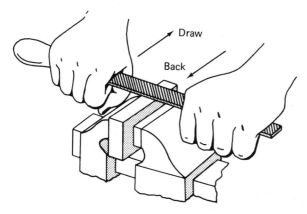

Fig. 4-30. Draw filing.

4-23. Care of Files

Break in a new file by first using it on brass, bronze, or smooth iron. Do not use a new file to remove fins or scales from cast iron or do not use it on a narrow surface such as sheet metal.

Chalk may be rubbed in the cuts of a file to help prevent particles of metal from clogging the teeth. When the teeth do become clogged, the file slips, scratches, and is inefficient. Clean a file frequently with a file card or brush and a scorer. The scorer cleans out individual rows and is used where the file card or brush cannot get the chips out. Use the file card or brush by drawing it in a direction parallel to the rows of teeth.

Use files properly and only for the intended purpose. Always use a firmly attached handle (except for Swiss pattern files). Do not strike a file to clean material from it; clean it properly and often with a file card or brush. Do not oil files as this causes the file to slip across the surface, preventing filing. Store files separately so that they do not rub against each other or other tools.

To install a handle, insert the tang into the handle socket. Hold the handle in your hand and tap the handle on the bench top until the file is seated. Do not hammer the handle.

To remove a handle, hold the file with one hand and the handle with the other. Pull the handle from the file while striking the handle (at the ferrule) on the edge of the workbench.

4-24. Resharpening Files with Acid

Files may be resharpened to a degree by the use of a sulfuric acid solution (three parts acid to two parts water). Always add acid to the water to prevent spattering. The file is placed in a container that is unaffected by acid. The solution is then placed in the container to a level just short of the top of the file teeth edges. The file is left in the acid solution for 1–2 hours. The acid cleans away particles of waste material in the teeth and also slightly eats away the metal of the file near the tops of the points. This results in cleaning and resharpening of the file teeth.

4-25. File Card and File Brush

A file card (Fig. 4-31) is a wire brush which is used for general cleaning of coarse file teeth. A file brush is a wire brush on one side and a hair brush

Fig. 4-31. (A) Combination file card and brush, and (B) file card. (Courtesy of Heller Tool Co.)

on the other; the hair brush is used for finer cut files. Clean files in accordance with the procedures described under the section on care of files.

4-26. Emery Cloth

Emery cloth is used dry or with oil primarily for light cleanup work of tools, rust removal, and for hand polishing nonplated metal surfaces. Emery cloth is made of grits of the abrasive mineral emery on a backing similar to sandpaper. Emery is actually a natural mixture of aluminum oxide and magnetite. Emery cloth is available in 9 × 10 inch sheets and grades as follows:

Grade	Mesh Size	Number
Fine	180	3/0
	150	2/0
	120	1/0
Medium	100	1/2
	80	1
	60	1 1/2
Course	50	2
	40	2 1/2
	36	3

4-27. Hacksaws

A hacksaw (Fig. 4-32) consists of a frame, a grip, and a blade. The hacksaw is used to cut metal sheets, iron pipe, tubing, screws and bolts, steel bars, brass, bronze, copper, and other metals.

The hacksaw frame provides tension on the blade. Most frames are adjustable, as for 10 and 12 inch blades, have notched studs to hold the blade which allow for positioning the blade in four positions, and unbreakable plastic grips. A wing nut adjusts the blade tension. The blade is mounted with the teeth forward to cut on the forward stroke.

Blades (Fig. 4-33) are available in 14, 18, 24, and 32 teeth per inch. The 14 teeth-per-inch blade is used for most metals with sections of one inch or more thickness; 18 teeth-per-inch blade for steel bars, brass, and copper $\frac{1}{4}$ to 1 inch; 24 teeth-per-inch blade for iron pipe and medium tubing $\frac{1}{8}$ to $\frac{1}{4}$ inch thick; and 32 teeth-per-inch blade for thin metal sheets and tubing up to $\frac{1}{8}$ inch. In general then, fine teeth are used for thin stock; course teeth for thick stock. A good rule to follow is that if the proper blade is chosen, two teeth should be able to rest simultaneously on the material (and the wall of a tube). If two teeth cannot be put on the material, tilt the hacksaw. Various

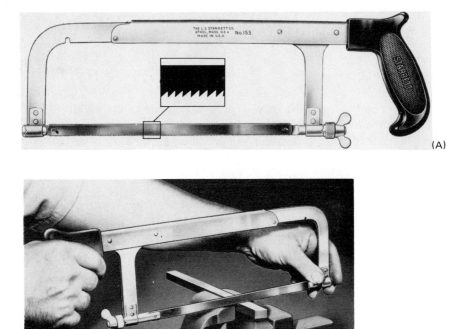

(A)

(B)

Fig. 4-32. (A) Hacksaw; (B) starting the hacksaw blade into the workpiece. (Courtesy of L.S. Starrett Co.)

Fig. 4-33. Hacksaw blades. (Courtesy of Simonds Saw and Steel Co.)

blade widths are available for cutting different size slots, as for slots in bolt heads.

There are high-speed and standard blades. Blades classed as high-speed, shatterproof, and flexible provide a balance between toughness and flexibility which assures safety and cutting efficiency. Blades made of molybdenum are hard, long wearing, straight-cutting, and preferred by tool makers for cutting rigidly clamped material. The blade is also ideal for general machine shop use. Tungsten blades cut the toughest materials. Tungsten blades are also used on tool and die steels and abrasive materials.

Standard flexible blades for general purpose cutting have only the teeth hardened; thus, the back remains flexible to minimize breakage. This type of blade is economical for the inexperienced or occasional user under a variety of conditions. Flexible blades are also used in cramped areas without a vise.

Special unbreakable standard blades are balanced, safe, tough, flexible, and are used for miscellaneous cutting. The blade is hardened throughout and is then made flexible. This results in hard, sharp teeth with a semi-hard back to withstand twisting, which reduces blade breakage and tooth strippage.

The standard all-hard blade is for the skilled technician. It is used for workpieces that can be held in a vise. The standard all-hard blade gives exceptional uniformity and cutting ability. It is important that the material being cut is held rigid to prevent breakage of the blade.

4-28. Use of Hacksaws

Place the workpiece firmly in a vise (see Fig. 4-32). Hold the saw firmly with one hand on the grip and the other hand on the top front of the frame. Use light pressure and slow strokes as the hacksaw is started on the guideline on the workpiece. Take short strokes of from 1 to 2 inches; a light downward pressure should be made on the forward stroke. There should not be any pressure on the rearward stroke as the blade does not cut in the rearward direction and pressure in this direction only causes dulling of the blade. After the cut has been started, increase the stroke length to from 6 to 7 inches. Do not twist the blade as this may cause the blade to break. When the cut is nearly through the workpiece, use light pressure and slow even strokes until the workpiece parts. Lubricants or coolants are generally not used. In cutting pipe or rod, the workpiece will sometimes close on the blade causing strain. Brace the cut open (as with a screwdriver) as required.

4-29. Care of Hacksaws

Use the proper blade for the workpiece being cut. Whenever possible, grip the workpiece in a vise; when this is not possible, use a flexible blade. Keep the blade from becoming excessively hot during use.

4-30. Nibbler

The hand nibbler (Fig. 4-34) is use to "nibble" holes, slots, or openings of any shape in steel up to 18 gauge and in plastic, copper, aluminum or other soft metal up to $\frac{1}{16}$ inch. The cut edges are safe, flat, and smooth. The "nibble" is a rectangle approximately $\frac{1}{16}$ by $\frac{1}{4}$ inch.

Fig. 4-34. Nibbler. (Courtesy of Adel Tool Co.)

In using the nibbler, first cut a $\frac{1}{2}$ inch square hole or drill a $\frac{1}{2}$ inch hole in the workpiece to allow the nibbler punch to pass through. Insert the nibbler punch and squeeze the handles together and follow the layout pattern around the workpiece. Make nibbler bites as close together as possible. File out any rough edges.

When the nibbler punch becomes dull, it may be replaced with a new punch. Remove and replace the punch according to the manufacturer's direction.

4-31. Tin Snips

Tin snips are hand tools that provide a shearing force for cutting various types of sheet metal, stainless steel, and Monel metal up to 18 gauge into various shaped pieces. Forged high carbon heat treated steel jaws cut smooth and evenly with a minimum effort. Tin snip lengths are 7 to 12 inches; jaws cut from $1\frac{3}{4}$ to 3 inches (in one closing of the jaws).

Tin snips are divided into four types (Fig. 4-35) according to the pattern cut: regular; straight or curved cuts; left cuts; and right cuts. Regular cut tin snips are used for cutting straight lines and also curves in locations that are easily accessible. Straight or curved cut tin snips cut straight lines, circles, squares, or any pattern. Left cut tin snips cut short straight lines or left cuts in locations where it is advantageous to keep the handles and the hand away from the metal stock. Right cut tin snips cut to the right.

To cut straight lines with tin snips, lay the workpiece on a bench with the guideline extending off the bench. Hold the workpiece with one hand and cut with the other. The top blade of the tin snips should cut along the scribed

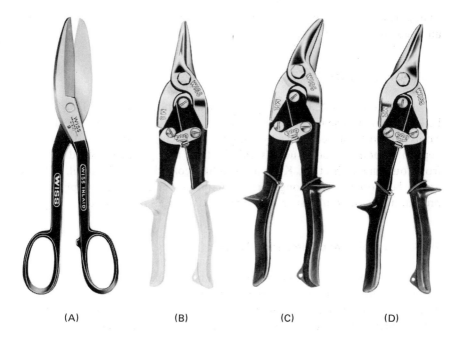

(A)   (B)   (C)   (D)

Fig. 4-35. Tin snips: (A) regular pattern, (B) straight or curved cuts, (C) cuts to the left, and (D) cuts to the right. (Courtesy of J. Wiss & Sons Co.)

line. The portion of the workpiece being cut off will bend below and out of the way of the tin snips. Keep the tin snips as far into the workpiece as possible so that the cutting is done from a point as near the pivot as possible. Keep the faces of the tin snip jaws perpendicular to the workpiece to prevent the cut edge from becoming bent or burred. Gloves should be worn to protect the hands from sharp edges and burrs.

When cutting curves, make the cut as continuous as possible. Stopping and starting at different points causes rough edges with sharp slivers of metal.

When a hole or opening is to be cut into a workpiece with tin snips, first cut a hole into the workpiece with a small cold chisel. Make the hole large enough to allow the tin snip tips to enter, open, and cut.

Do not attempt to cut material heavier than the tin snips can handle. Don't use the tin snips as a hammer.

Dull tin snips can be sharpened with an oilstone or a file. To sharpen, clamp the snips in a vise and file on the beveled edges only (never on the flat face). Stroke only in the direction shown across the full length of the edge and away from the edge. To remove the "wire edge" on the face of the snips (made from the filing on the beveled edge side), draw the face lightly across an oiled stone from the pivot to the tip.

4-32. General Tool Care

Hand tools should be used only for their intended function. They should be kept clean and sharp at all times. Burrs should be removed immediately. Tools should be stored either by hanging or placing in drawers. If tools are kept in a toolbox, tools such as chisels, files, and punches should be kept from touching each other. Keep tools lightly coated with a light oil. If tools are to be stored over a long period of time (over a month), apply a rust preventative compound, such as a light grease, to all metal parts.

4-33. Hardening and Tempering Tools

When tools such as screwdrivers, punches, chisels, and plier cutters, are ground to resharpen or reshape them, the tool should frequently be dipped in cold water to cool them. Too much heat from grinding can cause the tool metal temper to be lost; this condition is usually indicated by a blue color appearing on the metal.

To harden and temper a chisel or punch, first heat the tool to a cherry red color in a furnace or with a hand torch. Hold the tool in the center with a pair of tongs and dip the cutting end into cold water or oil to a depth of slightly over 1 inch. Dip the other end of the tool to the same depth. Using heat-resistant protective gloves, polish the hardened ends with an aluminum oxide abrasive cloth. Watch for the color to return to the ends from the hot center. Each time the cutting end turns purple, dip it. Everytime the head turns blue, redip it. When the red color disappears from the tool, dip the complete tool in water until it is cool. This checks further drawing of the temper. This procedure produces a hard cutting edge and head and a softer, tougher center section that withstands shock.

The same procedure is used on the cutting edges and heads of other types of tools when retempering or hardening is necessary.

4-34. Recommended Basic Hand Tools for the Mechanical Technician

This chapter has introduced the general hand tools used by the technician. Not all of these tools are used in all facets of work in which the technician may be employed. Not all of these tools are required by the beginning technician in any fields.

Table 4-3 lists a recommended starting set of general hand tools.

TABLE 4-3. RECOMMENDED STARTING SET OF HAND TOOLS

Conventional screwdrivers (set)
Phillips head screwdrivers (set)
Open end wrenches (set)
Adjustable open end wrench
Key setscrew wrenches (set)
Ball-peen hammer (4 or 8 ounce)
Plastic tip hammer
Prick punch
Center punch
Pin punch (optional)
Combination pliers
Needlenose pliers
Side cutting pliers
Locking plier wrench
Cold chisel
Flat file
Half-round file
Round file
Three square file
Square file
Pillar file
Mill file
File card
File handle
Hacksaw (and several blades)

SUMMARY

This chapter has provided the technician with descriptions, applications, and care of hand tools from a theoretical standpoint. Your job is to apply the knowledge gained from descriptions given here to the use of the hand tools. Begin now and practice with each tool. Try to be ambidextrous. Become proficient in the use of each tool. And by all means, use the proper tool required by the job.

REVIEW QUESTIONS

1. Explain the difference between the conventional, Phillips head, and Reed and Prince screwdriver tips.
2. Describe the procedures for removing a broken bolt from a metal plate.
3. When is an adjustable open end wrench used? How do you use it properly?

4. When is a strap wrench used? When is a pipe wrench used?

5. List the various parts in a basic socket wrench kit. What is the function of each part?

6. Explain the difference between a T handle, offset L, and striking socket wrench. What are the applications for each of these socket wrenches?

7. List the types of torque wrenches. What are the basic procedures for using and reading torque wrenches?

8. What are spanner wrenches? Where are spanner wrenches used?

9. How does a splined setscrew wrench differ from a key setscrew wrench? Draw an end view of each type.

10. List the types of punches and briefly describe the use of each type.

11. Which pliers are used for electronic assembly work?

12. Describe how to strip insulation from a wire with the cutting edges or notches of pliers. Use your own pliers and practice stripping wire.

13. What are the different types of chisels? Which chisel is used to cut V grooves?

14. List some safety precautions to follow when using chisels.

15. Briefly describe how to use a chisel.

16. What are the physical differences between:
    a. single and double cut files
    b. three square file and pillar file
    c. bastard and second cut grade file teeth

17. How are files cleaned? When are handles used?

18. When are fine teeth hacksaw blades used? What are the basic procedures for using a hacksaw?

19. What is a nibbler? What kind of materials can it be used on?

20. Refer to Table 4-3. Which of these tools are currently in your tool box? Which tools do you need to purchase next and in which order should you purchase them?

TOOLS FOR LAYOUT

All job skills have tools which are unique to the operations to be performed: the draftsman uses a T square and triangles, the carpenter uses a claw hammer and saw, and the plumber uses wrenches. Similarly, the mechanical technician has special tools to perform layout. This chapter provides the descriptions and uses of the layout tools needed to transfer a layout design, print, or sketch to a material. Chapter 6 describes layout in detail.

BASIC LAYOUT SET

A basic set of layout tools consists of the following tools:

1. layout fluid (dye)
2. scriber
3. dividers
4. trammel
5. prick punch and center punch
6. combination set (square head, bevel protractor, center head)
7. surface gauge
8. hermaphrodite calipers

The basic set of layout tools is used by the mechanical technician to transfer patterns from a designer's sketch or drawing to the workpiece.

5-1. Layout Fluid

If the technician is to lay out a pattern on a piece of cardboard or wood, he may use a sharp pencil to do an effective job of laying out the required lines, points, circles, and arcs. But, what happens when a pencil is used on a hard shiny surface or on a dark surface? Obviously, the pencil line is not plainly visible and the marking rubs off easily. How, then, can hard, shiny, or dark surfaces be marked?

Patterns can be drawn onto shiny, hard, or dark surfaces by first coating the surface with a layout fluid and then by scratching (scribing) the required lines, points, circles, and arcs onto the surface. The layout fluid prepares the surface so that the scratches will show.

5-2. Choosing a Layout Fluid Color

Layout fluid color may not immediately seem important, but consider the instance in which you may apply white fluid over an aluminum sheet. When thin scratches are drawn through the dried fluid, there is difficulty in seeing the aluminum scratched surface through the white fluid because both the aluminum and the white fluid reflect light at almost the same intensity. A poor choice of fluid color has been made! Hence, the technician should select a contrasting color from the selection of red, green, blue, black, and white fluids available. For example, a darker color would be easier for defining lines of layout on aluminum; a lighter color would be easier to define lines of layout on a black sheet of phenolic plastic, rust-coated castings, or dull-finish surfaces.

Layout fluid can be applied to the workpiece by one of two methods: by brushing on or by spraying on from an aerosol can. It should be pointed out however, that the solvent used in most of the commercial fluids is apt to soften or dissolve plastic materials and painted surfaces. To insure that this softening does not take place, try a small test area first. If the commercial fluid is not suitable, then a water base fluid made from whitewash may be substituted. The whitewash may be tinted with food color dye if a contrasting color is required. The whitewash is easily removed by flushing the surface of the workpiece with water.

Commercial layout fluid is removed by flushing or brushing the surface of the workpiece with a solvent recommended by the manufacturer on the label of the layout fluid container. Solvents such as acetone or alcohol may also be used to remove the layout fluid.

CAUTION

Layout fluid solvents (as well as most solvents) are highly inflammable and must not be used near open flames. Rags used to wipe solvents should be discarded in metal containers protected by lids.

5-3. Scriber

The scriber is the mechanical technician's "pencil"; it is used to draw (scribe) lines onto hard surfaces by making permanent scratches on the material. The scriber can be as simple as a hard-pointed metal rod or as sophisticated and expensive as a diamond chip scriber (Fig. 5-1). General purpose scribers are of tool steel and are carbide tipped. Tool-steel tips are generally longer lasting than carbide tips because carbide tips will shatter when dropped on hard surfaces and the carbide tips require special grinding wheels to resharpen.

Fig. 5-1. Scriber—straight and hook point. (Courtesy of L.S. Starrett Co.)

When scribing or "ruling" a line on a workpiece, a metal scale or bar is used as a guide. Proper technique in the use of the metal scale and scriber results in an accurately drawn line. The point of the scriber is placed in the "corner" formed by the scale and workpiece. The scribe is then held at a steady angle and is drawn across the workpiece using the scale as a guide. This technique allows for greater accuracy in aligning the scale with pre-scribed marks or lines. This technique also keeps the tip of the scriber from wandering, causing wavy irregular lines and inaccuracies.

5-4. Dividers

At times there is a need to draw circles or arcs, or to measure off lengths. The reader has, in the past, used a compass for drawing circles. As you recall, the compass is a hinged adjustable two-legged device. A metal point at the bottom of one leg is used as the pivot or center. A pencil or other marking device is placed in the other leg to mark the path of the required circle, arc, or measured length.

Remembering that pencil lines do not show very clearly and do not mark permanently, another tool must be substituted for the compass. If the pencil in the one leg is replaced by another metal point, the result is a divider.

Thus, with a metal point as a pivot and a metal point as a scriber, circles, arcs, and measured lengths can now be permanently marked on a hard surface.

Dividers are of two types: spring-joint and quick-release. Spring-joint dividers (Fig. 5-2) are so termed because spring-powered joints "spring" the legs apart. The dividers are adjusted by a knurled knob that opens or closes the tips of the dividers as the knob is rotated.

Fig. 5-2. Spring joint dividers. (Courtesy of L.S. Starrett Co.)

Quick-release dividers have both a coarse adjustment and a fine adjustment. By loosening the quick-release course adjustment lock screw, the legs of the dividers may be quickly spread apart with the hand to within $\frac{1}{8}$ inch of the required distance. The course locking nut is then tightened and the fine adjustment nut is rotated to set the required dimension (Fig. 5-3).

Fig. 5-3. Quick release dividers. (Courtesy of L.S. Starrett Co.)

In setting a divider in conjunction with a scale, several simple steps should be committed to memory. First, the scale should be placed on a firm surface having adequate overhead lighting. With the divider held at an angle of approximately 30–45° to the scale to protect the points, one point is held fixed on a *whole number* inch mark. The other leg is adjusted to the required dimension. It is of extreme importance to add the required distance to the setting of the *whole number* inch mark that was used as the starting point. For example, assume that a setting of $3\frac{5}{16}$ inches is required. If the whole number of one inch is chosen as the starting point, it is evident that the other leg must be adjusted to the setting of $4\frac{5}{16}$ inches, since the distance measured from the starting end of the scale to the first point is one inch and the distance from the starting end of the scale to the second point is $4\frac{5}{16}$ inches—the required setting. When setting accurate dimensions, it is important to keep the points of the dividers in such a location that a line drawn between them will be parallel to the edge of the scale. This is particularly important when using the $\frac{1}{64}$ inch graduations.

The size of dividers is determined by the maximum opening in inches between centers. Dividers are usually limited to a maximum of six inches of spacing between the tips. Thus, a twelve-inch diameter circle is the maximum circle that can be made. Larger than six-inch step offs, arcs, and circles greater than twelve inches in diameter are made with a trammel.

5-5. Trammel

The trammel (Fig. 5-4) is used to step off required lengths and to scribe arcs and circles that are of greater length or diameter than can be handled with dividers. A trammel consists of a rod or bar, to which a rigid pointed leg is attached on one end and another pointed leg is allowed to travel over the length of the rod or bar. In addition to the main knurled knob, which when loose allows the movable pointed leg to travel, there is a movable pointed leg having one of two types of adjustments (Fig. 5-4): an eccentric point or adjustment screw.

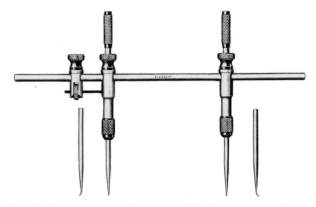

Fig. 5-4. Trammel set. (Courtesy of L.S. Starrett Co.)

The pointed leg of the eccentric pointed trammel is off center. When the knurled point lock nut is loosened, the off-center point can be rotated, thus setting the final desired dimension.

To set the fine adjustment on a trammel having an adjustment screw, the technician rotates the fixed adjustment screw, which increments the trammel movable point to the desired dimension.

5-6. Punches: Prick and Center

As the layout of scribed lines and intersections of circles is performed, a means of permanently marking them must be employed. It is for this need that center and prick punches are used. When a punch is struck with a hammer, the punch produces a small conical shaped indention in the surface. These indentions are permanent records or marks of intersections of lines and circles. The indentions also act as pivot points or centers of circles for drawing arcs or circles and as starting points for drilling holes.

The prick punch is preferred as a starting out tool in layout by some technicians for two main reasons: first, the small holes more clearly define a socket or pivot hole for use in conjunction with dividers or trammels in scribing arcs or circles; and second, it allows for a more accurate alignment of a punch point to a scribed line or to an intersection of lines.

After the layout, which contains punch points previously made with a prick punch, is completed, the centers of holes (after rechecking their accuracy) that are to be drilled are repunched with a center punch. The formation of this indention in the work surface allows for direct drilling with either a center drill or a drill proper without fear that the drill bit will "walk" over the surface causing an inaccurately drilled hole.

A simple step-by-step operation of punching is as follows:

1. Holding the punch in the hand, slant the anvil or striking end away from yourself. This allows for accurate alignment of the point with a line or line intersection in the layout.
2. Holding the punch point to the layout line, bring the punch body to a vertical position in respect to the work surface.
3. Firmly strike the punch anvil with a steel hammer to make an indention in the workpiece surface.
4. Inspect the punch mark to be sure that it has been properly placed on the layout.

A fifth step should be inserted at this point only if step 4 shows a large error in punch alignment. A large mispunch can be easily corrected by merely repunching. The punch mark that is off by $\frac{1}{64}$ or $\frac{1}{32}$ inch is sometimes a problem in repunching as the point of the punch will sometimes slide back into the misaligned indention.

To cope with this problem an operation known as "moving" the punched indention must be performed. This involves the physical shifting of the inaccurate punch indention by deforming the metal around it. The first step in moving the center is shown in Fig. 5-5. With the punch at approximately 45° to the workpiece surface, engage the point of the punch into the mispunched indention. Make sure that the punch is pointing in the direction of the correct location. Lightly tap the punch with the hammer; inspect the "moving" hole between blows. When its point is properly aligned with the layout mark, it should be straightened up. Raise the punch to a vertical position and give a final blow with the hammer; this gives the "moved" hole a normal shape for either drilling or for further layout.

Before moving to the next tool, a brief description of an automatic center

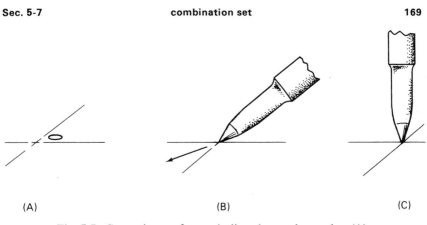

(A) (B) (C)

Fig. 5-5. Corrections of a misaligned punch mark: (A) misaligned punch mark, (B) moving punch mark, and (C) reforming moved punch mark.

punch is described. The purpose of an automatic center punch is to eliminate the use of a hammer to impart the force to punch an indention. The "hammer" in the case of this punch is built into the handle. The three parts of this punch (Fig. 5-6) are the hammer, anvil, and punch point.

Fig. 5-6. Automatic center punch. (Courtesy of L.S. Starrett Co.)

The amount of hitting force is variable by adjusting the spring pressure on the hammer. This is normally done by screwing the knurled handle either in or out (in for more striking force). As the punch is pushed to a surface, the punch point shaft and anvil are pushed back into the handle of the punch. When the preset punching or hitting pressure on the hammer is reached, the hammer strikes the anvil and allows the point of the punch to penetrate the workpiece surface. Tips for this type of punch are generally removable by unscrewing them from the punch point shaft.

5-7. Combination Set

A set of tools that is a necessity in the practice of layout is the combination set, which consists of a square head, bevel protractor, and center head. Each tool in the set performs a specific layout operation.

As a draftsman makes use of a T square, triangles, and a protractor in the process of drawing or drafting, the mechanical technician uses the tools of the combination set to layout patterns onto the workpiece. In place of the draftsman's T square is the square head (also referred to as a square butt) with the steel scale used for ruling lines. The draftsman's triangles and protractor are replaced by the bevel protractor for ruling lines at different angles. The center head is unique to the mechanical technician as its main purpose is to locate the center of round or square objects.

Square Head

The square head is a tool constantly used in layout work. The square head with a scale is used principally for two functions: spacing lines accurately from given edges of a plate, thus locating a point; and providing for drawing a vertical or perpendicular line to an existing edge or line. The normal accuracy of spacing with the square head is normally $\frac{1}{32}$ (0.031) inch, but with care, an accuracy of $\frac{1}{64}$ (0.0156) inch can easily be achieved. The four edges of the scale are graduated in $\frac{1}{8}$, $\frac{1}{16}$, $\frac{1}{32}$, and $\frac{1}{64}$ inch. Graduations of $\frac{1}{50}$ and $\frac{1}{100}$ inch and metric units are also available.

The square head is an iron or steel casting with three accurately machined faces. Two of the faces are machined perpendicular to each other. The third side is machined to 45° with respect to the other two faces. A 45° angle is machined rather than a 30 or 60° angle because the 45° angle is used in most problems encountered in layout. The square head is slotted perpendicular to its base to accept a grooved steel scale.

A draw bolt with an offset key is used to hold the square head to the scale (Fig. 5-7). To insert the scale into the square head, unscrew the square head's knurled locking nut. Hold the square head up to the light to determine which side of the slot the offset key is on. Grasp the required scale and with the scale groove on the same side as the square head offset key, start to insert the scale into the square head. With your thumb against the knurled nut, compress the spring, allowing for fast alignment of the scale and the offset key. When the offset key is properly aligned in the scale groove, tighten the knurled lock nut. Recheck that the scale is in the position required. The habit of rechecking the position or dimension should be developed, because there is always the possibility that the scale will slip during the tightening of the lock nut.

A spirit or bubble level is generally built into the square head. This level is parallel to the base and is used to level or to check that a job is level. Recalling that there is a perpendicular side of the square head, the level may be used to check that a surface is vertical. It should be pointed out that when using the square head without the scale to level an object, a bubble indication exactly in the center (between the marks) of the glass tube indicates that the surface immediately under the level (square head) is level, and nothing more.

From time to time the accuracy of the bubble level should be checked by the following fast and simple method. Use a known leveled surface as a reference, if possible, or construct a leveled surface from a smooth metal plate or bar as follows: place a smooth twelve-inch (approximate) surfaced plate or square bar on the top of a rigid table. Place the square head on this piece and

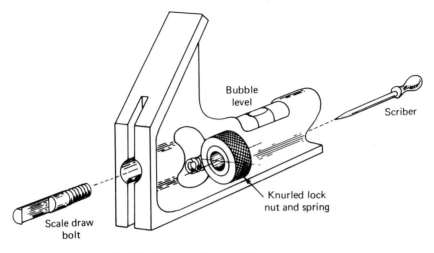

Fig. 5-7. Exploded view of the square head.

observe the bubble for levelness. Do not move the square head. Using metal or other similar shimming material, pack the plate or square bar until the bubble shows level between the marks. Now take the square head and turn it 180° and place it back in the same area as prior to the rotation. Observe the bubble. An accurate level will show the bubble still within the level marks, whereas an inaccurate level will show a shift of the bubble to the right or left of the level marks. Levels that are inaccurate can be repaired by the factory at the owner's expense.

One final but important tool is found in all quality square heads. A scriber is found stored in the square head casting near the base. (Refer to Fig. 5-7.) A split sleeve bushing places a restraint against the scriber, preventing it from falling out of the square head. The habit of storing the scriber in the square head immediately after use will prevent it from becoming lost or damaged.

Bevel Protractor

The bevel protractor (Fig. 5-8), is a tool having several different design shapes. It has two special layout functions: the measuring or measurement of an angle, and a means of providing a ruling edge for any required angle.

Fig. 5-8. Bevel protractor. (Courtesy of L.S. Starrett Co.)

Like the square head, the bevel protractor is an iron or steel casting with a machined base. Set into the base is a rotatable disc marked in degrees. The rotatable disc is used to hold a scale or straight edge (called the blade), and it also indicates the number of degrees of the angle formed by the base and the blade. The dial of the bevel protractor is graduated in one-degree increments over a range of 180°. This is done by a 0–90° segment and a 90–0° segment. The outer fixed rim of the protractor has positions which are exactly 90° apart and are noted by "0" marks. It is in relation to these marks that the angle is read. The graduated scale, when used with the bevel protractor, serves as a side of an angle; the scale is not used as a means of measuring a length. The scale is installed in the protractor head in the same fashion as described for the square head. Directly in back of the graduated face are two locking screws which are used to lock the protractor.

To increase the accuracy of angular measurement, some protractors are equipped with a vernier scale (Fig. 5-9). This vernier scale can at first be confusing, but with practice, its use will become easy.

Fig. 5-9. Vernier scale of a bevel protractor. (Courtesy of L.S. Starrett Co.)

The vernier scale segment is graduated with two groups of 12 divisions on either side of a zero mark. Each group covers a total space of 23° as shown on the outer rim. The total range of these 12 units (marked 0–60) is one degree. Therefore, 1 division on the vernier is $\frac{1}{12}$ of one degree, or 5 minutes— the ultimate accuracy of the protractor. Since the reading of a protractor from the left- or right-hand position is encountered from time to time, the vernier also has, as shown, a right and left hand scale about its zero mark.

For an example of reading the protractor, refer to Fig. 5-9. To properly read the angle indicated on the protractor, first count the number of whole degrees between 0° on the outer scale and 0 on the vernier. As shown, there are 50 whole degrees (50°) counting from right to left. The next step is the most important to remember: since the whole degrees have been read from right to left, then also read the vernier from right to left, starting at the zero point.

Looking along the left half of the vernier, it is seen that the line after the 15 numeral aligns with the degree division on the outer scale. Remembering that each division is 5 minutes (5'), the part of a degree is read to be 20'. Therefore, the total angle reads 50°20'. The divisions of 15, 30, 45, and 60 minutes are on the vernier to ease readability. In the above problem, the vernier was

1 division past 15′, so it was quickly determined that 1 division = 5′ and that by adding that increment to the 15′, the result is 20′.

The right-hand side of the vernier is used when reading whole angles from left to right starting with 0° on the outer rim. It is most important to remember that one only reads angles from 0 to 90° on the graduated disc. There are frequent times that it becomes necessary to read obtuse angles (obtuse angles are angles greater than 90 but less than 180°). The solution to this problem is to use the equation acute $\angle$ + obtuse $\angle$ = 180°.

Consider the problem that is shown in Fig. 5-10. The problem is to determine the included degree bevel on the workpiece.

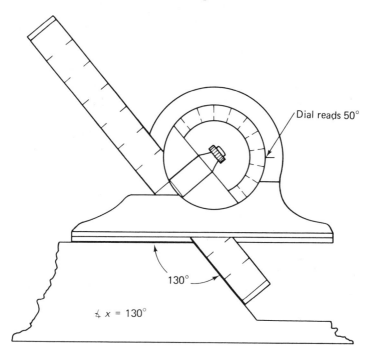

Figure 5-10.

PROBLEM

Determine obtuse $\angle$ x

Solution: Use bevel protractor.

By laying the protractor/scale on the workpiece and adjusting the two legs (protractor and scale), it is seen that an angle of 50° is indicated on the dial. By observation, one can see that the angle x is larger than 50°, but 50° is an acute angle. Referring to the equation, acute $\triangle$ + obtuse $\triangle$ = 180°, it is determined that the acute angle of 50° is known, but the obtuse angle is unknown.

Solving the equation for the obtuse angle and substituting all knowns into the equation results in obtuse $\triangle x$ = 180° − 50°. Obtuse $\triangle x$ or included $\triangle x$ = 130°.

With its built-in bubble level, the protractor head is also a useful tool for determining such things as slope or pitch lines of roofs, drainage systems, and the amount of angle that surfaces are out of level, out of horizontal or out of vertical planes. When determining the angle of conical items such as pictured in Fig. 5-11, a frequent source of error is the improper location of the protractor/scale on the center line of the workpiece. When confronted with this type of work problem, the first step is to locate the center of the workpiece and with a nonscratching scriber (pencil, chalk, etc.) and a scale, draw a center line on surface "A." With the protractor base edge on surface "A" aligned to the drawn centerline, as shown in Fig. 5-11, adjust both the base and the scale until the proper fit of the scale to the surface is achieved. Lock the protractor/scale and the angle. If, as shown by the dotted "offcenter line," the protractor/scale was aligned to it, an error would have resulted in determining $\triangle x$.

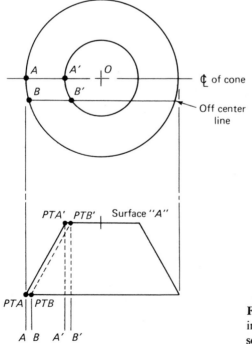

Fig. 5-11. Angle error due to improper measuring of conic sections.

Figure 5-11 illustrates the reasoning for this error. Points A and A^1, which are the bottom point and the top point on their respective circles, are intercepted by a true centerline and points B and B^1 are intercepted by an offcenter line. If these pairs of points are projected on a full half section of the cone, it is immediately obvious that the displacement between A^1 and B^1 is larger than A and B. The true accurate angle is bound by A to A' and A' to O. If the offcenter angle B to B' and B' to O were accurate, the angle would coincide with angle A $A'O$, which it does not.

Center Head

The center head tool is the third and final tool of the combination set. The center head pictured in Fig. 5-12 is an accurately machined tool, having two faces that are exactly 90° apart. The tail of the center head contains a machined slot and scale lock exactly as found in the square head and bevel protractor. In the case of the center head, the scale exactly bisects the 90° included angle when it (the scale) is installed in the machined slot of the center head. Although the actual uses of the center head are limited to only several applications, it is an invaluable tool in these applications.

Fig. 5-12. Center head of combination set. (Courtesy of L.S. Starrett Co.)

One of the center heads most important uses is in the location of centers of circular sections such as round bars or discs. The method of center location is a simple two step operation. With the machined registering faces held firmly against the workpiece and the scale locked tightly, scribe a line along the edge of the scale. Slide the head to a second position approximately 45–90° from the first position and scribe a second line. The intersection of the two scribed lines is the center of the workpiece.

Figure 5-13 shows a method of accurately measuring the diameter of a workpiece. Holding the center head firmly to the work, slide the scale until an even inch mark aligns over the edge of the workpiece closest to the center head; then lock the scale. Recheck the setting for accuracy. Read the second scale reading where the edge of the workpiece aligns to the closest $\frac{1}{64}$ inch increment of the scale. As shown in the example, the center head is initially lined up with the 5-inch mark and the second position aligns to the

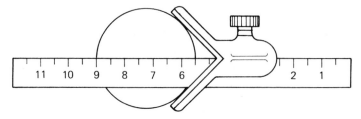

Fig. 5-13. Measuring round stock with a center head and scale.

9-inch mark. $9'' - 5'' = 4''$. Therefore, the diameter of the workpiece is $4 \pm \frac{1}{64}$ inches.

The final and least used application of the center head is the location of the center of square sections. While holding the machined faces of the center head firmly against one corner of the bar, scribe a line along the bisecting edge of a locked scale. Repeat this operation at a second position. Thus, the center of the bar is located.

5-8. Surface Gauge

The surface gauge is a tool that is used to perform a variety of scribing tasks demanded in layout practice. The surface gauge's basic function is to scribe a line or lines parallel to or concentric with existing edges on either a rectangular or a round workpiece. These parallel lines are made by adjusting a scriber point to a required diameter very accurately with respect to a surface or edge and then physically sliding the gauge to this reference edge to obtain the scribed line.

Figure 5-14 illustrates the surface gauge in detail. In describing its makeup, it is seen that the gauge consists of three main parts: the base, the main upright, and the scriber assembly.

The base of the surface gauge is the working heart of the assembly. The base is generally machined of cast iron or steel with the base and four edges mutually perpendicular. In the face of its machined base is a V groove. This machined groove allows for the base to be accurately set on a shaft and rotated (Fig. 5-15) for scribing concentric circles. Within the base, there are two or four hardened pins that are also mutually square to the base and parallel to the edges. These pins are friction fit into holes and can be slid out to extend past the base surface. This allows the base to be slid along the edge of a workpiece or a keyway. These pins are pushed back into the base when not in use.

On the top surface of the base block is a rod (upright), clamp, and adjusting bar. The unit serves to clamp a rod in a variety of positions. This clamping

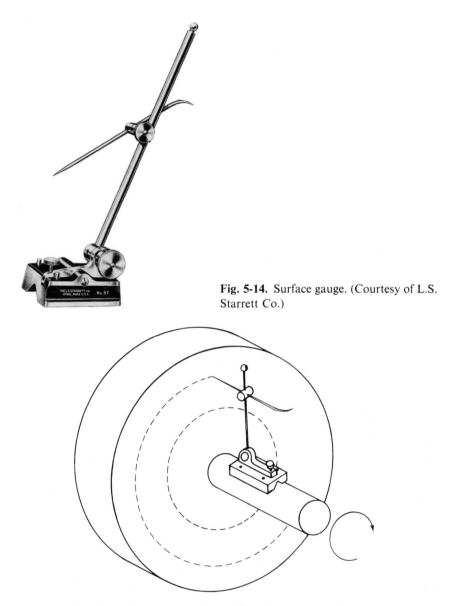

Fig. 5-14. Surface gauge. (Courtesy of L.S. Starrett Co.)

Fig. 5-15. Use of vee in base of surface gauge on a round shaft/flange layout.

arrangement, which is part of a rocking bracket, is fine adjusted by the knurled tipping adjustment nut. This nut will tip the clamping assembly up or down, adjusting the scriber point height. On most surface gauges, the rod clamp sleeve has a second smaller hole in it. This smaller hole is to hold the scriber rod itself when this situation is required. The main rod supplies a *means* for the scriber rod assembly to travel over for the different heights required. The base end of the main rod has a small thin washer or flange affixed very close to the rod end. The purpose of the washer or flange is to stop the rod from being pulled through the clamping joint. This allows the rod to rotate to any angle needed without interfering with the base. Short or long rod lengths can be either bought or made to meet any height required.

The scriber rod assembly or working end consists of an adjustable main rod clamp and a swivel locking clamp to hold and adjust the scriber. Both the main rod and the scriber locking is done by a single knurled nut. The scriber is a combination straight end and "hooked end" configuration.

The following discussion illustrates an exercise for setting an accurate height using the surface gauge, square butt, and scale. With the scale of the square head resting on a working surface, set a height of $8\frac{1}{4} \pm \frac{1}{64}$ inches (8.250 $\pm$.015 inches) on the gauge. This dimension is well within the limits of the scale accuracy. To perform this setup, first adjust the main rod to a vertical position and lock it. Loosen the scriber locking assembly and slide the assembly to a height where the scriber is approximately horizontal. Immediately adjust the point to the approximate dimensions required; in this case, $8\frac{1}{4}$ inches within $\frac{1}{16}$ inch either side of $8\frac{1}{4}$ inches. This is a rough setting of the height. Lock the scriber assembly. With the fine-adjustment screw in the gauge base, tip the pointer either up or down until it corresponds to the dimension of $8\frac{1}{4} \pm \frac{1}{64}$ inches as measured on the scale.

5-9. Hermaphrodite Calipers

Hermaphrodite calipers (Fig. 5-16) can be recognized by the fact that one leg has a bent and rounded end or pivot leg, and the other leg has a scribing point locked into it. The joint of the caliper is of the friction slip joint type which allows one leg to slip past the other. The hermaphrodite calipers are used for two main purposes: locating centers of irregular and regular rounds such as bars or holes, and stepping off lengths from edges or corners.

The method for locating the centers of irregular and regular rounds is discussed first. There are occasions when rough round stock, whether a forging, casting, or saw cut, is to be laid out prior to machining. For machining, the center location of the workpiece is required. To find the center, measure the approximate diameter. Using one-half of this measurement, set the

Fig. 5-16. Hermaphrodite calipers. (Courtesy of L.S. Starrett Co.)

gap between the caliper pivot leg and scriber point. The center point of the workpiece may be located by either a *three* or a *four arc* method, as shown in Fig. 5-17. In the three arc method, the pivot leg is held firmly against the outer edge of the workpiece and an arc is drawn. This is repeated two more times at approximately a one-third (120°) position on the outer edge. When properly scribed, a small triangle is produced and a center is found. In the four arc method, the same scribing is performed except at four positions around the edge. The scribed figure, in this case, is a square. By using a straight edge and scribe, diagonals are drawn from the corners of the square. The intersection of the diagonals determines the required center point. As can be seen, the four arc method is more accurate, as the center is not estimated as it is in the three arc method.

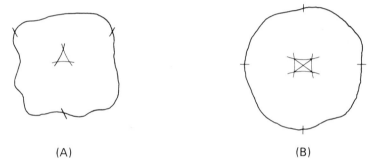

(A)   (B)

Fig. 5-17. Locating centers of irregular round stock with hermaphrodite calipers: (A) 3 arc method and (B) 4 arc method.

The second use of the hermaphrodite calipers is in the stepping off of lengths from edges or corners. In stepping off lengths from edges of workpieces, the thickness of it plays a part in the proper setting of the caliper. In the problem of Fig. 5-18, arcs are to be stepped off one inch in from the edge of a $\frac{1}{4}$-inch-thick plate. In setting the point, the thickness offset must be allowed for. By setting the pivot $\frac{1}{4}$ inch down from the scale, as shown, the point is then set to the one-inch mark. An error commonly found in this type of measuring has been eliminated. In working from a corner, this aforementioned problem does not appear, as the pivot leg and the point lie on the same plane. To work from a corner, simply reverse the legs by sliding them past one another as pictured in Fig. 5-18(B).

SUMMARY

For review, assume that it is desired to transfer the pattern shown in Fig. 5-19 to a piece of aluminum. First, the aluminum is coated with a colored layout fluid so that scribe marks will be easily visible on the workpiece. Then,

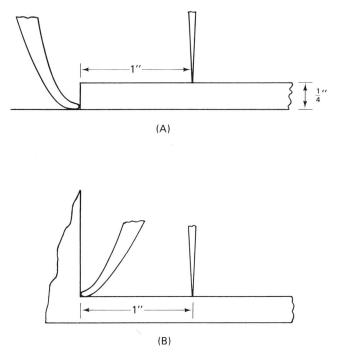

Fig. 5-18. Use of hermaphrodite calipers to (A) perform layout from edge of workpiece and (B) perform layout from corner of workpiece.

using the square head and scale, a scribe is used to draw a line AB perpendicular to one of the square sides of the aluminum workpiece. A scale is then used to measure line AC.

At point C, the bevel protractor is used to measure angle ACE to $23°$. A point is marked and a line is drawn, for example CJ. Point D is measured and an indenture is made with the prick punch. The dividers are then set to a radius of $3\frac{1}{2}$ inches and an arc is drawn from line CJ (intersection at point E) around to approximately point F.

The scale is used to measure GH. The surface gauge is then used to scribe line GF parallel with line AH. The layout is complete.

Had the radius of arc EF been greater than the capacity of the dividers, a trammel would have been used.

The center head is used for locating the centers of circular sections and sometimes for centers of square sections. The hermaphrodite calipers are used for locating centers of irregular rounds and for stepping off lengths from edges or corners.

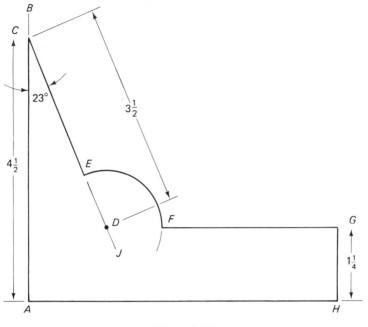

Figure 5-19.

REVIEW QUESTIONS

1. In using a bevel protractor with a vernier, what is the readable accuracy of any degree setting?

2. Looking at the section of the protractor pictured in Fig. 5-8 read the indicated angle to its best accuracy.

3. Name the three main parts of the automatic center punch.

4. Why is the divider held at an angle to the scale when setting its points?

5. What two methods are used to adjust the "floating" point end of a trammel?

6. List some uses of the guide pins in the base of a surface gauge.

7. What is the purpose of the V groove in the bottom of the surface gauge base?

8. What determines the selection of a layout fluid color?

9. Name the two methods used in locating the center of a piece of round stock with hermaphrodite calipers, and list the advantages of one over the other.

6

LAYOUT

The term *layout* means scribing or drawing required shapes, outlines, and intersections on a material as a guide prior to the actual machining operations on the material. Layout is an operation that is used extensively to eliminate errors and to graphically show the boundaries for machining operations.

Chapter 5 provided the descriptions and uses of the tools used in layout. This chapter describes layout aids and the four types of layout problems to be encountered by the technician. The types of layouts are: two-dimensional (flat), circular, three-dimensional, and shaft layouts.

LAYOUT AIDS

Prior to a discussion of layout, we will first discuss some special aids that are used in the art of layout. These aids are generally large and expensive and normally are supplied by the employer. These aids include: layout tables, parallels, angle plates, V blocks, and machine squares.

6-1. Layout Tables

The first special item used in layout is the layout table (Fig. 6-1). This table is generally a large cast iron plate that has a machined top work surface with its four edges also machined mutually square to each other and perpendicular to the top work surface. Cast iron is chosen for the reason that it has a natural slippery surface and is far less likely to distort or warp with age and use.

Fig. 6-1. Typical layout table with stand. (Courtesy of Challenge Machinery Co.)

The size of a layout table can run from one with a top surface of 18 by 24 inches to a large heavy industry table of approximately 30 by 100 feet in length. Small-to-medium size tables are generally mounted at waist height on a steel support frame. The sole purpose of the layout table is to provide an area to set up and layout workpieces with respect to a flat reference surface. Very often the top surface of a table will have mutually perpendicular reference lines ruled into it. They may be spaced on 4, 6, or 12-inch centers. These reference lines are very useful in aligning workpiece centers and for three dimensional layouts.

Layout tables are generally cared for by removing burrs, treating the surface, and removing rust. Raised burrs caused by dropping workpieces or tools are removed by filing or rubbing with a stone. The table surface is treated by rubbing fine graphite flakes into the top to provide a dry film lubrication that cuts down on scratching of the cast iron surface. Rust is removed with a crocus polishing cloth.

6-2. Parallels

Parallels are layout aids that are made of hardened and ground tool steel or cast iron bars and are normally supplied two to a set. The parallels may be square or rectangular in cross section with a wide range of sizes from approximately $\frac{1}{2} \times \frac{1}{2} \times 4$ inches to $12 \times 12 \times 24$ inches (Fig. 6-2). These bars provide an accurate spacing block to raise a workpiece from a machined

Fig. 6-2. Parallels. (Courtesy of L.S. Starrett Co.)

surface so that working on a layout is easier. This point will be made clearer in the discussion on layout.

The finish and the dimensional tolerances of parallels make them very costly. The width and thickness tolerances of these parallel bars are generally within a few ten thousandths of an inch and their flatness is also within the same range.

As with any precision tool, care must be exercised in maintaining the tool's accuracy. Nicks or burrs should be stoned off of the parallels with a fine grit stone. After use, parallels are oiled all over to prevent rust and should be stored in wooden cases separated from each other to prevent rubbing.

6-3. Angle Plates

Angle plates are layout aids that are generally L shaped castings made of iron. The angle plates form two outside surfaces that are perpendicular to each other, normally within a tolerance of 0.0005 inch in 12 inches. The sides and ends of the layout plate are machined but not to the closeness of the two work surfaces. The size of an angle plate is determined by the face length in inches of the two work faces followed by its face width (Fig. 6-3). Therefore, a 6 by 6 by 8 inch angle plate has a 6 inch base leg, a 6 inch vertical leg, and an 8 inch face width. This layout aid has wide use when used in conjunction with a layout table, as it adds a third dimensional reference surface where required. It also allows for workpieces to be worked on in a vertical position.

Fig. 6-3. Cast iron angle plate. (Courtesy of Challenge Machinery Co.)

Angle plates are commercially available up to a 24 by 24 by 24 inch model and can be procured in either a planer cut surface or the higher precision ground surface finish. The same care taken with the layout table should be extended to the angle plate.

6-4. V Blocks

V blocks are tools that are made of either steel or cast iron. V blocks, which are sold in pairs, are characterized by a 90° included notch of the same size in each block. Blocks having four different size V, one size on each face, are also available to cover a small-to-large range of work. (Fig. 6-4). Small V blocks start at 2 inches long by $1\frac{1}{2}$ inches square.

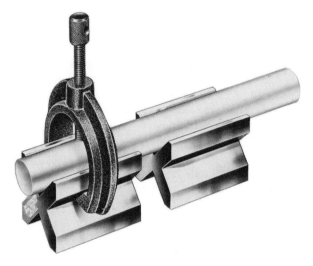

Fig. 6-4. Vee blocks shown holding a piece of round stock. (Courtesy of L.S. Starrett Co.)

V blocks, with their unique design, are used entirely for holding round workpieces, whether it be round discs or long round shafts. V blocks are not only used in layout practice but, as will become apparent, are also used to hold these same round workpieces for actual machining operations. Care is the same as for the layout table.

6-5. Machine Squares

The machine square (Fig. 6-5) is a tool used to align a surface of a part perpendicular with another surface. This is achieved because the outer edge of the square is machined or ground to within extremely close perpendicular tolerances, generally within a few ten thousandths of an inch. Basic blade lengths and beams begin at $1\frac{1}{2}$ inches.

The actual aligning procedure with a machine square is as follows:

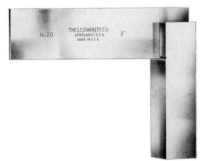

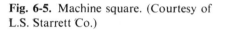

Fig. 6-5. Machine square. (Courtesy of L.S. Starrett Co.)

always wipe the edges of the machine square with an oil coated rag to remove any foreign material. Set the base of the machine square down on an equally cleaned section of the layout table. Slide the base of the square to the workpiece surface to be squared. After a light contact point has been made, the process of squaring the workpiece is begun.

Several trials must usually be made, each bringing the job surface more in parallel to the machine squares upright edge. When the surface is thought to be parallel with the square, one final check should be made. Using shimstock 0.0005 inch thick, check along the mating surface of the workpiece and straight edge. This procedure guarantees a squareness to within this tolerance (0.0005 inch).

The machine square is also used to check machined surfaces for flatness or warping. By using a variety of shims, the surface with respect to the straight edge can be inspected with very good accuracy.

The machine square is cared for the same as the layout table.

TYPES OF LAYOUT

There are four types of layout to be learned by the technician: two-dimensional, circular, three-dimensional, and shaft. These types of layout may each be performed by several methods using the layout tools described in Chap. 5 and the layout aids previously described in this chapter.

6-6. Two-Dimensional Layout

The most common form of layout is the flat or two-dimensional layout. This layout is used generally on sheet metal, flat sheet, and any job where the only measurements (two) required are width and depth. The third dimension (height–thickness) is ignored as it is usually either a round or other form hole through the material. Once the two-dimensional workpiece has its edges properly cut, it may be laid out by square head/scale and divider method, angle plate/surface gauge method, or the template method.

Selection of Reference Edges

In working with flat stock, two forms of cut workpiece edges are available: sheared and saw cut piece edges. In comparing the two forms, the sheared sheet generally has edges that are straighter and squarer to each other. Saw cut stock, particularly that done by a band saw, sometimes has wavy edges. Therefore, it is very important to inspect all, saw cut or sheared edges to select the straightest edge as the reference edge.

Prior to any layout, the first step is to inspect that the material is within the tolerance range of the engineering drawing dimension. Any undersize or oversize pieces should be reported for possible replacement. After inspection of the piece or pieces, they should be cleaned of grease or dirt, which could effect both the layout fluid to be used and the surface finish of the material. The next and most important step is the selection of the mutually perpendicular reference edges. By using either a machine square or the perpendicular edges on a layout table, locate the two adjacent edges that are the most perpendicular (90°) to each other. These two edges will serve as the only reference used during layout. Once the reference edges are determined, they should never be switched during layout, as errors will occur and could destroy the usefulness of the finished workpiece. This statement can be best demonstrated in the following problem (Fig. 6-6).

A plate, with two holes to be laid out on it, is sheared to an actual size

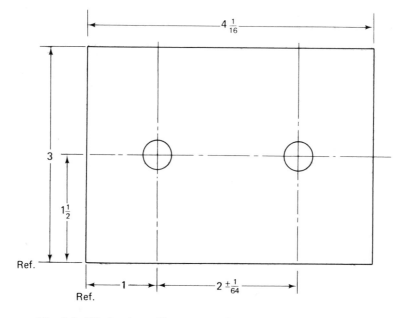

Fig. 6-6. Elimination of layout error by selection of only two reference edges.

of 3 by $4\frac{1}{16}$ inches. The engineering drawing dimension and tolerance on the sheet size is $3^{+1/16}_{-0}$ by $4^{+1/16}_{-0}$ inches and so the plate is within acceptable tolerances. The engineering drawing also shows two holes with a center to center tolerance allowance of plus or minus $\frac{1}{64}$ inch. The first hole is located 1 inch from the left edge and $1\frac{1}{2}$ inches from the base edge. This layout could be done very quickly with a square head and scale to lay out the hole centers. In looking at the engineering drawing, the right hand hole is 1 inch from its center to the right hand edge.

As stated, the plate is $\frac{1}{16}$ inch over 4 inches and if a scale were used to measure a hole 1 inch in from each end, an error of $\frac{3}{64}$ inch out of tolerance would be made and would void the workpiece. To prove this, a quick math check is used:

Plate (overall length)	$4\frac{1}{16}$ inches
Two 1-inch end spacings to holes	$-\,2$ inches
Center to center actual distance	$2\frac{1}{16}$ inches
Print range $1\frac{63}{64}$ through to $2\frac{1}{64}$ inches	
Actual range	$2\frac{4}{64}$ inches
Out of tolerance	$\frac{3}{64}$ inch

Having shown the main problem in changing reference edges, the discussion of two dimensional layout can continue.

When faced with a layout of a flat workpiece, there are three general approaches that can be used: square head and divider layout, surface gauge layout, and template layout.

Each of these layout methods possesses the capability of producing general tolerances of plus or minus $\frac{1}{64}$ of an inch. The only reason for choosing one over the other is the job application and the quantity and size of the workpiece.

Square Head/Scale and Divider Layout

In many cases of only one or two workpiece layouts, or in cases where a job cannot be moved to a layout table, the square head/scale and divider are used. The following approach is recommended for layout of the previously discussed plate (Fig. 6-6) and for similar layout problems. After locating the two best perpendicular edges of the workpiece, place them so that they appear on the bottom and left side as viewed by you when laid flat. Coat the surface with the pretested layout fluid as discussed in Chap. 5 and allow the fluid to dry.

Generally, all flat sheet drawings are laid out with the longest side running left to right (width) and the short side (depth) vertical (Fig. 6-7).

In laying out a series of long horizontal lines on a workpiece, it is a good practice *not* to lay out the horizontal lines by resting the square head on the

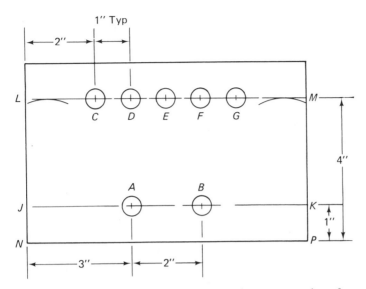

Fig. 6-7. Divider layout—correct approach to construction of parallel lines and elimination of error buildup.

one reference edge to draw them. The vertical edge, which may in general be straight, could be wavy in actuality. If this edge were used for the first horizontal line, it would be acceptable. Now if, as noted, the second horizontal line were drawn 3 inches up from the first line, the obvious next move would be to slide the square up 3 inches to define the second horizontal line.

In sliding the square head up 3 inches, the machine edge of the square head could lie within a wavy section of the workpiece edge. This could, and does, tip the horizontal edge out of parallelism with the first ruled line. If the layout were continued, the work would run out of tolerance so badly that the workpiece would be useless. The preferred method in performing this layout would be to scribe the first horizontal layout line using a scriber along the square. After scribing, refer to the drawing to determine the next horizontal line. This second line (*LM*) is located 4 inches up from the base line (*NP*).

Chapter 5 has shown the method of producing a line parallel to an existing line. This is our approach in producing the second line (*LM*) in our problem. The actual difference between our lines is 4 inches minus 1 inch, which equals 3 inches. We shall use the dividers set to 3 inches point-to-point, to measure the distance from the scribed line (*JK*) to the second line (*LM*). Using a prick punch, punch two small indentations at the far right and left end of the scribed line (*JK*). Using these indentations and the preset dividers (to 3 inches), draw arcs. Lay a scale tangent to the produced arcs, and scribe a line *LM* that connects the arcs. This line is parallel to the line *JK*.

This method of producing parallel lines can be extended to any amount of lines required but should, for accuracy, always be referenced to the original scribed line (as *JK*) whenever possible. Having produced the horizontal lines in layout (the *Y* dimensions), we are now ready to produce the required horizontal (or *X*) dimension.

Two methods are available to produce the horizontal dimensions: direct spacing off by use of a square head and scale; and use of dividers. In the case where several points or intersections lie along a common horizontal line (Fig. 6-7), the square head and scale are used. Since, the square head rests on essentially the same edge region, possibility of mistakes due to edge error is eliminated. Thus, spacing measured with a square head and scale is accurate as long as the reference edge that the square head rests on is the same for each measurement along that line. In the case of spacing the two rows of holes as per the engineering drawing (Fig. 6-7), it is a cut and dry method.

The second method of laying out the *X* dimensions is the divider method. This normally requires a reference point or intersection on a particular line to start from. Again, referring to the engineering drawing (Fig. 6-7), horizontal line *JK* has a hole (*A*) spaced in 3 inches from the left end. This hole (*A*) center can be located with a square head and scale set to the proper dimension.

After the intersection for the hole (*A*) center is located, the intersection should be checked for accuracy and then prick punched. With the tools still at hand, go to line *LM* to locate its first hole (*C*), which is located 2 inches from the left edge. Mark and prick punch it. Returning to the first line intersection (hole *A* center), we find the second hole (*B*) is spaced 2 inches away. Set the dividers for 2 inches and swing an arc using the punched hole center (*A*) as a pivot. Having laid out the first line of holes, we move to the second line *LM*. Looking at the number of holes required (five), the drawing calls out a 1 inch typical spacing. First impressions would be to set a divider at 1 inch and step the holes off from (*C*). This method can and does cause tolerance build up. This can be shown very quickly, if the error in setting the divider is, say, 0.010 inch over or under the 1 inch size. Stepping off four more spaces from hole (*C*) would give a total error of 0.040 inch build up (a 0.010 inch error per space). This error, as per the print tolerance, is not acceptable. The accepted method to layout this or other multi-hole layouts is to work both ends to the middle. This method does not allow for as great a tolerance build up. In our problem, the approach is as follows: having located the (*C*) hole, we set the divider to the outer hole (*G*) distance (4.000 inches) and draw an arc, using the prick punch indention at (*C*) as the pivot. Now set the divider to the one inch spacing and draw arcs for the adjacent hole (*D*) (from *C*) and (*F*) (from *G*). After this has been performed and

the prick punch indention made, use the set dimension of the divider for hole
(*E*) using both the (*D*) and (*F*) indentions as centers. In some cases, the two
arcs cross at the same point and in others a slight separation is noticed. In
the case of a separation the arcs have split the distance between the (*D*)
and (*F*) hole marks (Fig. 6-8). Mark the center of hole (*E*) with the prick
punch.

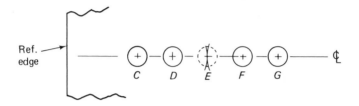

Fig. 6-8. Correct sequence for layout of equal spaces with
dividers: (1) *C* to *G*, (2) *C* to *D*, (3) *G* to *F*, (4) *D* to *E*, (5) *F*
to *E*.

In summary, this method keeps divider error from building by the fact
that the error is added, as in hole (*C*) and (*D*), and subtracted, as in holes (*G*)
and (*F*). Hole (*E*) is then located by splitting the distance between (*D*) and
(*F*). This procedure, which is used to eliminate tolerance build up, will be
extended to multi-hole layouts on circles (circular hole layout), to be discussed
in subsequent paragraphs.

Very often in panel layout, particularly in electronic panels, holes of
either square or rectangular configurations are required for mounting gauges,
lights, or switches. Figure 6-9 shows a typical rectangular hole layout problem.

When doing multi-hole layouts, always refer to the engineering drawing
or sketch for other holes that go with a particular hole such as mounting
holes for gauges, lights, etc. By grouping cutouts to holes, a much closer
working accuracy can be maintained.

In the problem of Fig. 6-9, we have a rectangular cutout with four
mounting holes associated with it. This grouping is ascertained quite often
by observing the dimensions that are normally referenced from the cutout
center lines.

Two methods of attack can be used in this type layout: divider/square
layout and angle plate/surface gauge layout. Angle plate/surface gauge
layout is described in a subsequent section of this chapter.

In the divider/square layout, the center lines are located and scribed
using the selected reference edges of the workpiece. With the two perpendicular center lines of the proposed cutout located, we are ready to begin the
layout. As shown, the horizontal limit of the cutout is 4 inches. Since we will

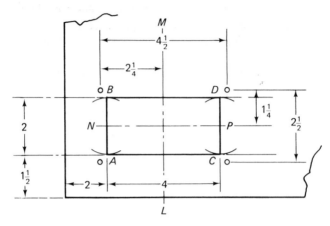

Fig. 6-9. Rectangular hole/bolt pattern layout with scale, square head, and dividers.

be working about the center lines, we set the dividers on 2 inches, which is $\frac{1}{2}$ of the total length of the horizontal cutout.

Locate one leg of the dividers in a prick punch mark on the vertical center line LM below the horizontal center line, and scribe arcs on either side (the arcs will be on lines AB and CD, which are not constructed at this point). Move the divider up on the same vertical center line LM 2 or 3 inches and repeat the construction of the arcs on either side, (again, the arcs will be on lines AB and CD). Taking a scale, lay it tangent to a pair of arcs on either side and scribe a connecting line (the line through AB). Repeat this on the other set (for line CD). These two parallel lines are now 4 inches apart and form the vertical sides of the rectangular layout hole $ABCD$.

To form the completed box layout $ABCD$, we look at the length of the vertical lines AB and CD; the drawing calls for a 2 inch O.A. (overall) length. Set the dividers at 1 inch, which is $\frac{1}{2}$ of the length. Where the two vertical lines (proposed AB and CD) cross the horizontal center line (NP), lightly prick punch indentions. With these indentions as centers, swing 1 inch arcs top and bottom at both ends. Lay a scale tangent to the top pair of arcs and scribe the line BD. Repeat for line AC. The rectangular cutout $ABCD$ is now defined.

The final layout requirement is the location of the four mounting holes. The use of the divider, as previously described in the rectangular layout, is used to produce arcs about the center lines for the mounting holes. It is not necessary to scribe long lines between these arcs, but rather short ones to locate the holes only. This method of layout, which is merely constructing parallel lines about a drawn center line, has two very good assets. The accuracy of this type layout is excellent, and the parallelism of the lines is very accurate. This method of layout should be practiced by the technician with a pencil compass and paper to develop the technique. Divider layout is used when only a few pieces are required and the use of a layout table is unavailable.

Angle Plate/Surface Gauge Layout

When many pieces of flat work layout are required, the use of the angle plate, surface gauge and layout table is recommended. This method allows for both speed and accuracy in the work.

As previously done for other layout methods, two square reference edges of the workpieces are chosen and marked. Select an angle plate to which the flat sheets can individually be C clamped in a vertical position to the layout table. Next, clamp a scale vertically to an angle plate or assemble a square head and scale and set it on the surface table. Either of these two methods provide a reference measurement for use during layout. Procure a surface gauge with a straight scriber point. In this method of layout, there is no need to scribe the center lines of a rectangular cutout. Select a reference edge on the workpiece, set the workpiece down on the table and clamp it to the angle plate. For example, say the horizontal edge of a workpiece (Fig. 6-9) is on the layout table. This means that we can scribe all of the horizontal lines of the layout parallel to the base with the surface gauge. This method always requires a paper and pencil to do additions or subtractions, since all of the measurements must be referenced to the edge of the plate resting on the surface table. The first horizontal line (AC), which is the bottom of the cutout, is $1\frac{1}{2}$ inches from the horizontal reference edge. Adjust the surface gauge scriber point to the $1\frac{1}{2}$ inch mark on the scale. Holding the base firmly down and with the scriber dragging on the workpiece, rule line AC (longer than required). This defines the base line of the cutout. The drawing shows the upper cutout edge (BD) 2 inches above the bottom cutout line; remember everything is referenced from the horizontal reference edge so that this line is located $1\frac{1}{2}$ inches plus 2 inches, which equals $3\frac{1}{2}$ inches above the reference edge. By setting the scriber at the $3\frac{1}{2}$ inch mark on the scale, a line can be scribed for the upper edge (BD) of the cutout.

The next item that can be scribed is the location of the four mounting holes located from the horizontal reference edge. The holes, it is noticed, are dimensioned off of the rectangular cutout center lines on this drawing. Since no center lines are used, a little addition and subtraction must be performed. The holes must be located from the horizontal reference edge. The horizontal center line is found from the fact that the bottom edge of the cutout is $1\frac{1}{2}$ inches up from the horizontal reference base and also that 1 inch must be added to it, one inch being the midpoint or the center line of the 2 inch width.

The center line is therefore $2\frac{1}{2}$ inches from the horizontal reference edge. From the drawing, the total distance between the holes is $2\frac{1}{2}$ inches, or $1\frac{1}{4}$ inch on either side of the horizontal center line.

To locate the top holes on the layout, add $1\frac{1}{4}$ inches to the calculated center line of $2\frac{1}{2}$ inches. The top holes are, therefore, located $3\frac{3}{4}$ inches from the horizontal reference edge. Set the surface gauge at $3\frac{3}{4}$ inches and scribe the line.

There is now only the bottom row of holes that need to be located. From the setting of $3\frac{3}{4}$ inches for the top row, subtract $2\frac{1}{2}$ inches. This answer of $1\frac{1}{4}$ inches is the location of the bottom holes to the horizontal reference edge and is also the setting of the surface gauge.

The next step in this layout is to rotate the vertical reference edge down onto the layout table and clamp the workpiece to the angle plate. Since the procedure is exactly the same as has just been mentioned, a quick run through for dimensional spacing will follow. The bottom edge of the cutout (AB) is located 2 inches from the vertical reference base. Set the gauge to 2 inches and scribe. Add 4 inches (the length of the cutout) to 2 inches. Set the gauge to 6 inches and scribe the upper end of the cutout (CD).

In the location of the mounting holes, the center line to vertical reference edge distance is 4 inches. Adding $2\frac{1}{4}$ inches to 4 inches, set the gauge to $6\frac{1}{4}$ inches and locate the intersection of the top row of mounting holes. Subtract $4\frac{1}{2}$ inches from $6\frac{1}{4}$ inches and a surface gauge setting of $1\frac{3}{4}$ inches is found. This is the location of the bottom row of holes. After checking the accuracy with a scale, remove the workpiece from the angle plate and lightly prick punch the corners of the cutout. This is done so that in case the layout fluid is wiped away by accident, the cutout could still be drawn using the prick punch marks. Center punch the four mounting hole centers of the layout.

At the beginning of this section, layout speed was mentioned as a requirement in multiple layouts. Using the layout method just described, it can be seen that if ten or more pieces were to be laid out, they could each have a line scribed on them at each setting of the surface gauge point. This would eliminate resetting the surface gauge for each individual piece.

Template Layout

Template layout is a process used on flat layout work where either many pieces are to be laid out or the layout is of a complex shape. (Refer to Fig. 6-10.) A template is an exact duplicate of the finished workpiece made in thin sheet metal (0.010 to 0.012 inch thick).

The actual process for making a template for template layout is rather simple. The outline of the piece is laid out with all hole intersections left on. All lines for joints or cutouts are also clearly marked. In Fig. 6-10, sides A and B of the workpiece (template) are to be bent 90° to side C, after putting

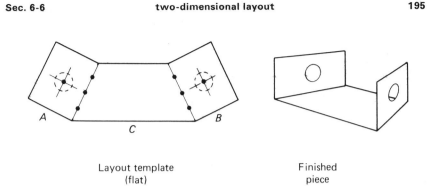

Layout template
(flat)

Finished
piece

Fig. 6-10. Use of a template to lay out a workpiece.

a hole in each side as shown. Therefore, both the holes and the bending line on the workpiece must be shown.

The method of marking these items on a layout template is simple: use small prick punch marks to show the centers of holes or the outlines of bend joints or other layouts on the template.

After the template has been laid out, a pair of metal hand shears are used to cut the outline. To use this template, simply lay the template on the pre-coated workpiece and scribe the outline. After scribing the outline, the second step is to place the prick punch through the punched template holes and punch the required hole centers, line points, etc.

After removing the template, a scale and scriber are used to scribe lines for cutouts between the punch marks on the workpiece. Holes can be scribed with a divider. A scriber should be used to mark the template with drawing number or to note the template's use.

Templates should always be retained for future use. Apply a thin coat of oil on the template prior to storage, and store the template flat to prevent bending.

Circular Layout (Chordal Method)

Circular layout is a process of locating evenly spaced angular increments around a scribed or drawn circle. These increments can be as few as 3 and as many as 150 or more increments. Circular layout is achieved with relatively few tools: a divider or trammel and scale.

The purpose of circular layout is to provide hole patterns on such items as flanges, discs, boiler heads, or any workpiece where evenly spaced holes on a specific diameter are needed.

Circular layout could be performed using a protractor, but this is difficult. For example, if the engineering drawing of a workpiece calls for five equal spaces to be located on a 4 inch diameter circle, this means that the 360° of the circle must be divided into five increments or angles (each 72°).

This layout is started by scribing a starting line from the center of the circle to any point on the 4 inch diameter circle. The next step is to locate the center of the protractor on the center of the circle, with the 0° line of the protractor lying on the previously scribed starting line. With a scriber, increments of 72° and 144° can be marked on one side of the starting line. Repeat this on the other side of the scribed starting line. Scribe lines to the circumference of the circle, using its center and the four 72° increments. These four lines plus the starting line form the five equal spaces required.

In circular layout by the chordal method, the solution to five equal spaces is as follows: multiply 0.588 × 4 inches = 2.352 inches, where 0.588 is a constant to be explained later and 4 inches is the diameter of the circle required.

Using a scale and divider, set the divider to 2.352 inches (very close to $2\frac{23}{64}$ inches). Prick punch a mark on the 4 inch diameter circle. Using this mark as a center, use dividers to scribe an arc on either side of center. Prick punch two indentions where the two arcs intersect the circle. Using these two points as new centers, scribe two more arcs. To check for accuracy, use the divider to see if the distance between the last two arcs is the same. If the arc distance is smaller than the divider, the original divider to scale setting was a little oversize; conversely, if the arc is bigger than the divider setting, the divider setting was a bit undersize in the scale setting. Correct the divider setting and start the layout again.

As can be seen, this method of circular layout has two distinct advantages over the protractor method: it is faster and its accuracy is easily checked.

In establishing the uniqueness of chordal circular layout, a discussion of its basic theory will be presented. In our previous problem, the divider was set to a $2\frac{23}{64}$ inch setting. This setting is in reality a chord that will be evenly spaced into the circumference of a 4 inch diameter circle. A chord, it will be recalled, is a straight line that connects the ends of an arc.

In a five equally spaced circumference, there are five angles each equal to 72° (5 × 72° = 360°).

Let us look at one of these five sectors in Fig. 6-11; but, instead of a 4 inch diameter circle, we will use a 1 inch diameter circle, sometimes called a *unit circle*.

Simply stated, the problem is to find the length of a chord sustained by a central angle of 72° in a 1 inch diameter circle. The knowns in this figure are the radius, which is equal to 0.500 inches, and the central angle, which is equal to 72°. The unknown is the length of the chord.

In any problem of this type, the first step is to break the figure into 90° triangles. By bisecting the 72° central angle into two 36° angles, the bisector line will meet the chord at its midpoint and will also be perpendicular to

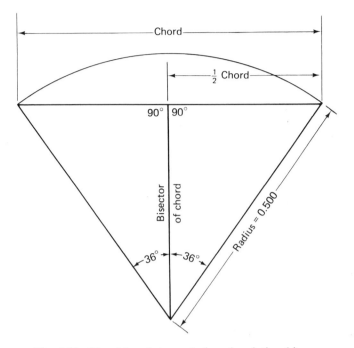

Fig. 6-11. Chord length to central angle relationship.

the chord. This bisection has now made two mutually equal 90° triangles, each having a side equal to $\frac{1}{2}$ the chord. By observation of one of the 36° right triangles, the solution to finding the length of the ($\frac{1}{2}$ chord) side is the use of the sine trigonometric function.

$$\text{sine } \theta = \frac{\text{opposite side}}{\text{hypotenuse}}$$

where $\theta = 36°$

hypotenuse $=$ radius $= 0.500$ inch

opposite side $= \frac{1}{2}$ hypotenuse $=$ unknown

From the trigonometric tables, the sine of 36° is 0.5878—substituting and solving for $\frac{1}{2}$ chord

(unknown) $\frac{1}{2}$ chord $= 0.5878 \times 0.500$ inch

(unknown) $\frac{1}{2}$ chord $= 0.2939$

The chord is therefore equal to 0.5878 inch.

This figure of 0.5878 inch is a constant that now can be used in any five equally spaced layouts regardless of the diameter of the circle. Since the above solution was performed on a unit diameter circle of 1 inch, a simple multiplication of the five space constant (0.5878) by the diameter of any circle is all that is required to determine the chord constant.

Solutions, typical of the above can be found for any range of equal spaces on a circle. For the technician's future use, Table 6-1 is included and lists the chord constants for a unit diameter circle.

Before leaving circular layout, a brief review will be given of some of the problems that will be encountered and their solutions.

Quite often a flange on the end of a pipe or casting will require a bolt circle layout. The first and obvious problem is how to scribe the required circle on the flange when there is no place to rest the point of the divider so that the required circle can be scribed.

A solution called "sticking the bore" is used. This entails the use of a square or rectangular wooden strip jammed into the bore or hole to be laid out. The face of the wood should be flush with the flange face. The wood near the center should now be covered with a dark colored wax crayon. Using a hermaphrodite caliper and the technique described in Chap. 5, locate the center of the flange by using the bore as a reference. With the center of the hole located, the next step will be the scribing of the proper diameter circle on the flange. Leave the stick in the bore until the layout is complete, since a recheck can be made easily with it in. The next step is to calculate the required chord length and start your spacing.

When stepping off even spaces, 6, 8, 10, 12, etc., it should be pointed out that a line from any one hole on the circle drawn to and through the center will locate a hole on the circle layout 180° opposite the starting hole. This fact is the basis for the next statement: always start your layout of even numbered holes from two points 180° apart on the circle (Fig. 6-12).

In the incorrect method of layout (Fig. 6-12), one starting point was used

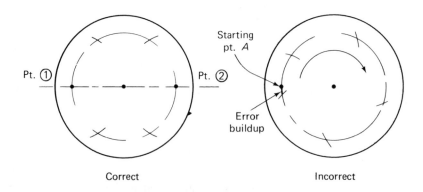

Correct   Incorrect

Fig. 6-12. Correct method of laying out even number spaces on a circle.

and the stepping off was performed consecutively in one direction. The result of this type procedure is always a poor layout, due to error buildup. The correct layout procedure where arcs are scribed on either side of point 1 and 2 has a cancelling effect on error buildup.

Odd number layouts, which are not too common, have a different approach. The odd number layouts all basically start off with only one starting point and by working to either side of it, eliminate error buildup. In the case of a three-space layout on, say, a 1 inch diameter circle, Table 6-1 shows a chord constant of 0.866 inch. Naturally this layout is completed when two arcs are swung on either side of a starting point.

TABLE 6-1. LENGTHS OF CHORDS FOR SPACING OFF THE CIRCUMFERENCE OF CIRCLES WITH A DIAMETER EQUAL TO I

(For Circles of Other Diameters, Multiply Length Given in Table by Diameter of Circle)

No. of Spaces	Length of Chord	No. of Spaces	Length of Chord	No. of Spaces	Length of Chord	No. of Spaces	Length of Chord
3	0.866025	51	0.061560	99	0.031727	147	0.021369
4	0.707106	52	0.060378	100	0.031410	148	0.021225
5	0.587785	53	0.059240	101	0.031099	149	0.021082
6	0.500000	54	0.058144	102	0.030795	150	0.020942
7	0.433883	55	0.057088	103	0.030496	151	0.020803
8	0.382683	56	0.056070	104	0.030202	152	0.020666
9	0.342020	57	0.055087	105	0.029915	153	0.020531
10	0.309017	58	0.054138	106	0.029633	154	0.020398
11	0.281732	59	0.053222	107	0.029356	155	0.020266
12	0.258819	60	0.052336	108	0.029084	156	0.020137
13	0.239315	61	0.051478	109	0.028817	157	0.020008
14	0.222520	62	0.050649	110	0.028556	158	0.019882
15	0.207911	63	0.049845	111	0.028296	159	0.019757
16	0.195090	64	0.049067	112	0.028046	160	0.019633
17	0.183749	65	0.048313	113	0.027798	161	0.019511
18	0.173648	66	0.047581	114	0.027554	162	0.019391
19	0.164594	67	0.046872	115	0.027314	163	0.019272
20	0.156434	68	0.046183	116	0.027079	164	0.019154
21	0.149042	69	0.045514	117	0.026847	165	0.019038
22	0.142314	70	0.044864	118	0.026620	166	0.018924
23	0.136166	71	0.044233	119	0.026396	167	0.018810
24	0.130526	72	0.043619	120	0.026176	168	0.018698
25	0.125333	73	0.043022	121	0.025960	169	0.018588
26	0.120536	74	0.042441	122	0.025747	170	0.018478
27	0.116092	75	0.041875	123	0.025538	171	0.018370
28	0.111964	76	0.041324	124	0.025332	172	0.018264
29	0.108118	77	0.040788	125	0.025130	173	0.018158
30	0.104528	78	0.040265	126	0.024930	174	0.018054
31	0.101168	79	0.039756	127	0.024734	175	0.017950
32	0.098017	80	0.039259	128	0.024541	176	0.017848
33	0.095056	81	0.038775	129	0.024350	177	0.017748
34	0.092268	82	0.038302	130	0.024163	178	0.017648

TABLE 6-1. (CONT'D.)

No. of Spaces	Length of Chord	No. of Spaces	Length of Chord	No. of Spaces	Length of Chord	No. of Spaces	Length of Chord
35	0.089639	83	0.037841	131	0.023979	179	0.017549
36	0.087155	84	0.037391	132	0.023797	180	0.017452
37	0.084805	85	0.036951	133	0.023618	181	0.017355
38	0.082579	86	0.036522	134	0.023442	182	0.017260
39	0.080466	87	0.036102	135	0.023268	183	0.017166
40	0.078459	88	0.035692	136	0.023097	184	0.017073
41	0.076549	89	0.035291	137	0.022929	185	0.016980
42	0.074730	90	0.034899	138	0.022763	186	0.016889
43	0.072995	91	0.034516	139	0.022599	187	0.016799
44	0.071339	92	0.034141	140	0.022438	188	0.016709
45	0.069756	93	0.033774	141	0.022278	189	0.016621
46	0.068242	94	0.033414	142	0.022122	190	0.016533
47	0.066792	95	0.033063	143	0.021967	191	0.016447
48	0.065403	96	0.032719	144	0.021814	192	0.016361
49	0.064070	97	0.032381	145	0.021664	193	0.016276
50	0.062790	98	0.032051	146	0.021516	194	0.016193

In larger number spacing layouts, both odd and even, the layout can sometimes be broken into smaller groups for ease and accuracy. For example, where 16 equal spaces are required, the 16 spaces could be divided into 8 spaces with each of the 8 spaces bisected, or into 4 spaces with 4 equal arcs in each space, all giving 16 spaces. This method should not be used until the technician is familiar with layout. The main purpose of subdividing is reduction of error in large number spacings.

6-7. Three-Dimensional Layout

Up to now we have discussed layout procedures involving only two dimensions, width and depth. We now come to a phase where we introduce height—the third dimension in layout. Three-dimensional layout is by far the most important as well as the most complicated in practice.

With today's variety of three-dimensional workpieces, many items, which include gear boxes, ship's rudders, cover plates, bearing blocks, etc., are produced in either a cast state or a welded state. In either case, a material allowance or "skin" is left so that the part can be machined to a finished size. Prelayout of these finished surfaces is one of the biggest single purposes of layout. This skin or allowance is normally added to all surfaces that must be machined. There are times that warpage or shrinkage of the metal may have occurred causing the casting or weldment to be useless. It is far less costly to find it out in layout than in actual machining, where several thousands of dollars may be lost.

In three-dimensional layout, the use of the machine square, angle plate, and surface gauge are used extensively. Before proceeding further, let's

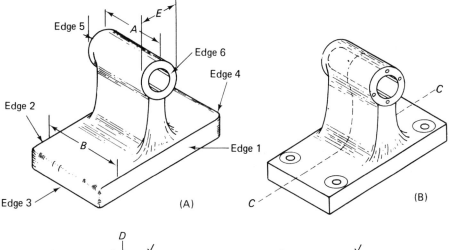

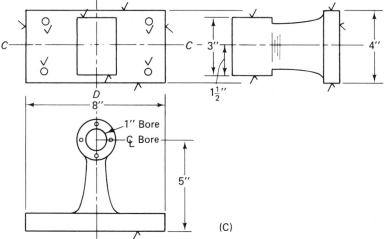

Fig. 6-13. (A) Rough casting, (B) finished casting, and (C) drawing.

introduce a rough casting, a finished casting and an engineering drawing (Fig. 6-13).

The layout problem is to take a casting that has no reference surface and lay out the finished edges according to an engineering drawing. As previously mentioned, this is performed for two reasons—to verify the usefulness of the casting or weldment, and to visually aid the technician in setting the workpiece on his machine.

Prior to putting the workpiece on the layout table, four metal parallels should be placed so as to rest the casting both above and somewhat square to the table top. The placement of the four support pads also allow for ease in shimming or packing of the workpiece. All cast holes should be "sticked" at each end and dark waxed near the centers.

The workpiece should then be somewhat squared up with the 90° lines in the layout table surface or the edge of the table. The casting can now be coated with layout fluid along all of the edges and surfaces to be machined, as well as the center lines pictured on the top view of the drawing. After the casting has been shimmed to approximately a level position with no rocking of the casting on the four pads, the center lines can then be defined. On either side of the raised bore, bisect the distance marked B with a hermaphrodite caliper. This is done to define the longest center line in the workpiece, thereby cutting down on errors.

Next, and very important, an angle plate face is aligned to a ruled table line or table edge with one of the workpiece edges roughly parallel to the angle plate. Take a scale or use a surface gauge on the vertical surface of the angle plate and align the bisected distance B pads of the workpiece parallel to the angle plate. Now, using the surface gauge base on the vertical surface plate face, scribe the center line (CC) on the workpiece.

Using the surface plate, all the lines parallel to it such as edges 1, 2, 5, and 6 and the location of the four mounting holes can be scribed. The casting at no time should be moved, as it is aligned to the table ruled line. If movement does happen, merely repeat the above alignment steps.

The start of the second center line (DD) is produced by bisecting the (E) dimension of the casting with a hermaphrodite caliper.

By sliding the angle plate to a table line now perpendicular to the first position, center line (DD) can be scribed by using a surface gauge and the bisected distance of E. From this second position of the angle plate, edges 3 and 4 can be scribed as well as the final location of the four mounting holes. The only remaining layout step is the location of the bore center. This is done by splitting the actual vertical height of the rough bore by use of a hermaphrodite caliper. Setting a surface gauge point at the center, horizontal lines are scribed on both front and back faces of the bore. By lowering the scriber point 5 inches, the bottom surface of the casting is located and scribed.

If a situation arises where either a surface or bore location will not clean up if machined, a compromise may be necessary. For example, if a bore scribed on the casting was found to have not enough material at the top of the bore to finish up, the solution would be to raise the center of the hole until a full 1 inch hole can be cut. This will mean, however, cutting more from the base to compensate. Whenever this situation is encountered, always check with your supervisor for a decision.

Before leaving this discussion, several points should be made concerning three dimensional layout. Whenever a surface is to be machined and have bolt or other holes put into it (Fig. 6-13), the hole pattern is never laid out since it would be machined off and lost. These hole patterns are always laid out after final machining of a part. All center lines, surface lines, and hole centers are to be punched after layout to ensure a more permanent line record. All "sticked" holes should also remain so that the technician can use them for reference points.

6-8. Shaft Layout

The layout of shafts is often required to locate such things as oil holes and keyways for machining. When working with long shafts, V blocks are the most important tools. These V blocks, as previously described, have a 90° included angle cutout; some are equipped with clamps to lock a round part firmly to the V block during layout or machining.

On shaft work, keyways or slots are the most often performed job task. These keyways are used with keys to drive gears, couplings, or other items that need high turning force. Keyway dimensions are listed as, for an example, $\frac{1}{2}$ inch (depth) by 1 inch (width) by a 6 inch length (Fig. 6-14).

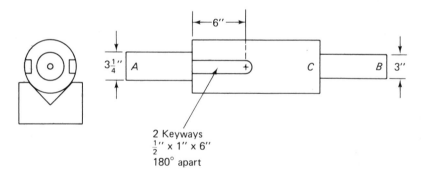

Fig. 6-14. Typical shaft-keyway layout problem with depth layout of keyways.

To perform the task of laying out two keyways 180° apart, the following approach is taken. Set the two ends of the shaft in V blocks. Since end A of the shaft is $\frac{1}{4}$ inch larger in diameter than end B, the shaft will not set level with the layout table. Most shafts have 60° centers cut in their ends that are used to hold and machine the workpiece and are the true and accurate centers of the workpiece.

A handy tool to transfer the cut center hole to a point center is a set of round thin metal discs with a light punch mark very accurately placed at the center. The discs should be in the range of $\frac{1}{64}$ inch or less in thickness.

A starter diameter group of $\frac{3}{8}$, $\frac{1}{2}$, and $\frac{3}{4}$ inch discs should carry through most of the work encountered. The center disc should bear on the inside faces of the cut center for best results. This center gauge with a surface gauge is then used to set the shaft center line parallel with the surface table. Shimming with thin metal sheets of the low V block is done until both shaft centers are the same distance to the table. With the shaft level and the gauge point set at the shaft center, a center line is scribed on the end of the shaft as a reference line and two center lines are scribed on either side for the two keyways.

Using a square head and scale, adjust the scale to the scriber point so that some even inch mark coincides with it. Note this reference inch mark down on paper. The required keyway width is 1 inch or $\frac{1}{2}$ inch on either side of the center line. By adding $\frac{1}{2}$ inch to the reference inch mark setting, the surface gauge point is adjusted to it. This setting is then scribed on both sides and faces of the C section.

Setting the scriber point 1 inch lower, the above scribing procedure is repeated. Next, the $\frac{1}{2}$ inch depth of the keyway must be laid out. This depth is never measured from the center line; rather, it is measured from either of the scribed key width lines, ½ inch above and below the center lines.

The final step is to define the 6-inch lengths of the two keyways. Simply lay a scale on the center lines and scribe a mark at 6 inches as measured from the end of section C. After prick punching this mark, scribe a semicircle with a radius equal to $\frac{1}{2}$ the keyway width, or in our case a $\frac{1}{2}$ inch radius. This radius sets the cutting limit of the keyway for the technician.

In cases where two keyways are located 90° apart, the second keyway can be located by rotating the shaft until the horizontal reference line on the end of the shaft, scribed there on the first keyway layout, is perpendicular to the layout table. This is achieved by using the machine square mentioned at the beginning of this chapter.

SUMMARY

This chapter defined layout as scribing or drawing required shapes, outlines, and intersections on a material as a guide prior to the actual machining operations on the material. Special aids used in layout including layout tables, parallels, angle plates, V blocks, and machine squares were also discussed. Finally, four types of layout—two-dimensional, circular, three-dimensional, and shaft—were discussed.

The student technician has now progressed through a large portion of this book. He can now read engineering drawings and sketches, can use mathematics to find unknowns, knows the proper application of hand tools, measures accurately, and can lay out designs from engineering drawings onto workpieces.

Subsequent chapters of this book are of a more practical nature. The student will begin to use drills, reamers, countersinks, counterbores, taps and dies, and power tools to work with the workpieces.

REVIEW QUESTIONS

1. What is a layout table? What are parallels? Angle plates? V blocks?
2. What considerations should be made in selecting a workpiece reference edge?
3. Once the workpiece reference edges have been selected, they are used as the layout reference edges for the completion of the job. Why is this true?
4. Name the three general approaches that can be used to lay out a flat workpiece.
5. Lay out a straight line *AB* about 6 inches long. Using dividers, lay out a line *CD* parallel to line *AB* and 1 inch away from it. Lay out *EF* parallel to and 4 inches from line *AB*.
6. From point *A* of question 5, use dividers to locate a point *K* which is on line *AB* and is $3\frac{1}{2}$ inches from point *A*.
7. Briefly explain how you can prevent error buildup when using dividers.
8. Using the angle plate/surface gauge method of layout, describe how to lay out the hole centers shown in Fig. 6-15.

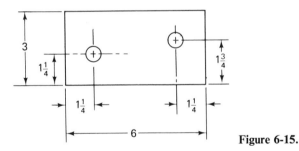

Figure 6-15.

9. Which of the layout methods described is used for laying out many workpieces of the same pattern?
10. When is a circular layout (chordal method) used? What are the advantages of circular layout over the protractor method?
11. How does three-dimensional layout differ from two-dimensional layout?
12. What are the basic steps in shaft layout?

7
DRILLS, COUNTERSINKS, REAMERS, AND COUNTERBORES

Thus far, the student technician has learned to measure using measuring tools, has learned to read engineering drawings, has seen and used general hand tools, has studied layout tools, and finally has learned to lay out the information from an engineering drawing onto a workpiece. This chapter, which covers the tools for drilling, countersinking, reaming, and counterboring, teaches the student technician how to perform the first cutting operations into a workpiece.

Drilling is the process of boring a circular hole of specific diameter into or through a workpiece. *Countersinking* is the process of enlarging the upper part of a hole, as by chamfering, so as to receive the cone-shaped head of a screw, bolt, etc. The process usually takes place at the surface of the workpiece. *Reaming* is the process of enlarging and smoothing a hole by a "shaving" action to a specified size. When a rough hole is reamed, a smoothly finished hole with an accurate diameter is produced. *Counterboring* is the process of enlarging the upper part of a hole into a flat bottomed cylindrical shape of greater diameter than the previously drilled hole (called a pilot hole). The counterbore may receive the head of a bolt, a nut, or other fastener. The tools—*drills, countersinks, reamers,* and *counterbores*—that perform these processes are described in this chapter.

This chapter is divided into four major areas: drills, countersinks, reamers, and counterbores. Each section describes the tool, how to use the tool, and the care of the tool.

DRILLS

The drill is by far the most commonly used tool in shop work. The drill is used to produce a range of holes from as small as 0.0059 inch in diameter to as large as a $3\frac{1}{2}$ inch diameter hole. Drills are made of hard high-speed tool steel alloys and other exotic metal compositions that allow the drills to cut a large range of materials that are much less hard than the drill. The basis of all cutting tools is that the tool be harder than the material to be cut.

Since the variety of drills is wide and sometimes confusing in application, this text limits discussion to those drills and their applications that are most commonly used. Drills to be discussed are the twist drill, the center drill, the combined drill and countersink, and the carbide drill. Recommended drilling procedures and speeds are also discussed. Finally, the drill gauge is described.

DRILL MATERIALS

Drills are made from three different materials that are used to cover a wide range of drilling applications. These materials are high-speed steel (HSS), cobalt high-speed steel, and carbide.

High-speed steel is a tool steel that is the most often used material in the manufacture of general shop drills. HSS will perform well in drilling aluminum, brass, copper and mild steels. It is the least expensive of the three drill materials.

The cobalt high-speed steel drill is, as the name implies, an alloy of cobalt and tool steel. Cobalt allows the drill to withstand higher heat buildup and abrasive action in the drilling of certain hard materials. Cobalt high-speed steel drills can also be run approximately 30% faster than regular high-speed steel drills; this allows for better production rates. Cobalt high-speed drills are recommended in applications such as drilling high-temperature alloys, hardened stainless steel, heat treated alloy steels, or any other tough material.

Carbide is a very hard, dense material that has recently been employed in drill applications. This material has the ability to drill the new and tougher materials found in today's shop practices, as well as perform quite well on other items such as glass, ceramic, or concrete. Carbide drills are generally run at speeds higher than those of cobalt alloy drills when employed in tough or hardened metals. Since this type drill material is far less susceptable to abrasion, it finds great use in drilling holes such as those required in printed circuit boards, where other type drills wear at a high rate.

7-1. Twist Drills

The foremost drill in use today is the twist drill. Its name was derived from the fact that the drill was originally made by twisting a flat ribbon of steel into a drill followed by the grinding of a cutting point at one end. The function of the twist is to lift out the material being cut by the cutting point, thereby allowing the drill to advance without jamming into the workpiece. In the manufacture of today's drills, the twist is either cut or ground from a solid round rod. This allows for stronger and longer-wearing drills.

Twist Drill Twists

Twist drills are available in three grades of spiral twist (Fig. 7-1): regular, fast, and slow. Regular spiral twist drills are used for general drilling operations in mild steels to high tensile strength steels. Fast spiral twist drills are recommended when drilling low tensile strength materials such as aluminum, copper alloys or die cast materials. The fast spiral design also allows for faster chip removal when drilling deep holes. The slow spiral twist drill is recommended when there is a need to drill holes in copper or other soft materials. Its slow twist prevents the drill from digging in or grabbing when used on soft or thin material.

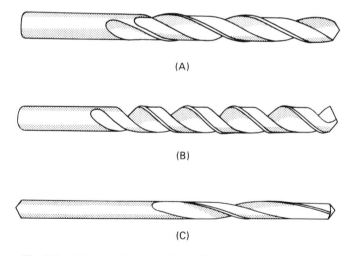

(A)

(B)

(C)

Fig. 7-1. Spiral grades of twist drills: (A) regular helix, (B) fast helix, and (C) slow helix.

Twist Drill Sizes (Jobbers Length)

Twist drills are available in three groups of diameter sizes which cover a range of 0.0059 to $3\frac{1}{2}$ inches in diameter. These three groups are fractional size drills, number size drills, and letter size drills. Each of these drills may be purchased individually or in boxed sets.

Twist drills are available in a wide range of lengths to cover special drilling applications. Special twist drills have been made in lengths exceeding 3 feet. The *standard* length drill in common use is known as the *Jobbers length*. This standard length, named for its originator, is used by manufacturers in both drill and reamer production.

Twist drills above a $\frac{1}{2}$ inch diameter are available with $\frac{1}{2}$ inch diameter shanks or in screw machine drills with shank diameters up to 2 inches. There are also drills available from $\frac{9}{32}$ to $\frac{1}{2}$ inch in diameter having $\frac{1}{4}$ inch diameter shanks. And finally, there are miniature drills and metric drills. Each of these drill size group variations—fractional, numbered, lettered, above $\frac{1}{2}$ inch diameter shanks, $\frac{1}{4}$ inch shanks, and miniature drills—are described in this section. Special operating considerations, where required, are included.

Fractional drills ranging from $\frac{1}{64}$ to $3\frac{1}{2}$ inches are available. Since supplying the number of drills by $\frac{1}{64}$ inch increments from $\frac{1}{64}$ to $3\frac{1}{2}$ inches would be extremely costly, a standard has been set up. Table 7-1 lists the increments in the fractional drill range.

TABLE 7-1. FRACTIONAL SIZED TWIST DRILLS

Drill Size Range (inches diameter)   *Drill Size Increments (inches)*

Drill Size Range (inches diameter)	Drill Size Increments (inches)
$\frac{1}{64}$ to 2	$\frac{1}{64}$
2 to $2\frac{1}{4}$	$\frac{1}{32}$
$2\frac{1}{4}$ to $3\frac{1}{2}$	$\frac{1}{16}$

By far, the most popular fractional drill package is the set of drills from $\frac{1}{16}$ to $\frac{1}{2}$ inch (by $\frac{1}{64}$ inch increments). This set can be purchased in the loose version or in a fold-out metal storage box. This type of set is recommended, as the case has marked size holes so that, after use, the drill can be returned to its proper hole for safe storage.

The standard set of number sized drills, also called wire gauge sized drills, consists of 80 drills starting with the no. 1 drill through the no. 80 drill. Table 7-2 shows the diameter decimal equivalent of each of these drills.

The sizes of these drills correspond very closely to Stub's steel wire gauge sizes, which are the only wire gauge sizes recognized by acts of Congress. These drills derive their beginning in the wire drawing industry, where holes had to be drilled in steel plates so the wire could then be sized or drawn through them. Today's applications of these drills find a vast use in the drilling of tap holes, clearance holes and pre-ream holes. Number drills, like the

TABLE 7-2. NUMBER SIZED TWIST DRILLS

Wire Gauge No.	Decimal Equiv.	Wire Gauge No.	Decimal Equiv.	Wire Gauge No.	Decimal Equiv.
1	0.2280	31	0.1200	61	0.0390
2	0.2210	32	0.1160	62	0.0380
3	0.2130	33	0.1130	63	0.0370
4	0.2090	34	0.1110	64	0.0360
5	0.2055	35	0.1100	65	0.0350
6	0.2040	36	0.1065	66	0.0330
7	0.2010	37	0.1040	67	0.0320
8	0.1990	38	0.1015	68	0.0310
9	0.1960	39	0.0995	69	0.0292
10	0.1935	40	0.0980	70	0.0280
11	0.1910	41	0.0960	71	0.0260
12	0.1890	42	0.0935	72	0.0250
13	0.1850	43	0.0890	73	0.0240
14	0.1820	44	0.0860	74	0.0225
15	0.1800	45	0.0820	75	0.0210
16	0.1770	46	0.0810	76	0.0200
17	0.1730	47	0.0785	77	0.0180
18	0.1695	48	0.0760	78	0.0160
19	0.1660	49	0.0730	79	0.0145
20	0.1610	50	0.0700	80	0.0135
21	0.1590	51	0.0670		
22	0.1570	52	0.0635		
23	0.1540	53	0.0595		
24	0.1520	54	0.0550		
25	0.1495	55	0.0520		
26	0.1470	56	0.0465		
27	0.1440	57	0.0430		
28	0.1405	58	0.0420		
29	0.1360	59	0.0410		
30	0.1285	60	0.0400		

fractional drills, can be procured either individually or in boxed sets. The boxed sets, however, are generally supplied only from a no. 1 drill through a no. 60 drill.

Letter sized drills, which further increase the range of drilling application, consist of 26 drills ranging from "A" (the smallest diameter of the set) to "Z" (the largest). The letter to decimal equivalent diameter of these drills is shown in Table 7-3. The use of these drills, along with the fractional and numbered drills allows for the total completion of such requirements

TABLE 7-3. LETTER SIZED TWIST DRILLS

Letter Size	Decimal Equivalent
A	0.234
B	0.238
C	0.242
D	0.246
E	0.250
F	0.257
G	0.261
H	0.266
I	0.272
J	0.277
K	0.281
L	0.290
M	0.295
N	0.302
O	0.316
P	0.323
Q	0.332
R	0.339
S	0.348
T	0.358
U	0.368
V	0.377
W	0.386
X	0.397
Y	0.404
Z	0.413

as tapping and reaming, as well as providing clearance and rivet holes, etc. in modern shop work. Boxed sets or individual drills are available from most machine tool supply houses.

Fractional, numbered, and lettered size twist drill shanks are available with two shank forms, straight and tapered. The straight shank, which is the most popular, has a cylindrical configuration and in some instances has three driving flats on it that keep the drill from slipping in the driving chuck. It should be noted that the straight shank diameter in some cases does not equal the diameter of the drill. Therefore, drill diameter measurements always refer to the diameter of the drill along the spiral, not the shank diameter. The diameter of the straight shank drill is not normally greater than 1 inch, as it is impractical to drill larger diameter holes due to the strain on the shank and also to slippage within the chuck.

The tapered shank, known also as the American Standard Taper and the Morse taper, is available in the fractional range from $\frac{1}{8}$ to $3\frac{1}{2}$ inches in diameter and also in the letter sized drills. The purpose of the tapered shank is that it gives the ability to transmit greater power for heavier drilling and it is adaptable to a variety of heavy machine tools. Table 7-4 shows the taper size to range of drills.

TABLE 7-4. STANDARD TAPER SHANK TWIST DRILLS

Drill Size Range (inches)	Taper Shank (ASA Taper–Morse)
$\frac{1}{8}$ to $\frac{15}{32}$	no. 1
$\frac{31}{64}$ to $\frac{25}{32}$	no. 2
$\frac{51}{64}$ to $1\frac{1}{16}$	no. 3
$1\frac{5}{64}$ to $1\frac{1}{2}$	no. 4
$1\frac{33}{64}$ to 3	no. 5
$3\frac{1}{8}$ to $3\frac{1}{4}$	no. 6

As previously stated, tapered shank drills allow for greater power transmission during drilling and are recommended for drilling holes above $1\frac{1}{4}$ inches in diameter. Drills above 1 inch diameter, when used in portable drilling machines, should not be handled by one person as the torque involved in most drilling operations tends to turn or buck the drill machine and can cause the person drilling to lose his balance and possibly be injured.

There are two types of straight shanked drills above $\frac{1}{2}$ inch diameter to choose from: $\frac{1}{2}$ inch shank drills (available only up to $1\frac{1}{4}$ inch drill diameter); and screw machine twist drills up to 2 inches in diameter. Quite often, when working in the field, there is a requirement to drill holes larger than $\frac{1}{2}$ inch in diameter. This drilling is usually performed by portable electric or air-powered drill units. Generally, these portable units have a small chucking capacity from $\frac{1}{16}$ to $\frac{1}{2}$ inch and for this reason, a range of drills to $1\frac{1}{4}$ inch drill diameter with $\frac{1}{2}$ inch maximum shank diameters is available. These drills are known under the originator's name of Silver and Deming; they are also very popular in light-to-medium drill press operations where alumimun and mild steels are encountered. Half-inch shank drills are not recommended for heavy work above the $\frac{3}{4}$ inch diameter size.

A screw machine drill has an upper fractional limit of 2 inches in diameter and a lower limit of $\frac{3}{64}$ inch in diameter. The screw machine drill is generally shorter in overall length and flute length than the Jobbers drill, which makes this drill very rigid and excellent for portable drilling. Screw machine drills are also available in wire gauge and letter sizes.

The shank diameters of fractional drills are the same as the drill size up to the 1 inch size. In drills from $1\frac{1}{16}$ through $1\frac{1}{4}$ inches, the shank is 1 inch in diameter; from $1\frac{5}{16}$ through $1\frac{1}{2}$ inches in diameter, the shank is $1\frac{1}{4}$ inches in diameter; and from $1\frac{9}{16}$ through 2 inches in diameter, the shank is $1\frac{1}{2}$ inches in diameter. Drills with larger shanks require special split driving collets.

In $\frac{1}{4}$ inch diameter shanks, there exists another important group of drills. This group has a size range of from $\frac{9}{32}$ to $\frac{1}{2}$ inch diameter drill sizes in $\frac{1}{32}$ inch increments, but the drills have $\frac{1}{4}$ inch diameter shanks. This group is very popular in sheet metal and duct work where a $\frac{1}{4}$ inch chuck capacity electric drill is used extensively because of its light weight and convenience.

Where extremely fine holes (above no. 80) are required for such items as fuel jets for aircraft and diesel engines or precision instruments, the miniature twist drill is used. This group of drills is really an extension of the previously mentioned number (or wire gauge) drills (no. 1 through no. 80). The miniature group continues with a no. 81 drill (0.013 inch) through a no. 97 drill (0.0059 inch). Refer to Table 7-5.

TABLE 7-5. MINIATURE PRECISION TWIST DRILL SIZES

Number Sizes	Decimal Equivalent	mm Equivalent
81	0.0130	
82	0.0125	
83	0.0120	
84	0.0115	
85	0.0110	0.28
86	0.0105	
87	0.0100	
88	0.0095	
89	0.0091	0.23
90	0.0087	0.22
91	0.0083	0.21
92	0.0079	0.20
93	0.0075	0.19
94	0.0071	0.18
95	0.0067	0.17
96	0.0063	0.16
97	0.0059	0.15

Miniature twist drills are normally driven at high speeds (10,000 to 20,000 RPM, depending on material) and are used only in miniature precision drill presses equipped to handle them. When using miniature drills, the most gentle tool pressure along with a good lubricant must be used to prevent breakage. The first sign of too much pressure or dulling is the bowing or bending of the drill between the workpiece and the driving chuck. Frequent withdrawal of the drill to clean both the drill and the workpiece should be performed. At the high speed that these drills are driven, their life span is short and the drills should be replaced if poor cutting is noted. These drills are generally available in high-speed steel (HSS) and are sold separately or in packs of 12 each per number size.

Metric drills are available in a wide range of sizes. The shanks of these drills, however, are produced in the United States in the same range and styles as the fractional twist drills, straight, and AST shanks. A listing from the major manufacturers will show the diameter range available from 0.35 mm (0.0138 inch) to 12.5 mm (0.4921 inch). Metric diameters larger than 12.5 mm are available from most manufacturers.

7-2. Center Drills

As mentioned in Chapter 6 on layout, the beginning operation prior to drilling any job is the center drilling (predrilling) of the center punched holes of the laid out workpiece. This center drilling allows for better alignment of the drill to the layout lines and also allows the drill to start without walking.

Center drills are twist drills which have a very short overall length with corresponding short flutes. A diameter range is normally available from $\frac{1}{16}$ inch to 1 inch in diameter. Although generally listed as fractional in size, some manufacturers make them available in a limited range of number gauge sizes. The shortness of these drills makes them very useful in predrilling center punched layouts prior to finish drilling. In predrilling, the center drill is normally drilled just deep enough to spot or guide the finish drill. Centering drills are generally made of high-speed steel.

7-3. Combined Drill and Countersink

By far the most popular type predrill in use today is the combined drill and countersink (it is now also referred to as a center drill). This drill combination is available in a variety of sizes and in two distinct types, and is used very often to spot drill a layout prior to drilling. The two types of combined drill and countersink (Fig. 7-2) are the plain type and the bell type. They are both basically of the same design in the fact that they are made in the same diameter sizes and are double ended.

Tables 7-6 and 7-7 show the size number vs. body and drill tip diameter for both the plain and bell type combined drill and countersink. The main

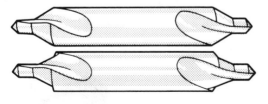

Fig. 7-2. Combined drill and countersink. (Courtesy of Greenfield Tap and Die)

difference between the plain and bell type is in their countersink design. The plain type is available in only a 60° included angle countersink. The bell type is available in a 60° as well as a 120° included angle countersink, but with one added special feature. This feature is an extra chamfer near the place where the countersink meets the drill body diameter, known as the bell diameter. The purpose of this extra chamfer in the bell type center drill is to produce a sunken protected center hole. The main use of the bell type drill and countersink is in machining operations such as lathe work where a plain center hole is apt to be hit or burred if not recessed.

When drilling with the plain type combined drill and countersink, it is generally limited in drilling depth to the point where body diameter and countersink meet. When using the bell type, the chamfer generally is not drilled further than $\frac{1}{16}$ inch in depth.

TABLE 7-6. PLAIN TYPE COMBINED DRILL AND COUNTERSINK

Size No.	Body Diam. (Inches)	Drill Diam. (Inches)	Drill Lgth. (Inches)	O.A. Lgth. (Inches)
00	$\frac{1}{8}$	0.025	0.040	$1\frac{7}{32}$
0	$\frac{1}{8}$	$\frac{1}{32}$	$\frac{3}{64}$	$1\frac{7}{32}$
1	$\frac{1}{8}$	$\frac{3}{64}$	$\frac{3}{64}$	$1\frac{1}{4}$
2	$\frac{3}{16}$	$\frac{5}{64}$	$\frac{5}{64}$	$1\frac{7}{8}$
3	$\frac{1}{4}$	$\frac{7}{64}$	$\frac{7}{64}$	2
4	$\frac{5}{16}$	$\frac{1}{8}$	$\frac{1}{8}$	$2\frac{1}{4}$
5	$\frac{7}{16}$	$\frac{3}{16}$	$\frac{3}{16}$	$2\frac{3}{4}$
6	$\frac{1}{2}$	$\frac{7}{32}$	$\frac{7}{32}$	3
7	$\frac{5}{8}$	$\frac{1}{4}$	$\frac{1}{4}$	$3\frac{1}{4}$
8	$\frac{3}{4}$	$\frac{5}{16}$	$\frac{5}{16}$	$3\frac{1}{2}$

TABLE 7-7. BELL TYPE COMBINED DRILL AND COUNTERSINK

Size No.	Body Diam. (Inches)	Drill Diam. (Inches)	Drill Lgth. (Inches)	O.A. Lgth. (Inches)	Bell Diam. (Inches)
11	$\frac{1}{8}$	$\frac{3}{64}$	$\frac{3}{64}$	$1\frac{1}{4}$	0.100
12	$\frac{3}{16}$	$\frac{1}{16}$	$\frac{1}{16}$	$1\frac{7}{8}$	0.150
13	$\frac{1}{4}$	$\frac{3}{32}$	$\frac{3}{32}$	2	0.200
14	$\frac{5}{16}$	$\frac{7}{64}$	$\frac{7}{64}$	$2\frac{1}{8}$	0.250
15	$\frac{7}{16}$	$\frac{5}{32}$	$\frac{5}{32}$	$2\frac{3}{4}$	0.350
16	$\frac{1}{2}$	$\frac{3}{16}$	$\frac{3}{16}$	3	0.400
17	$\frac{5}{8}$	$\frac{7}{32}$	$\frac{7}{32}$	$3\frac{1}{4}$	0.500
18	$\frac{3}{4}$	$\frac{1}{4}$	$\frac{1}{4}$	$3\frac{1}{2}$	0.600

One very useful side application of the combined drill and countersink is in drilling holes in thin sheet metal or shimstock. Twist drills tend to produce noncircular holes, but the combined drill and countersink design produces very clean, sharp, and round holes in thin metal stock.

7-4. Carbide Drills

In recent years, the carbide drill has been introduced in the metal working industry. The development of this drill was prompted by the use of newer and tougher alloys in job applications. A carbide drill has the ability to cut through hardened metals such as leaf springs or similar metals of equal hardness. The cutting action of most carbide tools is really a pushing action, which is achieved in two ways. First, all carbide drills used in metal work are run at speeds higher than the speeds used for cobalt drills. Second, heavy tool pressure is employed in the cutting action. The combination of higher speed and heavier tool pressure acts on hardened metals in the following ways. First, the high speeds build up extreme amounts of heat where the cutting point and metal are in contact. This frictional heat tends to anneal or soften the metal being drilled. Second, the application of heavy pressure to the drill plows or pushes the softened metal away and out of the hole. In some dry drilling operations using carbide tools, the removed metal may be ejected as blue hot chips, which can vary from 900 to 1000° F. If this is encountered, the drilling area must be shielded to eliminate operator burns.

Carbide drills are available in two basic forms: cemented tip, and solid carbide. Cemented drills are really tool or carbon steel twist drills with carbide cutting tips silver brazed to them. This combination is relatively low in cost and is in widespread use where required. One disadvantage of the cemented tip drill is that on some occasions, temperatures developed in drilling have loosened the silver braze holding the carbide cutting tip, which results in drill failure. This problem is solved by using a solid carbide twist drill, which is designed to take the heat buildup during drilling and also allows for multiple resharpenings. In all carbide tools, one must remember that sharp shock, blows, or stops during drilling can crack the carbide cutting edge.

One final note that should be mentioned here is the use of carbide tipped masonry drills. These drills are used in drilling holes into such materials as concrete, stone, brick and cinder block.

7-5. Use and Speed of Drills

Twist Drills

In drilling operations, determination of the proper speed and feed is very important to both job and drill. Two factors determine the selection of proper speed and feed: drill diameter and the material to be drilled. Speeds that are too high on hard metals will tend to either dull the drill or in some instances work harden the job material (especially stainless steels). Lower than normal drill speeds tend to cause the smaller drills (below $\frac{1}{4}$ inch diameter) to bow or bend, thereby causing runout and oversized holes.

A general rule of thumb is to drill at a high speed with a light to medium feed for small drills (below $\frac{1}{2}$ inch diameter) and medium to low speed with medium to heavy feed on the larger drills.

Drill feeds (inches advance per revolution) are mainly effected by the material hardness. In soft material (copper, lead) or very brittle material (hard brass, plastic) the feed should be light to medium to prevent the drill from digging in and ruining the workpiece. In the harder alloys, high carbon steels and stainless steels, the drill feed should be toward the heavy side. The heavier feeds are needed to keep the drill cutting and to prevent the metal of the workpiece from hardening. Included as a reference is Table 7-8, which

TABLE 7-8. DRILLING SPEEDS AND FEEDS FOR HIGH-SPEED AND CARBON STEEL DRILLS[1, 2]

High-Speed Steel Drills

Drill Diam. (Inches)	Aluminum, Bronze, Brass	Cast Iron, Annealed	Mild Steel	Drop Forgings	Mal. Iron	Cast Steel
		Revolutions Per Minute				
$\frac{1}{16}$		9170	6111	3660		2440
$\frac{1}{8}$	9170	4584	3056	1830	2745	1220
$\frac{3}{16}$	6112	3056	2037	1210	1830	807
$\frac{1}{4}$	4585	2292	1528	915	1375	610
$\frac{5}{16}$	3660	1833	1222	732	1138	490
$\frac{3}{8}$	3056	1528	1019	610	915	407
$\frac{7}{16}$	2614	1310	873	522	784	348
$\frac{1}{2}$	2287	1146	764	458	688	305
$\frac{5}{8}$	1830	917	611	366	569	245
$\frac{3}{4}$	1525	764	509	305	458	203
$\frac{7}{8}$	1307	655	436	261	392	174
1	1143	573	382	229	349	153
$1\frac{1}{4}$	915	458	306	183	275	122
$1\frac{1}{2}$	762	382	255	153	212	102
$1\frac{3}{4}$	654	327	218	131	196	87
2	571	287	191	115	172	77

[1] The drilling feeds per revolution are, for $\frac{1}{8}$ inch diameter, 0.002 inch; $\frac{1}{4}$ inch diameter, 0.006 inch; $\frac{1}{2}$ inch diameter, 0.007 inch; $\frac{3}{4}$ inch diameter, 0.010 inch. The feeds for rough drilling should be all that machine and work will withstand. Finishing feed depends upon finish required and kind of material and tool.

[2] Carbon steel bits are run at speeds of approximately one-half the speeds of HSS drills.

lists speeds in RPM for a range of drill diameters and materials. As time and experience increase, the technician will soon judge for himself the speed/feed of drills for particular jobs.

In the process of drilling, the material being cut will form two coils or *spinners* of chips. These chips, if allowed to continue in length, can and often do cause serious injury to the technician. A simple, effective method of removing the coils is to momentarily stop feeding the drill into the workpiece. This stops the feed shears or breaks the chip coils and allows the coils to fall from the drill.

Center Drills

When using center drills, speed selection normally is the same for high-speed drill diameters of comparable size (Table 7-8). Due to the short rugged design of center drills, they can take heavy feeds.

Combined Drill and Countersinks

The speed selection of this tool is basically governed by the diameter of its drill point, not by its body diameter. Referring to Tables 7-6 and 7-7, the drill point diameter for any particular size can then be referenced to Table 7-8 for its proper speed. Feeding of this type of tool is usually with a very light pressure.

Carbide Drills

As previously mentioned, carbide drills perform well under a higher speed range. The speeds recommended on carbide drills are normally 30% faster than a comparable sized HSS drill. With the added speed of the carbide drill, heavier feeds are possible. In some applications, feeds two to three times those of an HSS drill are attainable.

7-6. Care of Drills

Twist drills (Jobbers and center) generally require little maintenance other than point grinding. Normally, minor problems of burrs or gouge marks on the drills' shanks occur. These burrs and marks are readily removed by using a smooth stone. In the case where large raised burrs are to be removed, hand grinding followed by stoning is recommended. Drills should be stored separately in either storage boxes or cardboard tubes.

Twist Drill Cutting Point Requirements

Figure 7-3 shows the cutting point of a drill. For an efficient cutting point, three critical angular requirements must be met at the cutting point: point angle; lip angle (clearance); and chisel angle. The point angle (118, 135, or 90°) that is defined for most general drilling applications is the 118° point. This point is used in drilling mild steels, aluminum and other material where work hardening does not take place.

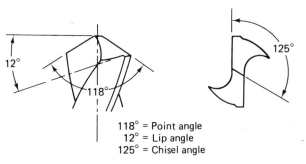

118° = Point angle
12° = Lip angle
125° = Chisel angle

Fig. 7-3. Twist drill cutting point requirements.

For extremely tough or hard materials such as alloy steels, the 135° point is recommended as it will withstand both the increased heat and pressure encountered in drilling. In drilling materials such as soft copper, plastics, lead, or other soft and ductile material, the 90° point is recommended for two very important reasons. First, with a large area cutting edge, as in the 135° point, there would be a tendency for the drill pressure to push the material forward without cutting it. Second, there is a tendency for larger point angles (above 90°) to dig in and tear the material being drilled.

The lip angle (clearance angle), which varies with drill diameter, can also vary as to material application. The main purpose of the lip angle is to impart the slicing/cutting action to the drill point and not allow the area immediately in back of the cutting edge to rub or come in contact with the material being drilled. A second variation of the lip angle is determined by whether the material is hard or soft. If it is a hard material, the lip angle is reduced by approximately two degrees of its defined angle (refer to Table 7-9) so as to impart more strength to the cutting edge. In the case of soft materials (plastic,

TABLE 7-9. LIP RELIEF ANGLES

Drill Diameter Range		Lip Relief Angle (degrees)[1]
#80 to #61	(0.0135 to 0.0390)	24
#60 to #41	(0.0400 to 0.0960)	21
#40 to #31	(0.0980 to 0.1200)	18
$\frac{1}{8}$ to $\frac{1}{4}$	(0.1250 to 0.2500)	16
F to $\frac{11}{32}$	(0.2570 to 0.3438)	14
S to $\frac{1}{2}$	(0.3480 to 0.5000)	12
$\frac{33}{64}$ to 1	(0.5156 to 1.0000)	10
$1\frac{1}{64}$ to $1\frac{3}{4}$	(1.0156 to 1.7500)	8
$1\frac{49}{64}$ and larger (1.7656 up)		6

[1] For harder materials reduce these values by approximately 2°, and for soft and nonferrous materials increase these values approximately 2°.

copper, etc.), the drill will tend to advance faster per revolution and therefore may require extra lip angle relief to prevent back rubbing of the drill and the material.

Where the two cutting faces of a drill meet, a common edge is formed called the chisel angle (Fig. 7-3). This obtuse angle is 125° for most applications, as referenced from the flat cutting edge of the drill. Gross deviations below this angle cause the drill to demand high thrust to cut and thereby create high heat. Too great a chisel angle causes chipping of the cutting edges and causes an oversized hole to be drilled. It should be noted that the correct chisel angle is automatically formed by the proper grinding of the point angle and the lip angle.

Hand Grinding Twist Drills

In all phases of cutting, there is a time to resharpen a dulled cutting edge. Drills, when possible, should be machine ground to ensure perfect cutting form and clearance. Unfortunately, the situation does arise where the technician is forced to hand sharpen a drill. Here again, the technician will become proficient in hand sharpening drills only after several tries. In drilling, when the pressure to advance the drill has to be increased and excessive heat builds up, the drill generally is dull. Stop the drilling and inspect the cutting edge for dullness and heat marks (bluish burn marks). If blue heat marks are present, the drill must be reground past the marks since the marks are unhardened drill stock. In field sharpening, bench or pedestal grinders are used (Chap. 9) with a medium to fine grit wheel. During the grinding operation, it is necessary to frequently dip the drill into water to retain the drill's temper.

In this free-hand drill resharpening operation, only one motion is used (Fig. 7-4). This motion is the arc (swing) required to grind from the back clearance edge to the front cutting edge on each flute of the drill. The pivot point of this swing is the technician's finger, which is resting on the tool rest near the wheel. Extreme care should be exercised during this operation to prevent injuries. Practice grinding as follows:

1. With the grinder turned off and with the finger under the drill as a pivot, align the drill cutting edge and the wheel face to establish the rotation angle.

2. Swing the drill shank down with the other hand to the back surface of the cutting flute without touching the wheel. Now, slowly pull the shank of the drill up until you come to the tip of the cutting edge. Do this several times until you develop the procedures.

In actual grinding, the following procedure is recommended:

1. Use the finger/fingers as a pivot point of drill rotation on the tool rest of the grinder.

2. Align the drill cutting edge to the wheel face (parallel). This establishes the drill to wheel rotation angle.

3. Dry run several passes of back-to-front edge rotation on the clearance angle.

4. After several trial runs, feed the drill back edge into the wheel until a light grind is noticed. Immediately, but slowly, rotate the drill shank upward about the rotation angle in step 2, so that the grinding action is toward the cutting edge.

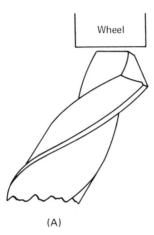

Wheel

(A)

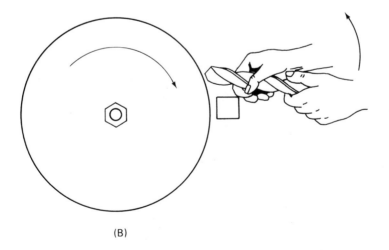

(B)

Fig. 7-4. Hand grinding twist drills.

5. When the cutting edge is reached, sparks will fly over the edge. Immediately pull the drill from the wheel.

6. Repeat the above on both flutes until a sharp edge is achieved.

7. Using a bevel protractor, check that both cutting edges are of the same angle and also use a scale to measure and verify that the length of both cutting edges are the same.

There are times when there is a requirement to grind a drill point flat. These drills are used to flatten the drill point left by previous conventional point drilling operations. The first procedure is to hold the drill point flat on the tool rest of the grinder. Then, with a rotary motion, the end of the drill is ground flat (perpendicular) to the drill axis. After this flat is produced; the previous steps of grinding can be performed on each flute. But in this case, one must grind a clearance relief for the cutting edges. The rotation angle is perpendicular to the wheel face.

Drills should be stored separately in either storage boxes or cardboard tubes.

Grinding Carbide Drills

Carbide drills suffer the most damage by sudden shock, either in drilling or by accidental dropping. Regrinding of carbide drills must be done by a tool room grinder as they require special wheels to cut them and special point geometry to make them cut properly.

7-7. Drill Gauges

An important item that is used quickly and easily in sizing drills is the drill gauge. Drill gauges are made of hardened steel and are available in three types to cover the full range of drills up to and including the $\frac{1}{2}$ inch size; the three types of drill gauges cover fractional, number, and letter size drill sets. The true worth of these gauges is in the identification of drills where the size has been rubbed or worn away.

Wire drills, which have the smallest diameter increments from drill-to-drill size, should be gauged for accurate sizing prior to any drilling. For example, in the case of a no. 34 drill, the difference between it and a no. 35 drill is only one thousandth of an inch (0.001). In gauging this drill, it is found that it will fit loosely in a no. 33 hole (0.113), snugly in its true size, a no. 34, and not at all in a no. 35 hole. This procedure of loose, snug, or no-go should be used in gauging all drills. Never gauge the drill shank, as errors due to burrs and swelled sections will be encountered. *Always* gauge the cutting end of the drill.

COUNTERSINKS

Countersinks (Fig. 7-5) are cutting tools that are used to produce conical shaped holes and to deburr drilled holes. They are also used for drilling holes in thin sheet metal. The general range of included angles of countersinks is 60, 82, 90, and 100°. These angles have been selected to meet the requirements in the machining and fastening field.

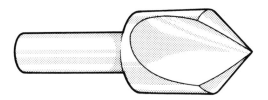

Fig. 7-5. Countersink. (Courtesy of Greenfield Tap and Die)

7-8. Countersink Included Angles

The 60° countersink is used quite often to cut or put a center in a shaft to be turned in a lathe. The 82° countersink is by far the most often used countersink in shop work. This angle is the general standard angle of flat head screws in the United States. A 100° flat head standard also exists, but is not in general use. Quite often, the use of a countersink rivet is required and generally this is where the 90° countersink is used.

7-9. Countersink Flutes and Major Diameters

In addition to classifying countersinks by their cutting angles, countersinks are also classified by the number of cutting flutes and by the major (largest) diameter. Countersinks are available in either three flute, four flute, or single flute versions. The three flute countersink is preferred in machine use as it produces a smoother countersink hole and has less of a tendency to chatter while cutting. It should be pointed out that any odd number of cutting flutes on a countersink tends to eliminate chatter while cutting and produces smoother cut surfaces.

The single flute countersink has been developed for light portable drilling applications. With its one cutting edge, less pressure is required in cutting. It produces excellent finishes. One important point should be made concerning single flute countersinks; the manufacturer recommends that a predrilled hole of at least 10% of the countersink diameter be drilled prior to countersinking.

The major diameter of countersinks normally run from $\frac{1}{4}$ to $1\frac{1}{2}$ inches in diameter. Shank diameters generally run from $\frac{3}{16}$ inch for the small $\frac{1}{4}$ inch countersink to $\frac{1}{2}$ inch shanks for countersinks above the 1 inch diameter range.

7-10. Use and Speed of the Countersink

When using countersinks on metal jobs, a general rule of thumb is that the speed be run in the low range (300 to 500 RPM) with good lubrication applied to the cutting surfaces. Quite often, shops supply four flute countersinks, which may start to chatter badly during some cutting operations. If the chatter doesn't stop in slowing the speed down further, there is a simple solution to stop the chatter. Take a small bit of shop waste or an oily waste rag and place it between the countersink and the chattered countersunk hole. Lubricate the rag and again bring the countersink to the work with the waste in between. The rag acts to dampen out vibrations, which eliminates the chattering during the cutting operation.

7-11. Care of the Countersink

Countersinks should be inspected regularly for dullness of the cutting edges and burring of the shanks. Generally, the countersink is considered by many as an expendable tool. This means that the cost of resharpening it is as costly as purchasing a new one. In the larger diameter types, however, resharpening is economical and can and must be performed on a tool room grinder.

REAMERS

There are times when very accurately sized diameter holes with an extremely smooth surface finish are required. Drills generally do not produce smooth high tolerance diameter holes. When holes that require an exacting tolerance, such as 0.0001 inch in diameter, are called for, two methods of attaining this tolerance are available. By the first method, high tolerance holes are cut by a machinist using such machine tools as lathes, milling machines and boring mills. The second method is to size a drilled hole by using either a hand or a machine reamer.

In drilling, as previously discussed, the only cutting action is performed by the cutting point itself. The flutes of a drill, while sharp to the touch, do not perform any cutting action. The resulting drilled hole is generally rough and scraped. In reaming, the process is reversed in that the flutes do the cutting (shaving). Since the flutes cut along their entire length, the finished surface is devoid of feed or scrape marks.

7-12. Reamer Flutes

The cutting flutes on reamers are of two types: straight flutes and spiral flutes. Straight fluted reamers are recommended for most general purposes. In areas where reaming a hole with interrupted cuts is required, such as when a pulley with a keyway or slot is encountered, the spiral fluted reamer is used. The reason is that in reaming a keywayed hole with a straight reamer, the reamer would allow its edges to drop down and catch on the edge of the keyway; the spiral fluted reamer would not.

Reamers are generally made of high-speed steel, but in recent years have been made in both carbide tipped and solid carbide to meet the tougher alloys being used. The three most often used types of reamers are straight, taper pin, and expansion reamers.

7-13. Straight Reamers

Straight reamers, which produce holes that are cylindrical in shape, are available in a range of sizes from 0.040 to $1\frac{1}{2}$ inches in diameter. The total range of the straight reamer can be broken into four groups: fractional sized reamers, wire gauge sized reamers, letter sized reamers, and over or undersized reamers.

Fractional reamers are available in $\frac{1}{64}$ inch increments in a range of from $\frac{1}{64}$ to $1\frac{1}{2}$ inches in diameter. Wire gauge reamers are to the exact diameters of the number (wire gauge) twist drill diameters and likewise the same for the letter size reamers. In over and undersized reamers, which are too numerous to list, a plus and minus diameter reamer size range for any given fractional, wire gauge, or letter reamer is available. For example, in the case of a letter R size reamer, which has a diameter of 0.3390 inch, the closest minus reamer is 0.3380 inch (one thousandth of an inch less), while the closest plus reamer is 0.3395 inch (or 0.0005 inch) larger than the R size reamer.

The purpose of the over/undersize reamers is to provide holes where either a pin of a particulr diameter is to be driven into a hole or where there is to be clearance in a hole for a pin to slide into. For example, dowel pins are used to align machine parts and normally one end of the dowel is driven or held into one piece, while the mating piece has a hole which slides smoothly onto and off of the dowel pin. Thus, in using $\frac{1}{8}$ dowels (0.1250 inch), the technician must choose a reamer to hold the dowel in the one plate while picking another size reamer which allows the dowel to slide freely in the other plate. A quick check of the available sizes shows that a 0.1247 inch diameter reamer exists. Since this is 0.0003 inch smaller than the 0.125 inch dowel, the dowel will have to be tapped or pressed into the plate. This will lock the dowel to the plate. In looking for a dowel pin clearance hole for the other plate, a range of reamers (plus) from 0.1252 inch to 0.1253 inch to 0.1255 inch to 0.126 inch are available. This oversize range allows for clearance from 0.0002 inch to 0.001 inch above the 0.125 inch dowel, and any of the four oversize reams would allow for a smooth slide fit.

Drill Selection for Straight Reamers

Selection of a proper size drill that will allow sufficient material for a smooth and accurate straight reamer hole is at times difficult. One main factor that governs the selection is the size of the reamer—whether it is in the small, medium, or large range. The small range will be defined as a no. 60 wire drill reamer up to and including the $\frac{1}{4}$ inch (0.250) reamer. The amount of material to be left in a hole for reaming within this group should be 0.002–0.004 inch at the small end with 0.004–0.005 inch left at the upper end.

The medium size reamers with a range of from $\frac{1}{4}$ to $\frac{3}{4}$ inch in diameter should have drill sizes of from 0.004 inch undersize for the small reamers to 0.015 inch for the larger $\frac{3}{4}$ inch reamer. The larger range reamer covering from $\frac{3}{4}$ to $1\frac{1}{2}$ inches in diameter generally requires a 0.010 to 0.015 inch undersize drill for the low end to $\frac{1}{32}$ inch for the high end. It should be remembered that most reaming problems exist in the small range, where a dull drill or excessive drilling pressure can sometimes cause a drill to "run out of size" and cut out what little material was left for the reamer. Extreme care should be exercised during drilling in the small reamer range to eliminate this problem.

Straight Reamers Used by Hand

Hand straight reamers are important for field use where power tools are not available. The available size range in the hand reamer class is limited generally to the fractional sizes. These reamers are basically the same configuration as the chucking reamer except for the square driving end at the end of the shank, which is fit into either a "tee" handled tapper or the straight tap wrench. The tee handled tapper and the tap wrench are discussed in Chap. 8.

7-14. Taper Pin Reamers

While there are several forms of taper reamers in use, this text will discuss only the *taper pin reamer* (Fig. 7-6). These reamers, which are available in both hand and machine forms, are manufactured to cut a standard taper of $\frac{1}{4}$ inch per foot. As shown in Table 7-10, a range of 17 taper pin reamers are listed from the smallest, a 7/0, to the largest, a no. 10. The purpose of the taper pin reamer is to produce a tapered hole to receive a tapered pin. Tapered pins are used to secure and to mate rods and shafts to such items as collars, gears, sleeves and couplings.

Fig. 7-6. Taper pin reamer. (Courtesy of Greenfield Tap and Die)

TABLE 7-10. TAPER PIN REAMER SIZES

Size No.	Shank Diameter (Inches)	Diam. of Small End	Diam. of Large End	Flute Length (Inches)	Overall Length (Inches)
7/0	$\frac{5}{64}$	0.0497	0.0666	$\frac{13}{16}$	$1\frac{13}{16}$
6/0	$\frac{3}{32}$	0.0611	0.0806	$\frac{15}{16}$	$1\frac{15}{16}$
5/0	$\frac{7}{64}$	0.0719	0.0966	$1\frac{3}{16}$	$2\frac{3}{16}$
4/0	$\frac{1}{8}$	0.0869	0.1142	$1\frac{5}{16}$	$2\frac{5}{16}$
3/0	$\frac{9}{64}$	0.1029	0.1302	$1\frac{5}{16}$	$2\frac{5}{16}$
2/0	$\frac{5}{32}$	0.1137	0.1462	$1\frac{9}{16}$	$2\frac{9}{16}$
0	$\frac{11}{64}$	0.1287	0.1638	$1\frac{11}{16}$	$2\frac{15}{16}$
1	$\frac{3}{16}$	0.1447	0.1798	$1\frac{11}{16}$	$2\frac{15}{16}$
2	$\frac{13}{64}$	0.1605	0.2008	$1\frac{15}{16}$	$3\frac{3}{16}$
3	$\frac{15}{64}$	0.1813	0.2294	$2\frac{5}{16}$	$3\frac{11}{16}$
4	$\frac{17}{64}$	0.2071	0.2604	$2\frac{9}{16}$	$4\frac{1}{16}$
5	$\frac{5}{16}$	0.2409	0.2994	$2\frac{13}{16}$	$5\frac{3}{16}$
6	$\frac{23}{64}$	0.2773	0.354	$3\frac{11}{16}$	$5\frac{7}{16}$
7	$\frac{13}{32}$	0.3297	0.422	$4\frac{7}{16}$	$6\frac{5}{16}$
8	$\frac{7}{16}$	0.3971	0.505	$5\frac{3}{16}$	$7\frac{3}{16}$
9	$\frac{9}{16}$	0.4805	0.6066	$6\frac{1}{16}$	$8\frac{5}{16}$
10	$\frac{5}{8}$	0.5799	0.7216	$6\frac{13}{16}$	$9\frac{5}{16}$

Since the taper is so slight and therefore self holding, pins put into rotating assemblies are not loosened during use. In small gear box assemblies where no looseness between gear and shaft units is allowable, the taper pin is used to hold, drive, and locate these assemblies. Taper pin reamers, as shown in the Table 7-10, have an effective flute length that limits their working reaming length. For example a no. 3/0 reamer could not be used to ream an assembly whose thickness is greater than $1\frac{1}{4}$ inches, since its effective cutting length is only $1\frac{5}{16}$ inches.

Drill Selection for Taper Pin Reamers

Predrilling of an assembly for taper reaming can be difficult because of proper drill selection. It is obvious from Table 7-10 that the diameter of the small end of the reamer can be used as the approximate drill size. Generally, on the small size reamers, drills that are several thousandths of an inch larger than the small diameter of the reamer are used. This method of a single drill size is always positive, but also involves a large amount of stock removal by reaming. Quite often in the larger sizes of reamers, holes are *stepped* into the workpiece so that reaming becomes easier. For example, in putting a no. 10 taper pin in a 3 inch diameter collar and shaft assembly, it is required that the large diameter, small (exit) diameter, and the midsection diameter of the reamer be known so that two drill sizes may be selected. The actual diameter of the large end of a no. 10 pin is listed in Table 7-10.

Determination of the diameter of the pin at the point where it comes through the assembly (exit) can then be found by the following equation:

$$d_{sm} = D_{lg} - L(0.0208) \qquad (7\text{-}1)$$

where

$$D_{lg} = \text{large diameter of pin}$$

$$d_{sm} = \text{small diameter (exit) of pin}$$

$$L = \text{length of work to be pinned}$$

$$0.0208 = \text{taper per inch of pin (constant)}$$

In the original problem, the large diameter of the no. 10 taper pin (Table 7-10) is 0.7216 inch. Therefore, the small (d_{sm}) or exit diameter of the reamer at a length L equal to 3 inches is

$$d_{sm} = D_{lg} - L(0.0208)$$

$$d_{sm} = 0.7216 - 3(0.0208)$$

$$d_{sm} = 0.7216 - 0.0624$$

$$d_{sm} = 0.6592$$

Having determined the exit diameter of the tapered pin, we can now determine its diameter halfway through the 3 inch assembly. Referring back to Eq. 7-1, we calculate for a length L equal to $1\frac{1}{2}$ inches, which is the mid point of the assembly.

$$d_{sm} = 0.7216 - 1.5(0.0208)$$

$$d_{sm} = 0.7216 - 0.0312$$

$$d_{sm} = 0.6904 \text{ inch}$$

We can now put two drills, one halfway and the second completely through the assembly prior to reaming. Remember that the above two calculations are for the actual finish diameters, so allowance for stock to be removed by reaming must be made. In the case of the no. 10 pin, the small or exit diameter was calculated at 0.6592 inch. Therefore, a $\frac{41}{64}$ inch drill can be used. This size will allow for 0.018 inch of stock excess to be reamed out. For the mid range, or $1\frac{1}{2}$ inch section of the 3-inch assembly, the calculated diameter was 0.6904 inch. Allowing for stock to be left, an undersize drill of $\frac{43}{64}$ inch, or 0.6719, can be used to counter drill the hole to a depth of

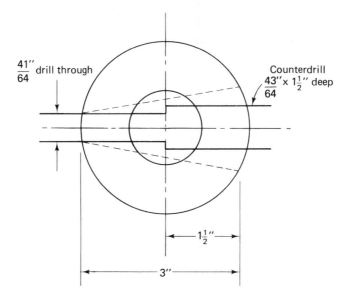

Fig. 7-7. Counterdrilling an assembly for taper reaming.

$1\frac{1}{2}$ inches. Refer to Fig. 7-7. This procedure of counter drilling allows for faster taper pin reaming with less wear imparted to the reamer.

Taper Pin Reamer Flutes

Cutting flutes are available in both straight and spiral groove for the hand driven type of taper pin reamer and with helical flutes in the machine type. The helical grooves reduce tool breakage during machine reaming by preventing chips from clogging the reamer.

7-15. Expansion Reamers

Expansion reamers are normally available in fractional sizes only. The range, as in the AST straight reamers, runs from $\frac{1}{4}$ inch (0.250) to $1\frac{1}{2}$ inch (1.500 inch). The one unique difference of expansion reamers is the fact that the diameter can be adjusted to a very accurate "over" size of a standard fraction. The expansion reamer is very useful in close fitting of shaft-to-hole situations where only a few thousandths of an inch over a standard size is needed.

On looking at the construction of the expansion reamer (Fig. 7-8), the first thing that is noticed is that the slots are machined along the length of the reamer between the flutes. These slots give the spring or flexibility to the cutting flutes. The oversizing is achieved by adjusting a screw thread located at the end of the reamer. This thread acts on tapered inserts within the reamer to force the cutting flutes out to the required dimension.

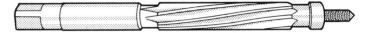

Fig. 7-8. Expansion reamer. (Courtesy of Greenfield Tap and Die)

Whenever the expansion reamer is to be adjusted, a micrometer should be used to check the required size. The size should be checked for accuracy in at least two positions of the expanded size. Over expanding of the reamer can cause cracking of the cutting edges and destruction of the tool. Recommended ranges of expansion for the hand expansion reamer family are:

$$\tfrac{1}{4} \text{ to } \tfrac{15}{32} \text{ inch—0.006 inch}$$

$$\tfrac{1}{2} \text{ to } \tfrac{31}{32} \text{ inch—0.010 inch}$$

$$1 \text{ to } 1\tfrac{1}{2} \text{ inch—0.012 inch}$$

Hand expansion reamers are available in both straight and spiral flute designs.

7-16. Types of Reamer Shanks

All reamers, regardless of their type, are available in three basic shank styles: straight shank (chucking reamer), taper shank (AST), and square end (hand reamers). Straight shank (also called chucking) reamers are the most popular shank style reamer used in the shop. They are designed to be used in the full range of chucks, from the small hand drill to large machine tools such as lathes or milling machines.

On reamers with the taper AST shank, the size range is available only in the fractional sizes. Generally the size starts at 0.250 inch and extends to the $1\tfrac{1}{2}$ inch reamer, with increments in size of $\tfrac{1}{32}$ inch. The use of these taper shank reamers is generally directed toward the heavy machine tools that are equipped with the AST drive.

The square end hand reamer is held and driven by means of the "tee" handled taper or the straight tap wrench. These tools are discussed in Chap. 8.

7-17. Use and Speed of the Machine Reamer

When using reamers in power tools, the question of proper speed and feed becomes very important. Proper speed and feed effect both smoothness of the finished hole and also the life and accuracy of the reamer.

Generally, reamers are run at low speeds with light, medium, and heavy feeds, depending on the material. As in any cutting operation, coolant in the form of oil or cutting fluids should always be used. A discussion on coolants during cutting operations is discussed in Chap. 8. Table 7-11 lists the recommended speeds in RPM for reamers in the small, medium and large range.

TABLE 7-11. MACHINE REAMER SPEEDS[1, 2]

Reamer Size (inches)	RPM Mild Steel	RPM Stainless Steel	RPM Aluminum and Brass
no. 60 to $\frac{1}{4}$	500	450	580
$\frac{1}{4}$ to $\frac{3}{4}$	350	300	425
$\frac{3}{4}$ to $1\frac{1}{2}$	250	200	300

[1] The above speeds are merely starting points for machine reaming. Speeds above or below any reamer size group can and will be encountered.

[2] Carbide reamers can generally be run at higher speeds in the range of feeds used by reamers.

A light-to-medium reamer feed is recommended for best results. Table 7-12 lists reamer feeds.

TABLE 7-12. REAMER FEEDS

Size (inches)	Inch/Revol. Light	Inch/Revol. Medium
no. 60 to $\frac{1}{16}$	0.0003 to 0.001	0.0005 to 0.002
$\frac{1}{16}$ to $\frac{1}{8}$	0.001 to 0.002	0.002 to 0.005
$\frac{1}{8}$ to $\frac{1}{4}$	0.002 to 0.005	0.005 to 0.007
$\frac{1}{4}$ to $\frac{1}{2}$	0.005 to 0.007	0.007 to 0.010
$\frac{1}{2}$ to 1	0.007 to 0.010	0.010 to 0.020
1 and up	0.010 to 0.020	0.020 to 0.040

For applications in brass and aluminum, a medium feed is recommended. Reaming in stainless steel also calls for medium reamer feeds. It must be pointed out that in any reaming process, the reamer should be backed out of the work while still turning. Failure to do this will result in flute edge scratches being formed in the hole due to tool edge drag.

7-18. Care of the Reamer

Reamers normally require special care in storage. Since their cutting surfaces are on the outer surface of the tool, care must be taken not to dull them. When dull, reamers can be reground to a smaller or undersize configuration. This regrinding must be performed by special tool room grinders. During long periods of storage, a thin coat of oil should be applied to the overall tool. Taper reamers, when dull, are replaced.

COUNTERBORES

Counterbores (Fig. 7-9) are tools of high-speed steel used to produce flat bottomed holes in line with previously drilled holes. The main purpose of counterbores is to provide a space or pocket for mounting nuts or bolt heads below a surface and to also provide a machined surface for bolt heads or nuts to bear on, particularly in cast pieces where surface flatness is poor. Counterbores can be procured in diameters of from $\frac{1}{4}$ to 3 inches in diameter. From the $\frac{1}{4}$ inch size to the 1 inch size, increments in $\frac{1}{32}$ of an inch are available. From a 1 inch diameter to a $1\frac{3}{8}$ inch diameter, increments are by $\frac{1}{16}$ inch. From $1\frac{1}{2}$ inch to 3 inch diameters, $\frac{1}{8}$ inch increments are available.

(A)   (B)

Fig. 7-9. (A) Counterbore and (B) counterbore pilot.

7-19. Counterbore Cutting Flutes

Cutting flutes on counterbores are made in the odd number range to eliminate chatter during the cutting process. Counterbores, like drills, only cut on their points while the flutes carry chips away from the workpiece. Three flutes are used in the diameter range of from $\frac{1}{4}$ to $1\frac{3}{16}$ inches in diameter and five flutes on the $1\frac{1}{4}$ to 3 inches in diameter sizes.

7-20. Counterbore Pilots

As the name implies (counterbore), a predrilled or finished hole must be available before counterboring can take place. There is a hole and set screw in the cutting end of the counterbore that is used to locate and lock pilot hole inserts to the counterbore. These pilots or pilot hole inserts are made in a fractional range of diameters from $\frac{1}{8}$ to 3 inches in diameter, but can also be obtained in letter and number (wire gauge) diameters. The single purpose of these pilots is to guide the counterbore in the predrilled hole and act as a bearing during cutting.

7-21. Counterbore Shanks

Counterbores are available in both straight and AST (taper) shank styles. In the straight shank style, shank diameters of $\frac{3}{8}$ and $\frac{1}{2}$ inch are most often used.

AST shank counterbores, like drills, have a range of diameters assigned to a particular taper shank number. For example, in the $\frac{3}{4}$ to the $1\frac{1}{16}$ inch diameter counterbore, the AST taper shank is made in either a number 1 or a number 2. Manufacturer's lists are available to give the range and variations of both straight and taper shank counterbores.

7-22. Use of Counterbores and Counterbore Pilots

Counterbores should never be used to perform any work without a pilot, and any attempt to do so will cause work and tool damage and possible injury to the technician. The correct approach procedure is to first measure the existing hole diameter that is to have a counterbore. After determining its size, a pilot of the same size is inserted into the proper diameter counterbore and locked with the set screw. It is always wise to hand check that the pilot is not tight in the drilled hole. A pilot that is too tight will heat up and gall during the machining operation. The counterbore is used at a low to medium speed in relation to drills. The feeds are less than or equal to the feeds of drills.

7-23. Care of Counterbores and Counterbore Pilots

During all cutting operations, some form of coolant should be used to both cool the cutting and lubricate the pilot. When the counterbore appears to be dull from overheating or use, it should be reground. Regrinding must be done by machine (not hand ground). Either the factory or a tool shop can regrind, maintaining both squareness and equal cutting of the cutting edges. Pilots at times will heat up and gall. They should be inspected and whenever a burr or rub mark is found, they should be either smoothed with a stone or replaced.

SUMMARY

Drills, countersinks, reamers, and counterbores are hole cutting tools which are used by the technician in his daily work. Drilling is the process of boring a circular hole through a workpiece; countersinking is the process of enlarging the upper part of a hole to receive a cone-shaped fastener; reaming is performed to enlarge and smooth a hole; and counterboring is the method of enlarging the upper part of a hole into a flat bottomed cylindrical shape with a diameter larger than the original hole.

In using drills, countersinks, reamers, and counterbores, the technician must be aware of the correct drilling speed and correct feeding speed for the workpiece materials. Accuracy comes through repeated practice.

REVIEW QUESTIONS

1. Define: drilling, countersinking, reaming, and counterboring.

2. What three size groupings of drills are available to the technician?

3. Why is such a size range needed?

4. What two types of shanks are used to hold and drive drills?

5. List the two types of combined drill and countersink drills, and state briefly an application for each particular style.

6. What are the three angular requirements to produce a drill cutting point?

7. What is the purpose of a countersink? What is the most popular included angle in common use?

8. The cutting flutes on reamers are of two types. List them.

9. What is the purpose of over and undersize straight reamers?

10. Taper pin reamers have a standard taper of _____?

11. List the three basic shank styles available in all reamer types.

12. What is the purpose of an expansion type reamer? Explain how it is adjusted and checked for a particular size.

13. Describe the correct procedure used in the selection of a pilot for use in a hole.

8

TAPS
AND
DIES

This chapter covers two processes that are frequently used by the mechanical technician—use of the tap, and use of the thread die. The tap is a tool that is used to produce a continuous helical groove (thread) in a hole, whereas the thread die is a tool used to produce a continuous helical groove on a rod. There are three reasons for producing threaded holes and rods: to clamp or screw two or more pieces of material together, to fasten tubes or pipes together to carry air, gas, or liquids, and to pry or jack apart two items, such as jacking a car off the ground with a screw jack.

The chapter begins with a discussion of thread characteristics. Machine taps, pipe taps, and removal of broken taps are presented next. The discussion of taps is followed by a discussion of machine thread dies and pipe thread dies. Finally, lubricants/coolants used in the process of cutting threads with taps and dies are discussed.

THREAD CHARACTERISTICS

There are six basic characteristics required to describe a helical thread. The characteristics are *thread form, thread size, threads per inch or pitch, thread series, helix,* and *class of fit.*

8-1. Thread Forms

The thread forms in most common use today in the United States are the unified thread, the acme thread, and the buttress thread.

Unified Thread

On November 18, 1948, an international accordance was signed by representatives of the United States, the United Kingdom, and Canada for a common thread form. This thread, known as the unified screw thread form or UN, allows for threads from these countries to be interchanged. Figure 8-1 shows the British standard form thread (Whitworth) with its rounded peaks and bottoms, and the American thread form with its flat peaks and bottoms. The accordance on thread form had concentrated on two areas, the 60° included angle thread and adjustment of the round tops and bottoms of the British thread and the flat tops and bottoms of the American thread form to allow interchangeability. The British made the greatest contribution to the agreement on the unified thread form. They had adjusted their 55° included angle thread form to the 60° included angle thread form. The final phase was the mutual adjustment and acceptance of allowances to allow rounded British threads and flattened American threads to be interchanged.

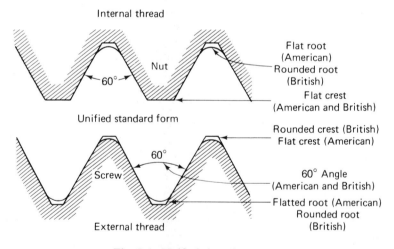

Fig. 8-1. Unified thread system.

The unified thread form is used extensively on all bolt, nut, and machine screw hardware in the United States. Its geometric form is a 60° equilateral triangle with its top or peak left flat. Between any two threads, there is an equal flat at their bottom. This flat is a specific dimension and is equal to $\frac{1}{8}$ (0.125) of the distance between two successive threads.

The dimension of these flats has been carefully chosen for two reasons.

The first reason is that the outer flats act as a protective surface to absorb hits or bumps which could damage the thread. The second reason is that the flats allow for longer life of the cutting tap or die. It can be seen that if a pointed tool was used, it would be more prone to wear during heavy use.

Acme Thread

The acme thread form (Fig. 8-2) is found in areas where heavy work must be performed by the thread without damaging itself. A car screw jack is a good example of an acme thread application. The strength of the acme thread can be seen by the fact that the cross section of the thread physically has more material than the UN and is thus able to transmit more push or pull without distorting its form. The acme thread, while important, finds very little use in day to day general shop use.

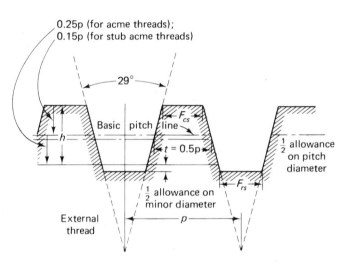

Fig. 8-2. American standard acme thread.

Buttress Thread (American Standard)

The buttress thread has a special and unique shape and application. The shape has one of its faces almost perpendicular to the center line of the screw, while the other side of the thread has a slope of 45° to the center line position. Anyone who has noticed the design of large dams can immediately see the strength of this thread. With the side of the dam adjacent to the water almost vertical, the sloped or buttressed outer face acts to hold back the tremendous pressure exerted by the water. The buttress thread is designed for this same principle: for the task of holding or exerting great pressure. This thread is used in the breech loading mechanisms of large cannons.

The pressure on the buttress thread must be applied in the direction of the "almost" vertical face of the thread. Its general abbreviated notation is written as N. Butt. (National Buttress).

8-2. Thread Sizes

When the word *size* is used in reference to machine screw and fractional taps, it refers to the actual diameter of these taps. The diameters of these two groups of taps are listed in Tables 8-1 and 8-2. In the left column of Table 8-1, the size runs from no. 0 to no. 14 for the screw gauge inclusive. The range of fractional taps (Table 8-2) generally runs from $\frac{1}{4}$ to 2 inches.

In the column marked major diameter (inches) of Table 8-1, it is noted that a difference of thirteen thousandths (0.013) of an inch separates any two successive numbers; i.e. screw gauge no. 1 major diameter minus screw gauge no. 0 major diameter equals 0.013 inch: 0.073 − 0.060 = 0.013 inch.

TABLE 8-1. UNIFIED THREAD MACHINE SCREW CHART

Size	Major Diameter (Inches)	Threads per Inch Series		General Tap Drill*		Minimum Clearance Drill
		Coarse, Unc.	Fine, Unf.			
0	0.060		80	$\frac{3}{64}$''		#52
1	0.073	64	72	#53	C	#48
				#53	F	
2	0.086	56	64	#50	C	#43
				#50	F	
3	0.099	48	56	#47	C	#37
				#45	F	
4	0.112	40	48	#43	C	#32
				#42	F	
5	0.125	40	44	#38	C	#30
				#37	F	
6	0.138	32	40	#36	C	#27
				#33	F	
8	0.164	32	36	#29	C	#18
				#29	F	
10	0.190	24	32	#25	C	#9
				#21	F	
12	0.216	24	28	#16	C	#2
				#14	F	
14†	0.242	20†	24†	#10	C	"D"
				#7	F	

* Where F is fine thread and C is coarse thread.
† Not of the Amer. standard form.

TABLE 8-2. AMERICAN NATIONAL FRACTIONAL SCREW CHART

Size (Inches)	Major Diameter (Inches)	Series (Threads per Inch)		General Tap Drill*		Minimum Clearance Drill
		Coarse N.C.	Fine N.F.			
$\frac{1}{4}$	0.250	20	28	#7	NC	$\frac{17}{64}$
				#3	NF	
$\frac{5}{16}$	0.3125	18	24	F	NC	$\frac{21}{64}$
				I	NF	
$\frac{3}{8}$	0.375	16	24	$\frac{5}{16}''$	NC	$\frac{25}{64}$
				Q	NF	
$\frac{7}{16}$	0.4375	14	20	U	NC	$\frac{29}{64}$
				$\frac{25}{64}''$	NF	
$\frac{1}{2}$	0.5000	13	20	$\frac{27}{64}''$	NC	$\frac{33}{64}$
				$\frac{29}{64}''$	NF	
$\frac{9}{16}$	0.5625	12	18	$\frac{31}{64}''$	NC	$\frac{37}{64}$
				$\frac{33}{64}''$	NF	
$\frac{5}{8}$	0.625	11	18	$\frac{17}{32}''$	NC	$\frac{41}{64}$
				$\frac{37}{64}''$	NF	
$\frac{3}{4}$	0.750	10	16	$\frac{21}{32}''$	NC	$\frac{49}{64}$
				$\frac{11}{16}''$	NF	
$\frac{7}{8}$	0.875	9	14	$\frac{49}{64}''$	NC	$\frac{57}{64}$
				$\frac{13}{16}''$	NF	
1	1.000	8	14	$\frac{7}{8}''$	NC	$1\frac{1}{64}$
				$\frac{15}{16}''$	NF	
$1\frac{1}{8}$	1.125	7	12	$\frac{63}{64}''$	NC	$1\frac{9}{64}$
				$1\frac{3}{64}''$	NF	
$1\frac{1}{4}$	1.250	7	12	$1\frac{7}{64}''$	NC	$1\frac{17}{64}$
				$1\frac{11}{64}''$	NF	
$1\frac{3}{8}$	1.375	6	12	$1\frac{7}{32}''$	NC	$1\frac{25}{64}$
				$1\frac{19}{64}''$	NF	
$1\frac{1}{2}$	1.500	6	12	$1\frac{11}{32}''$	NC	$1\frac{33}{64}$
				$1\frac{27}{64}''$	NF	
$1\frac{3}{4}$	1.750	5		$1\frac{9}{16}''$		$1\frac{49}{64}$
2	2.000	$4\frac{1}{2}$		$1\frac{25}{32}''$		$2\frac{1}{64}$

* Where NC is national course and NF is national fine.

Therefore, to obtain any major diameter of a unified screw gauge number, the technician only has to remember the screw gauge number 0 diameter, which is 0.060 inch. To obtain any other major diameter multiply the screw number by 0.013 and add the base diameter of a no. 0 screw (0.060).

(Unified Screw thread no.) $\times (0.013) + (0.060) =$ major dia. of screw thread (8-1)

Example:

Determine the major diameter of a no. 8 unified screw. (8 × 0.013) + (0.060) equals diameter of no. 8 screw or 0.164 inch.

By looking at the list of fractional taps in Table 8-2, it is evident that the fractional size of the tap is the major diameter. A $\frac{5}{8}$ inch tap therefore has a major diameter of 0.625 inch. Taps are sometimes made in sizes larger than $1\frac{1}{2}$ inches. This is not usually done on a production basis as the demand for them is limited and their cost is rather high.

8-3. Threads Per Inch and Pitch

In the English system of thread notation, the number of threads per inch is used exclusively. It simply denotes the number of threads or fractional threads per one inch (linear). The third column of Tables 8-1 and 8-2, marked threads per inch, shows that the range is from 80 to $4\frac{1}{2}$ threads per inch for the thread system.

Another term of importance is *pitch*. It will be recalled from Chap. 1 that one full turn of an English micrometer screw is equal to 25 thousandths (0.025) of an inch. This travel or displacement due to one revolution of a thread is called its pitch. An equation for determination of a thread's pitch is as follows:

$$\text{pitch} = \frac{1}{\text{no. of threads per inch}} \qquad (8\text{-}2)$$

Pitch is the distance in inches between any two successive threads. It is the distance in inches that a thread advances in one revolution. As previously stated, the thread or anvil of the micrometer was displaced 0.025 inch for one revolution; this is the pitch of the thread. Using Eq. (8-2), the number of threads on the micrometer screw can be calculated:

$$\text{no. of threads per inch} = \frac{1}{\text{pitch in inches}} = \frac{1}{0.025} = 40 \text{ threads per inch}$$

8-4. Thread Series (Coarse, Fine, Extra Fine)

Consider a workpiece consisting of a $\frac{1}{4}$ inch thick aluminum plate that is to have a series of 1 inch tap holes put into it. If the standard 1 inch tap of 8 threads per inch were to be used, it is quite evident that only two threads would be cut into the plate [refer to pitch Eq. (8-2)]. The pitch of this thread is $\frac{1}{8}$ inch; therefore, only two threads could go into the $\frac{1}{4}$ inch thick plate. Two threads do not allow for any strength or stability from "rocking" when a bolt is inserted. It would be far more beneficial to put more threads into the plate. Therefore, a need for a range of threads per inch within a tap size is necessary. This range is classified into three groups:

1. coarse—NC (national coarse)
2. fine—NF (national fine)
3. extra fine—NEF (national extra fine)

As for the threaded holes in the aluminum plate, a better selection is a 1 inch —14 thread tap since it would put 3.5 threads into the plate.

Another important factor in choosing a particular thread series is the brittleness of the material to be tapped. In the case of cast iron, which is very brittle, a coarse thread is always a wise choice as a fine thread series tends to chip and break the thread during use. Certain other materials, such as plastics and phenolics, react the same as cast iron and should also be tapped with the coarse series of thread.

8-5. Helix (Right Hand or Left Hand)

When looking at a bolt or a thread, one fact that should be noticed immediately is whether the thread is sloping to the right or to the left when held in front of you. This simple, but important point, determines if the helix spiral is a right- or left-hand thread. Remember, a right-hand thread advances or pulls itself into a nut when turned in a clockwise direction (to the right) and a left-hand thread advances when turned counterclockwise (to the left).

The importance of a left-hand thread can be shown in the following examples. Some car manufacturers put left-hand threads on the wheels on one side of the car, while the other side has right-hand threads. The reason for this is that if a right-hand lug bolt was very loose on the left side of the car, sudden stops could completely unscrew the lug, whereas a left-hand lug on the left side tends to retighten itself on sudden stops. Turnbuckles, used to tighten wires to ship masts or to towers, use both a right- and left-hand thread in combination so that both threads tighten or loosen when the turnbuckle is rotated.

8-6. Class of Fit (Tolerance)

In the manufacture of a standard ground thread tap, a tolerance or range of size deviation has been set up for above or below the basic pitch diameter size (refer to Fig. 8-1). In ground thread taps from size no. 0 to 1 inch in diameter, seven levels or limits of tolerance are defined (Table 8-3). On ground thread taps which are above the basic pitch diameter, the letter H is used to denote high and the letter L to denote low or minus.

TABLE 8-3. LIMITS OF TOLERANCE OF GROUND THREAD TAPS (INCHES)

L1-Basic to basic minus 0.0005 inches
H1-Basic to basic plus 0.0005 inches
H2-Basic plus 0.0005 to basic plus 0.001 inches
H3-Basic plus 0.001 to basic plus 0.0015 inches
H4-Basic plus 0.0015 to basic plus 0.002 inches
H5-Basic plus 0.002 to basic plus 0.0025 inches
H6-Basic plus 0.0025 to basic plus 0.003 inches

It should be noted that an L1 or H1 class tap is rather expensive compared to the others (H2 or H3) due to the close tolerance range of basic to plus 0.0005 inch. In most general work applications, H2 or H3 class taps are used.

MACHINE TAPS

Having learned the six thread characteristics, let us proceed to discuss taps that are used almost daily in shop work. This section discusses machine tap forms, shank nomenclature, hand tap driver wrenches, and tapping procedures.

Very often, prints or sketches note a threaded hole merely as "$\frac{10}{32}$ thru" or "10–32 thru." When such a threaded hole is called out, there are several facts that are assumed unless otherwise noted: (1) the thread is right handed; (2) the thread form is American unified; and (3) the class fit is H2 or H3 (general).

The only remaining unknown is the tap drill size—the diameter of the hole to be drilled, so that the threads can be cut or formed into the material. The tap drills shown in Tables 8-1 and 8-2 are for producing a cut thread form that is 75° of the true thread form. This is a general practice so that commercial bolts and threads enter the tap holes easily, even if small burrs or bumps occur on the bolt. It also makes tapping much easier, since the material being tapped is not in full area contact with the tap during the cutting or forming operation. Full contact would cause extreme friction and quite often breakage of the tap.

8-7. Forms of Machine Taps

Of the three thread forms discussed at the beginning of this chapter, only taps utilizing the unified thread form are discussed, because the unified thread covers the bulk of threading applications. The unified tap varieties fall into two catagories for discussions: (1) hand machine screw and fractional taps that include cut taps (ground taps) and roll form taps; and (2) pipe taps that include regular and dryseal (discussed in Sec. 8-11).

All taps are generally made of high speed tool steel and consist of a threaded section for the cutting or forming of the thread, a cylindrical shank, and a square on the end of the shank (Fig. 8-3) for a driving tool.

In cut taps and ground taps, room is required for lubricant and chips during tapping. Therefore, grooves are machined into the tap leaving flutes with cutting thread on them. Of the types available, machine taps (Fig. 8-3) are described as two-flute taps, three-flute taps, and four-flute taps. Ground taps are of a closer tolerance and quality than the cut tap.

Fig. 8-3. Thread cutting tap. (Courtesy of Greenfield Tap and Die)

In a roll forming tap, which has little or no flutes, the thread is not cut but is squeezed or formed by displacing the material into the tap and also by the tap forming into the material.

It should be noted that Table 8-4 shows the tap drill sizes for roll forms only and should never be used in conjuction with the cutting tap drill charts in Tables 8-1 and 8-2. For example, a comparison between tap drills of a $\frac{1}{4}''$—20 thread cut tap (Table 8-2) and a $\frac{1}{4}''$—20 thread roll form tap (Table 8-4) shows a no. 7 drill 0.201 for the cut tap and a 0.224 diameter drill for the roll form tap: a difference of 23 thousandths (0.023) inch. If the roll form tap was tried in the cut tap no. 7 hole, it would jam and break, but the cutting tap would fit in the roll form tap hole and would produce a thread that would be so shallow that it would be unsafe to use.

Machine or fractional taps, except for the forming taps, are available in a set consisting of three taps to any particular size. The purpose of having three taps to a set is severalfold: it allows for easier starting in the tapping of hard material; it allows for aligning the tap "squarely" to the workpiece; and it allows for threading a "blind" hole (a hole that is not all the way through a workpiece).

The name and purpose of each tap in a cutting tap set (Fig. 8-3) is:

1. TAPER (or start tap)—used for starting a threaded hole generally in a hard material.
2. PLUG—used after the taper tap. Also sometimes used in place of the taper tap where material "tapability" allows.
3. BOTTOMING—used after the plug tap to cut threads to the bottom of a drilled hole (blind hole). This tap is never used to start a tap hole.

The taper tap has its threads slightly ground down or chamfered at

TABLE 8-4. DRILL SIZES FOR ROLL FORM TAPS

Tap Size	Threads per Inch	Theor. Hole Core Size	Nearest Drill Size	Decimal Equiv.
0	80	0.0536	1.35 mm	0.0531
1	64	0.0650	1.65 mm	0.0650
1	72	0.0659	1.65 mm	0.0650
2	56	0.0769	1.95 mm	0.0768
2	64	0.0780	$\frac{5}{64}$	0.0781
3	48	0.0884	2.25 mm	0.0886
3	56	0.0899	43	0.089
4	40	0.0993	2.5 mm	0.0984
4	48	0.1014	38	0.1015
5	40	0.1123	34	0.1110
5	44	0.1134	33	0.113
6	32	0.1221	3.1 mm	0.1220
6	40	0.1253	$\frac{1}{8}$	0.1250
8	32	0.1481	3.75 mm	0.1476
8	36	0.1498	25	0.1495
10	24	0.1688		
10	32	0.1741	17	0.1730
12	24	0.1948	10	0.1935
12	28	0.1978	5.0 mm	0.1968
$\frac{1}{4}''$	20	0.2245	5.7 mm	0.2244
$\frac{1}{4}''$	28	0.2318		
$\frac{5}{16}''$	18	0.2842	7.2 mm	0.2835
$\frac{5}{16}''$	24	0.2912	7.4 mm	0.2913
$\frac{3}{8}''$	16	0.3431	$\frac{11}{32}$	0.3437
$\frac{3}{8}''$	24	0.3537	9.0 mm	0.3543
$\frac{7}{16}''$	14	0.4011		
$\frac{7}{16}''$	20	0.4120	Z	0.4130
$\frac{1}{2}''$	13	0.4608		
$\frac{1}{2}''$	20	0.4745		

the starting end. This taper allows the tap to start easily in hard material and to be adjusted so that it starts "square" to the workpiece.

The plug tap is used after the taper tap has been run into a hole to increase the depth of the threads. The plug tap is recognized in that it has just a few threads at the end which are chamfered. The plug tap is the most popular in the set. It can very often be used to tap a hole directly without using a taper tap.

The bottoming tap is used to continue the threads to the bottom of a blind hole. It is recognized by the fact that its thread runs untouched except for one thread or so that is beveled on its end.

The roll form tap is designed to start into a tap hole and deform the material surrounding it into a finished thread. This tap is available in both the plug and bottoming types. There are no chips since there is no cutting action. The only requirement is the use of a lubricant to allow for smoother forming of the material. Application of this tap is limited to metals and materials that are not brittle (ductile).

Several advantages of the roll form tap over the cut tap are: the thread is stronger, the thread is smoother (burnished), and the threads last longer.

8-8. Machine Tap Shank Nomenclature

On the shank of every machine tap of American origin there is a concise description of the tap. A listing in the order of appearance of the nomenclature is as follows:

1. Size—screw no. or fractional size
2. Threads per inch
3. Thread series—coarse (NC), fine (NF), extra fine (NEF)
4. Class of fit—(Refer to Table 8-3)

For example, the following may appear on a tap shank.

$$\tfrac{1}{4} - 20NC - GH_3$$

This is interpreted as a $\tfrac{1}{4}$ inch fractional tap of 20 (threads per inch) NC (national coarse) and G (ground tap) of H3 class fit. Refering to Table 8-3, H is "high" or above basic with a range of 0.001–0.0015 inch above basic.

8-9. Hand Tap Driver Wrenches

In hand tapping, two tools are used to drive a tap—the T handled wrench and the straight tap wrench (Fig. 8-4). The T handled wrench generally is available in two sizes, small and large capacity, which accomodate taps from no. 0-80 to $\tfrac{1}{2}$ inch. The collet end of the wrench is designed to lock tightly

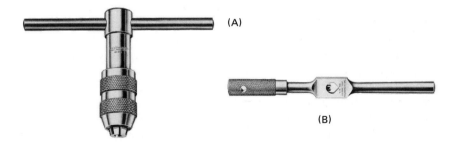

Fig. 8-4. Driving wrenches for taps: (A) tee and (B) straight. (Courtesy of L.S. Starrett Co.)

to the square end of the tap. In driving the tap, torque is applied by the bar on the wrench. The bar is slid through to either side of the wrench for use in cramped quarters.

The straight tap wrench is used when more torque or turning power is required. The straight tap wrench has both a fixed jaw and an adjustable jaw at the center of the wrench. Each jaw has a 90° notch. By adjusting the rotating handle, the movable jaw moves in or out to the required size. The straight tap wrench generally requires less torque in tapping due to its longer moment arm. It is normally available in three sizes covering a tap range of from 0-80 to $2\frac{1}{2}$ inches. The straight tap wrench is recommended for tapping pipe threads, since it has the ability to deliver the additional needed torque. Both the T handled wrench and the straight tap wrench are also used to hold square ended reamers, drills, and countersinks for performing other hand operations.

8-10. Tapping Procedures

After a properly selected tap drill hole has been drilled in a workpiece, the hole is deburred with either a burring knife or a countersink. Next, the hole is cleaned of chips or pieces from the drilling/burring operation.

If the tap hole is a deep hole through the workpiece or is a blind deep hole, a taper tap also called a start tap is used first, whereas a plug tap can be used for thin section tapping. With the tap secured in a tap driver wrench, coat the tap with the proper lubricating fluid selected from Table 8-7. Also put a little lubricating fluid inside the drilled hole. Next, bring the tap to the hole and "square" to the work surface. With a firm downward pressure, turn the tap until it starts to cut into the material. Turn the tap approximately one to one-and-one-half turns more and stop. Using a small square butt or a scale blade against the tap and workpiece, check the squareness of the tap to the workpiece at two places, 90° apart.

If an out-of-square situation exists, mark the workpiece with an arrow showing the direction in which the tap should slant to return it to a square position. "Back" the tap off (revolve in a direction opposite to the original turning) one turn or so and start the tapping again, this time with slight tipping pressure in the direction of the arrow. As the tap begins to cut, maintain the pressure directed by the arrow until approximately two turns are made; stop and recheck for squareness. With the tap now entering squarely into the material, continue the turning. As the tap is turned (in a cutting tap), a chip is formed which is packed into the fluted area. If tapping is continued, the chip will eventually jam the tap. It is for this reason, as well as for supplying new lubricating fluid to the edges, that a tap must be "backed off" one or two turns for every three or four cutting turns. The action of reversing the tap one or two turns causes the long chips being produced by the tapping operation to break off, thereby eliminating jamming or breaking of the tap. In the case of deep holes, the tap should be removed completely from time to time to remove chip material from the hole and tap flutes.

The procedure of "backing off" is extended to all hand taps including the forming tap. Even though there are no chips formed using the forming tap, backing off allows lubricating fluid to flow to the working edges of the tap.

PIPE TAPS

8-11. Forms of Pipe Taps

The two forms of tapered pipe taps that are used in job applications in the order of most common use are American Standard Taper Pipe Tap (regular), designated as NPT, and American Standard Dryseal Taper Pipe Tap, designated as NPTF. Both thread forms and tapers are basically the same except that the dryseal thread has truncations of crests and roots that when assembled leave no mating clearance, since the flats on the external and internal thread meet metal to metal. The dryseal thread requires no sealer or "dope." This dryseal joint is very essential when assembling piping where extreme cleanliness is required, such as in food packaging lines or beverage-filling equipment.

In using the standard taper pipe tap (NPT), a sealer is used in all assemblies to prevent leakage. A general sealer in most pipe work is a mixture of white lead and linseed oil which is coated on the thread prior to assembly. Recently, a much more sophisticated sealing method has been introduced; the male thread is wrapped with thin Teflon tape 0.001–0.002 inch in thickness prior to assembly. Teflon is a very inert slippery material that allows for smooth assembly. It seals extremely well in temperatures below 500°F.

In the case of commercial pipe taps (Table 8-5), the most often used range is from $\frac{1}{16}$ to 2 inches NPT. The number of flutes start with four flutes for the $\frac{1}{16}$ to the $\frac{1}{2}$ inch regular tap, and five flutes for the $\frac{3}{4}$ to $1\frac{1}{4}$ inch range. The remaining group of $1\frac{1}{2}$ to 2 inch taps generally have seven flutes.

Since the pipe tap is already tapered, $\frac{3}{4}$ inch to the foot, there is only one tap required to start and finish the thread. One form of pipe tap available for use is the "interrupted" thread type (Fig. 8-5). Since regular pipe tapping requires large amounts of power, whether hand or machine, there was a need for an easier operating tap. A method of grinding away every other tooth along a flute without interfering with its helix was developed. It is from this "interrupted" thread tap that much of the friction is eliminated in the tapping operation and a smoother thread is produced.

TABLE 8-5. PIPE TAPS

Nominal Pipe Size	Outside Diam. of Pipe	Threads per Inch	Handtight Engagement	
			Length (*Inches*)	Diam. (*Inches*)
$\frac{1}{16}$	0.3125	27	0.160	0.28118
$\frac{1}{8}$	0.405	27	0.1615	0.37360
$\frac{1}{4}$	0.540	18	0.2278	0.49163
$\frac{3}{8}$	0.675	18	0.240	0.62701
$\frac{1}{2}$	0.840	14	0.320	0.77843
$\frac{3}{4}$	1.050	14	0.339	0.98887
1	1.315	$11\frac{1}{2}$	0.400	1.23863
$1\frac{1}{4}$	1.660	$11\frac{1}{2}$	0.420	1.58338
$1\frac{1}{2}$	1.900	$11\frac{1}{2}$	0.420	1.82234
2	2.375	$11\frac{1}{2}$	0.436	2.29627

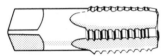

Fig. 8-5. Interrupted pipe thread tap.

Unlike machine tapping, where the tapping is either completely through or to the bottom of a blind hole, pipe tapping is more of a hand fitting operation.

8-12. Pipe Tapping Procedures

Generally, a problem arises in tapping a hole to match a male pipe thread. On all male pipe threads, there exists a section of imperfect threads that is the result of the chamfer of the dies used to produce the thread. In column 4 of Table 8-5, the length is the length of a proper hand tight engagement between a particular pipe assembly prior to a final sealing operation.

When using NPT thread joints, the final sealing of the pipe threads usually occurs within two to four turns after a handtight position. But here again is a case where a good "feel" is developed with experience. An approach that eliminates a loose, leaking assembly is as follows: first drill the tap hole (Table 8-6) and then tap the hole until approximately one-half of the tap is inserted. Remove and hand assemble both cleaned pipe joints (male and female). Mark the position of engagement on the male thread. Unscrew and visually inspect the advance mark made on the male thread. Inspect the number of full unused threads left on the pipe that are past the mark. Insert the tap until cutting pressure is felt. Remembering that at least two to four turns are needed for sealing, subtract three from the number of full unused threads counted. In the above case, let us say that nine full threads were counted; by subtracting three from nine, there are six remaining, the number of full turns that the tap is to be turned. After retapping the hole, again remove and clean the thread and hand assemble. There should be at least three full threads left for the final tightening seal. Remember that all NPT joints on final assembly should have either pipe dope or other thread sealant used on them. The same procedure of tapping the NPTF is followed except that in the final assembly, no sealant is used at all.

TABLE 8-6. TAP DRILLS FOR PIPE TAPS

Size of Tap	Tap Drill
$\frac{1}{8}$	$\frac{11}{32}$
$\frac{1}{4}$	$\frac{7}{16}$
$\frac{3}{8}$	$\frac{19}{32}$
$\frac{1}{2}$	$\frac{23}{32}$
$\frac{5}{8}$	
$\frac{3}{4}$	$\frac{15}{16}$
$\frac{7}{8}$	
1	$1\frac{5}{32}$
$1\frac{1}{4}$	$1\frac{1}{2}$
$1\frac{1}{2}$	$1\frac{23}{32}$
$1\frac{3}{4}$	
2	$2\frac{3}{16}$
$2\frac{1}{4}$	
$2\frac{1}{2}$	$2\frac{5}{8}$
$2\frac{3}{4}$	
3	$3\frac{1}{4}$
$3\frac{1}{4}$	
$3\frac{1}{2}$	$3\frac{3}{4}$
$3\frac{3}{4}$	
4	$4\frac{1}{4}$
$4\frac{1}{2}$	$4\frac{3}{4}$
5	$5\frac{5}{16}$
$5\frac{1}{2}$	
6	$6\frac{3}{8}$

REMOVING BROKEN TAPS

There are times when through either misuse or the tapping of very hard material, a tap will break, leaving part of the tap protruding out of the work. The first step to perform in all cases is to remove all chips from around the hole by using either an air hose or small metal tweezers. When using an air hose, always wear safety goggles. Next, using a pair of pliers or locking plier-wrench with a piece of soft metal sheet such as copper or aluminum to protect the jaw faces, grasp the protruding part of the tap and with a back and forth rotation, loosen it and eventually unscrew it from the hole. Apply lubricant to the hole when removing broken taps.

When a tap is broken off flush to or below a surface, there are five methods of removing the tap (listed in the preferred order): tap extractor, nitric acid solution, annealing (and drilling), tap buildup, and electrical discharge machining (EDM).

8-13. Tap Extractor

A tap extractor (Fig. 8-6) is used only on fluted broken taps. It consists of a metal bar with a series of slots running lengthwise. Two, three, or four (depending on the number of flutes on the tap) hardened metal rods or fingers are inserted into the slots and are locked in place by a collar. The fingers are then slid into the flutes of the broken tap. A metal sleeve is adjusted down to the tap and locked. Using a wrench with a slight back and forth action, the broken tap is removed. Continual cleaning of the hole with an air hose is recommended to eliminate smaller chips that are sometimes produced in the tap removal.

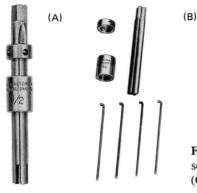

(A)   (B)

Fig. 8-6. Tap extractor: (A) completed assembly and (B) subassembly of extractor. (Courtesy of Walton Co.)

8-14. Nitric Acid Solution

When small taps break off in metals such as steel or stainless steel, a weak solution of nitric acid can sometimes be used to aid in removing the tap. The acid solution eats away some of the metal thread in the hole, thereby loosening the tap. Generally a solution of one part acid to four or five parts water is effective. One important note must be made here: when mixing any acid/water solutions, remember the three A's (A-A-A) of safety. *Always add acid to water*; never add water to acid. The mixing of water to some acids produces violent heating and spattering of the mix which could result in injury or blindness. After a tap has been removed from a workpiece, the remaining acid solution within the hole is diluted with a thorough water washing. Failure to dilute the acid could result in destruction of the thread by the acid.

8-15. Annealing (And Drilling)

After the above methods of tap removal have been tried, a third method that is more time consuming can sometimes be used. Tap annealing is a process of heating a broken tap to cherry red and allowing it to slowly cool to an annealed state. This procedure can be utilized, provided that the job material will stand the heat and that heat warpage will not ruin the workpiece. Small workpieces can be placed into a furnace and brought to an annealing temperature along with the tap. After annealing, a drill bushing the same size as or smaller than the tap drill of the broken tap is located and clamped over the tap. A drill bushing is a hardened sleeve that guides and locates a drill. The process of drilling out the tap is begun. As soon as a hardened part of a tap is encountered, the process of reannealing starts over again.

8-16. Tap Build Up

In the case of large broken taps, a method of adding metal by welding can sometimes be used to remove the tap. It is very important that no welds occur between the tap and the thread hole, as this locks the tap in tighter within the hole. The weld material is built up on the tap until it is possible to weld a hex nut on the end. With the nut secured, the tapped hole area is then cooled, cleaned and lubricated before any attempt is made to remove the tap. When cool, a wrench is used to back the tap out of the hole.

8-17. Electrical Discharge Machining

E.D.M., which is a final but positive method for tap removal, is effective but costly. The process involves the use of large electrical power directed to the tap from an electrode associated with the machine. This power discharge creates a guided arc that breaks down the tap with little or no effect on the threaded hole. This process is normally performed under a liquid flow to both carry material away and also to cool the workpiece.

MACHINE THREAD DIES

Having discussed taps in the beginning of this chapter, the taps counterpart, the die is now discussed. Dies are hardened metal tools used to cut thread forms on round workpieces such as rods and pipe. There are normally four cutting edges or flutes, which contain the thread form. Between each flute lies an area for chip storage and lubrication during the threading operation.

The starting end of the die is the most important part of the die. This end, which first comes into contact on threading, is recognized by the fact that several of the threads at one end have been ground down to form a tapered end. The taper allows for easier starting and squaring of the die to the workpiece and, generally, for easier threading.

As the die is started on a workpiece, the "mouth" or start end of the die starts to cut a slight (not full) thread, which, as the die is turned, comes to the full cutting section. This progressive cutting action allows the die to pull itself onto the rod without undue stress on the die or on the workpiece.

8-18. Forms of Machine Thread Dies

There are five basic types of machine/fraction dies that are in common shop use (listed in the order of most popular use): round split adjustable die, insert die, round solid die (truing die), hexagonal rethreading die, and square die.

These dies offer the full range of thread sizes of no. 0 machine to $1\frac{1}{2}$ inch in both national coarse (NC) and national fine (NF) threads. The size of a die is normally stamped into the "start" face. A typical notation is as follows:

$\frac{1}{4}$—20 N.C. This notation means that the nominal size of the thread is $\frac{1}{4}$ inch. There is no reference to a class thread, since most dies are either adjustable or made fixed for cutting a loose fit of bolt thread.

Round Split Adjustable Die

The round split adjustable die (Fig. 8-7) has a small screw that is used to adjust the gap of the die. This adjustable gap allows for two important threading features. First, by opening the die, thread cutting is made easier. A rough thread can be cut; then by closing the die, a smoother thread is produced by a light shaving cut. Second, the adjustable gap allows a male thread to be matched to a female thread for a precision class of fit.

Round die bodies are available in a range of diameters. The diameter generally runs from $\frac{5}{8}$ for small thread ranges to 3 inches O.D. for the larger

Fig. 8-7. Round split adjustable die. (Courtesy of Greenfield Tap and Die)

thread sizes. When ordering a round die, such as a $\frac{1}{4}$—20 thread die, its diameter must also be specified.

There is a conical indentation located on the periphery of the adjustable die. This indentation is used to secure, adjust, and hold the die in its holder, which is discussed under stocks in Sec. 8-19.

Split dies are of the same configuration as the adjustable die except for four differences: there is no adjusting screw in the die, there are two conical points on the die periphery instead of one and they are each 90° apart from the split (180° apart from each other), they require a different holder (stock) from the adjustable type, and they can be opened or closed without removing them from their holder. Other than this last item of adjustment, which is somewhat of a time-saving feature, the quality of the thread produced by the split die is the same as produced by the round die.

Insert Dies

The insert die (Fig. 8-8) generally ranks second in popularity because of its convenience and lack of adjustment during threading. In the insert die arrangement, a set of two matched cutter dies are used in a holding cap. Each of the two dies has two thread cutting flutes with a chip clearance between them. When assembled and locked into the collet head with the guide, the dies form a combination of four cutting flutes, the same as the round split adjustable die. Care should be taken in assembly so as not to install the dies improperly, causing die breakage or thread distortion.

(A)   (B)   (C)   (D)

Fig. 8-8. Insert die and basic parts: (A) complete die, (B) collet, (C) dies, and (D) loading cap. (Courtesy of Greenfield Tap and Die)

Round Solid Die (Truing Die)

In the area of thread skimming or truing, three dies are used: the round solid die, the hexagonal rethreading die, and the square nut die. Their main purpose is to retrue existing thread only, and not to cut a full thread. The first die is the solid round die (not shown). This die, with its four cutting flutes, is not split and therefore is fixed in its cutting size. It normally is manufactured for the "bolt" class of fit and is intended in the truing of bolts. This die has one conical locking and driving hole on its periphery and is used in either of two types of holders discussed in Sec. 8-19. The round solid die has the manufacturer's name, size, threads per inch and series, NC or NF on its starting face. The size range of this die is generally no. 0 through 1 inch.

Hexagonal Rethreading And Square Dies

Situations arise, for example in the case of a damaged stud or threaded rod, when field correcting of the threads is necessary. This problem is solved by using a hexagonal rethreading or square nut die (Fig. 8-9). These dies are made in the form of either a hexagon or a square nut so that they can be driven with a wrench. This method of retruing a bruised or rusty thread is also useful in very close areas where conventional die stocks can not be used due to their size. Thus, the hexagonal rethreading and square nut dies are only truing dies. They may not be used for cutting new threads.

(A)   (B)

Fig. 8-9. Rethreading dies: (A) hexagon and (B) square. (Courtesy of Greenfield Tap and Die)

8-19. Machine Thread Die Drives (Stocks)

Two types of tools, called single screw and three screw round die stocks, are used to hold and drive round dies. Hand power is transmitted through two handles of a stock, which are attached to a holder (Fig. 8-10). The single screw stock is designed to accommodate all round dies of the adjustable split type and the solid truing die. Its single setscrew enters the indentation on the die's periphery. The three screw type stock is designed to hold round adjustable dies of any design. Both stocks are made to accept the wide range of die diameters that range from $\frac{5}{8}$ to 3 inches in diameter.

(A)   (B)

Fig. 8-10. Die stocks: (A) single point and (B) 3 point. (Courtesy of Greenfield Tap and Die)

8-20. Procedures for Threading with Machine Thread Dies

Cutting threads on rods to produce bolts or special threaded products is a task that the technician must perform quite often. This operation is explained in detail, but first some ground work facts are presented.

Dies produce threads of high quality in both finish and class of fit. All threading, especially a coarse series thread, is done in several steps by opening or adjusting the thread die to take light cuts instead of one full cut. Failure to take several cuts usually results in very poor finish of the threads. This poor finish is usually in the form of burrs or tears sometimes called "fish scaling" (due to its appearance). Lubrication is required during all die threading operations—absence of a lubricant also results in "fish scaling" and weak threads as well as having a dulling effect on the dies.

Material preparation, before threading, falls into two areas: cleaning of the rod material of all paint or other surface coatings, especially rust which is very abrasive to dies; and filing, grinding, or machine cutting of a bevel of from 45 to 60° included angle on the end of the rod. This bevel allows the die to start easily. Whenever possible, the rod should be held securely in a vertical position in a vise.

The first step in using the die is to obtain the proper size split or adjustable split die and a stock: a three screw stock for the split die and the single screw stock for the adjustable split die. If using the split die, insert the die in the three screw stock with the die face out. Locate it so that the middle screw is at the die split and the other two screws engage the indentations on the die. Turn and adjust the middle screw inward to open or spring the die apart. After adjusting, lock the other two screws into the die. If using the adjustable split die, open the die by adjusting its built in screw. Install the die face out in the single screw stock and tighten the screw.

After lubricating both the die and the rod, engage the two, chamfer of rod to chamfer of die. Exert a medium to heavy pressure on the stock and die and apply a turning force until the die begins to cut two to three threads on the rod. Using either a square butt or a scale end, check at two places on the rod (90° apart) to be sure that the die has started squarely on the rod. The same squaring procedure, as described in tapping, is extended to the stock and die. During threading, the die should be backed off to clean both it and the thread being cut.

After the first or second roughing cut has been made, this final fitting procedure is recommended. Use a nut or the tapped workpiece as a gauge. Adjust the die smaller and begin a cut of several threads. Try the nut or workpiece for a start. Adjust the die "down" until the nut or workpiece starts with a smooth turn. The rod thread can now be cut to its full length and size.

PIPE THREAD DIES

8-21. Forms of Pipe Thread Dies

Several hand pipe thread dies are available for cutting male pipe threads. The first die discussed is the round solid die, which is basically of the same exterior configuration as the solid machine screw die and can be driven by the stocks previously described. Hand pipe dies are generally available in sizes from $\frac{1}{16}$ to $\frac{1}{2}$ inch NPT. Sizes above $\frac{1}{2}$ inch in the round dies are somewhat impractical due to the difficulty in starting them on the pipe. Whenever pipe dies are used, it must be remembered to never allow the workpiece being threaded to extend past the die end by more than one thread as the sealability of the thread could be affected. This is because as the die is run past the workpiece end, the small part of the die starts to cut straight thread rather than tapered thread.

8-22. Pipe Ratchet Dies and Stocks

In most common use today are the ratchet type inserted pipe dies (Fig. 8-11). The cutting dies are held in a collet head by a cover plate very similar to the insert dies previously mentioned (Fig. 8-8). In this case, the start end of the dies face down into a hole through the collet assembly. The purpose of the hole is to guide the pipe to be threaded into the dies for easier starting. Near the middle of the die assembly and on the outside of the collet is an area with ratchet teeth set around the periphery. The ratchet teeth are used for a ratcheted stock to drive the die.

Fig. 8-11. Pipe ratchet set. (Courtesy of RIDGID)

A ratchet stock wrench is used to drive a pipe ratchet die. The wrench is designed to fit a set of dies and ratchets to the left or right as selected by a lever. Since the hole through each die head is for a specific pipe, only the dies for that particular pipe size can be used. The normal range for a set of pipe ratchet dies is $\frac{1}{16}, \frac{1}{8}, \frac{1}{4}, \frac{3}{8}, \frac{1}{2}, \frac{3}{4}$, and 1 inch.

Unlike machine or fractional thread dies, which note the diameter of the material to be used in cutting a thread, pipe die notation does not tell the diameter of the workpiece. Table 8-7 gives the O.D. of a tube or rod required to cut a specific pipe thread size. It should be remembered that a tube size is normally given as its actual O.D. A $\frac{1}{2}$ inch tube has an O.D. of $\frac{1}{2}$ inch, whereas in pipe notation the size refers very roughly to the I.D. of a standard pipe. Table 8-7 shows the various wall thicknesses for any given pipe size. ($\frac{1}{2}$ inch pipe has an O.D. of 0.840 inch).

TABLE 8-7. STANDARD DIMENSIONS OF WELDED AND SEAMLESS STEEL PIPE

Size (Nom. Inside Diam.) (Inches)	Outside Diam. (Inches)	"Standard Weight" (Standard Wall) Pipe Wall Thickness, (Inches)	"Extra Strong" Pipe Wall Thickness, (Inches)	"Double Extra Strong" Pipe Wall Thickness, (Inches)
$\frac{1}{8}$	0.405	0.068	0.095	
$\frac{1}{4}$	0.540	0.088	0.119	
$\frac{3}{8}$	0.675	0.091	0.126	
$\frac{1}{2}$	0.840	0.109	0.147	0.294
$\frac{3}{4}$	1.050	0.113	0.154	0.308
1	1.315	0.133	0.179	0.358
$1\frac{1}{4}$	1.660	0.140	0.191	0.382
$1\frac{1}{2}$	1.900	0.145	0.200	0.400
2	2.375	0.154	0.218	0.436
$2\frac{1}{2}$	2.875	0.203	0.276	0.552
3	3.500	0.216	0.300	0.600
$3\frac{1}{2}$	4.000	0.226	0.318	
4	4.500	0.237	0.337	0.674
5	5.563	0.258	0.375	0.750
6	6.625	0.280	0.432	0.864
8	8.625	0.322	0.500	0.875
10	10.750	0.365	0.500	
12	12.750	0.375	0.500	
14	14.000	0.375	0.500	
16	16.000	0.375	0.500	
18	18.000	0.375	0.500	
20	20.000	0.375	0.500	
24	24.000	0.375	0.500	

Since applications of some pipe require stronger walls to contain high pressures, the walls are increased in thickness. Since the O.D. of a particular pipe is fixed, the extra wall thickness results in a smaller bore in the pipe. Most pipe work used for low pressures uses standard (STD) pipe.

8-23. Procedures for Cutting Pipe Thread

In using a round pipe die and stock, the procedure is the same as in cutting machine/fractional thread. Here again, lubrication is required during the cutting operation.

In using the ratchet stock and pipe die, a pipe vise or other holding device is used to firmly hold the pipe. The die collet is inserted onto the ratchet stock by holding the reversing knob in a neutral position. When the head is inserted, flip the knob into either a left or right position. This secures the die head to the ratchet wrench. Slide the pipe into the die collet hole until it bottoms. Hold the die head and check to see that the ratchet is driving in a cutting or right hand direction. Lubricate the die and pipe with a sulphur base oil. Put pressure on the die in a direction toward the pipe with one hand. The ratchet is then turned until cutting pressure is felt. The die has now started and both hands can be used on the ratchet stock handle. After cutting about half of the thread length, reverse the ratchet and back the die off the thread. After cleaning and inspecting the threads, rescrew the die onto the thread and lubricate. Gauge the proper length of the thread being cut by trying its mating part or by counting its threads as previously described. Finish cutting the thread and then clean the chips and cutting oil from the workpiece and the die.

THREADING LUBRICANTS/COOLANTS

A lubricant is anything that reduces friction and increases workability. In general, during all tapping or machining operations, whether cutting or forming, a lubricant is used (except in the case of several special materials). Lubrication acts in two distinct ways: it removes a great part of the friction in thread forming with the end result being a longer tool life and a smoother stronger thread, and it acts as a coolant by carrying away heat built up by friction. The range of lubricants runs from common water and oil to new chemically blended fluids.

Lubricants in general use are broken into four groups: sulphur base oils, water soluble oils, mineral oils, and chemical cutting fluids.

8-24. Sulphurated Oil

The most commonly used oil is sulphur based. It is characterized by its somewhat strong noxious odor. It has excellent coolant and cutting qualities that allow its use in most hard metal applications. When applied to some metals, such as brass or copper, however, there may be a darkening or discoloration of the metal due to the reaction between the metal and the sulphur, but this does not affect its purpose. Sulphurated oil is most often used in cutting operations on steel, aluminum, and some stainless steels. It can be obtained in pints, quarts, and 1, 5, 30, and 55 gallon quantities. In cost, this oil is by far the least expensive of the commercial cutting fluids. It is not recommended for use on plastics, phenolics, or asbestos board.

8-25. Water Soluble Oil

As its name implies, this oil as supplied by the manufacturer, has to be thinned with water prior to its use. Originally it was designed to be used in machines where fast flow rates and runoff of coolant could be achieved. When thinned, its consistency is very much like water, and in high heat producing operations, it can cool faster than if a slow flowing oil is used. Water soluble oil is chiefly used in machine operations such as drilling, turning, grinding, and milling. It is of little or no use in hand forming of threads, but is mentioned here for future use in these areas.

8-26. Mineral Oils

Oils that are obtained from the distillation of petroleum (namely paraffin, kerosene and mineral seal oil) are known as mineral oils. They are generally blended with the "base" oils to obtain a consistency to suit a particular job. Use of straight mineral oils for tapping free cutting steel or nonferrous metals (aluminum, brass, copper) can be used, but check for ease of tapping as well as the finish of the thread. Kerosene can be used uncut on copper and aluminum.

8-27. Chemical Cutting Fluids

Within recent years, a variety of chemically compounded cutting fluids have been introduced. The purpose of the fluid is to impart longer tool life and to ease cutting operations, especially in some of the harder steels of the common and stainless variety. The fluid's main function is to chemically break down the metal-to-metal bond at the cutting edge of the tool, thereby allowing for a cleaner and far more effortless operation. The cutting fluid prevents glazing or hardspots from developing in the workpiece. Glazing is caused by the tool edge rubbing over a work area instead of cutting. This cutting fluid is not used on roll form taps.

CAUTION

Some chemical tapping fluids cannot be used on certain metals due to adverse chemical reactions. Always refer to manufacturer's suggestions.

Table 8-8 lists the recommended lubricants/coolants for different types of materials to be tapped.

TABLE 8-8. LUBRICANTS/COOLANTS

Material to be Cut	Drilling	Tapping
Aluminum	Soluble oil (75–90% water) (or) 10% lard oil with 90% mineral oil	Lard oil (or) Sperm oil (or) Wool Grease (or) 25% sulphur-base oil mixed with mineral oil
Alloy steels	Soluble oil	30% lard oil with 70% mineral oil
Brass	Soluble oil (75–90% water) (or) 30% lard oil with 70% mineral oil	10–20% lard oil with mineral oil
Tool steels and low-carbon steels	Soluble oil	25–40% lard oil with mineral oil (or) 25% sulphur-base oil with 75% mineral oil
Copper	Soluble oil	Soluble oil
Monel metal	Soluble oil	25–40% lard oil mixed with mineral oil (or) sulphur-base oil mixed with mineral oil
Cast iron	Dry	Dry (or) 25% lard oil with 75% mineral oil
Malleable iron	Soluble oil	Soluble oil
Bronze	Soluble oil	20% lard oil with 80% mineral oil

SUMMARY

The tap is a tool that is used to produce a continuous helical groove (thread) in a hole. A die is a tool that is used to produce a continuous helical groove on a rod. The threads produced by these tools have been discussed as well as the procedures to be used in performing the cutting.

The practical knowledge and applications discussed in this chapter should be combined with those discussed in Chap. 7 on drilling. The technician should practice all of the procedures from both chapters until he can successfully use drills, taps, and dies.

REVIEW QUESTIONS

1. It was recorded that a rotation of 120° advanced an adjusting screw 0.075 inches. Calculate the pitch and number of threads per inch of the screw.

2. Suppose you have been asked to tap several holes already drilled in a workpiece. The size tap is to be 6—40. When you measure the holes, they are found to be 0.125 inch in diameter. What type tap would you use?

3. In obtaining a set of taps for general tapping from a distributor, it is noted on the shank that a "GL$_1$" is listed. Would you use this tap, and why or why not?

4. Why is "backing off" performed while tapping a hole?

5. What is the main purpose of pipe threads? How is this result obtained?

6. Approximately how many turns should be allowed for sealing a pipe thread assembly?

7. In threading a rod with a $\frac{5}{8}$ inch—11 thread, an adjustable die is to be used. What would be the best approach to cut the thread with ease and obtain a good finish?

8. What is the most diverse and economical type of cutting lubricant available?

9. What are the chemical cutting fluids main actions in cutting operations?

10. Since a forming tap does not cut, should a cutting lubricant be used?

POWER TOOLS

A power tool is one of a group of tools that increases the work range and speed of a man in such areas as drilling, sawing, grinding, and sanding. Using either electrical or compressed air as a means of power, work can be performed with very little energy being supplied by the technician.

Like any tool, the proper and safe use of a power tool comes from an understanding of its basic functions, capabilities, and limitations, from proper care, and from knowing what special accessories are available and how to use them to do a particular job.

In this discussion of power tools, the electrical power tool is the principal type discussed, as this is the most popular type in use. The air powered type, however, has two advantages over its counterpart—it runs cooler (electrical powered tools heat up under load and time); and it is safer since shock and sparking are eliminated (in an explosive atmosphere, air tools are the safest).

Air tools have a vaned air turbine as the power unit. Air pressure supplied from a compressor by a hose is run through an on/off regulating valve. The air, upon entering the turbine area, spins the turbine at speeds up to 30,000 rpm. The speed of the turbine rotor is then reduced through gears to its operating speed range. Reduction of a high speed rotation, as in the case of the rotor, to a lower final speed output always results in a high turning torque output.

The speed output of air powered tools is controlled through the on/off valve. This valve regulates the flow of air into the turbine area and, by doing so, regulates the output speed.

Proper care of air powered tools generally falls into two areas: oiling vital parts and cleaning the on/off valve. At the high speeds that the turbine is rotating, special lubrication is needed for both the turbine bearings and the gear reducer. The manufacturer's recommendations should be adhered to strictly. If the tool encounters heavy use over long periods of time, the frequency of oiling should be increased.

In all air systems, from compressor to hand tool, water and rust form and unless thoroughly filtered, enter the hand tool. Rust generally collects at the on/off switch, affecting its speed regulation. Frequent cleaning of the valving system should be performed, using manufacturer's recommended procedures.

Having briefly described the basics of air powered tools, the remainder of this chapter describes electrical powered tools. The chapter is divided into three sections: hand held power tools, bench and floor mounted power tools, and grinding wheels and polishing compounds.

HAND HELD POWER TOOLS

Hand held power tools aid the technician in completing a job more efficiently, with less fatigue, and often more accurately. Those tools most frequently used by the technician are drills, jigsaws, hand grinders, and orbital sanders.

9-1. Hand Drills

Probably the most popular rotary power tool in use today is the hand drill. This tool with its many attachments performs many other jobs besides drilling, such as buffing, grinding, driving screws (in or out), and others. These tools are available in two basic styles (Fig. 9-1), the standard pistol design and the 90° offset design.

The standard pistol grip style is the most popular, since its design lends itself to the majority of job situations. On pistol grip drills of larger sizes, extra hand holding power is provided to keep the drill from torquing or twisting during operation. A bar located at 90° to the chuck end is available and is installed or removed as the situation requires. The need for more forward push by the operator during heavy drilling has necessitated the installation of a permanent handle grip located at the back of the larger-sized drills (see Fig. 9-1). This back grip gives the operator more control in both holding and drilling, especially in hard materials.

The 90° angle drill was developed to meet one basic need. This need was

Fig. 9-1. Standard pistol (top) and 90° (bottom) offset drills. (Courtesy of SKIL Corp.)

the ability to drill holes in areas where there is not enough room for the standard drill because of its length. In fieldwork, where sheet metal work is being done, this drill allows holes to be drilled from within sheet metal ducts and also from within recesses such as wood floor beams.

Basic Parts of the Drill

The basic parts of an electric drill can be broken into four units: an electrical on/off control switch, an air cooled motor, a gear reducer, and a chuck to hold tools.

The electrical on/off control switch is a means of both stopping and starting the drill as well as controlling the speed of the drill motor. This switch is called the trigger and is so designed that it operates on being compressed only. When pressure is released, the electrical power that is supplied to the motor is switched off.

The motor of an electric drill is generally of a high speed, air cooled variety. The motor shaft is run with bronze bearings at each end for both smoothness and long life. The sole purpose of the motor is to generate a high-speed rotational force. This rotational force is shaft coupled into the third unit of the drill, the gear reducer.

The gear reducer takes the high speed revolutions supplied by the motor and reduces them. In the reduction, performed by the gears, the final exiting speed is lower, but the torque or turning power is increased. This is the same principle as described for the air power motor.

With the output speed now at a high torque, the final unit of the drill, the chuck, is reached. The chuck (Fig. 9-2) is a three jaw device used to grip an infinite range of round shafts with diameters up to a particular maximum size or capacity.

Fig. 9-2. Drill chuck and key. (Courtesy of Jacobs Manufacturing Co.)

The three jaw gripping design has been chosen for its ability to grasp round pieces and hold them always within the centerline of the chuck. A tee handled wrench called the chuck key is inserted into the side of the chuck and is used to both tighten the jaws on the tool or to unclamp and remove the tool.

Hand Drill Sizes and Speeds

Hand drills are normally graded into three sizes: $\frac{1}{4}$, $\frac{3}{8}$, and $\frac{1}{2}$ inch. The drill size is graded by the maximum diameter drill or tool shank that will fit into the drill's chuck. The chuck size or drill size is selected by the manufacturer to correspond to the design strength of the motor and gears of the drill. Overloading the designed capacity of the drill chuck is not recommended as this undue strain shortens the working life of the drill.

Most drills are available in any of several output speeds—fixed, two speed, variable speed—and some drills are able to reverse their direction of rotation.

Originally, the fixed speed drill was the only type available. Its speed, which may have been perfect on materials such as wood or plastics, sometimes proved to be too fast when drilling into metals. This high speed drilling of hard metals tends to dull tools at a high rate. Drills of single speed generally have speed ranges suited to the drill size. The $\frac{1}{4}$ inch drill has a speed in the range of from 2000 to 2250 rpm. The $\frac{3}{8}$ inch fixed speed drill, whose use is intended for larger drilling applications, has a speed of about 1000 rpm. The $\frac{1}{2}$ inch fixed speed drill, used in drilling both steels and masonry, has an output speed in the 500 rpm range.

The two speed drill is mostly found in the $\frac{1}{4}$ inch drill. The two speeds are controlled by a switch located on the handle. This switch electrically controls the speed in either mode selected. This type drill, with its two speeds, is able to drill both soft and hard materials without undue strain on the drill bit. The speeds in the $\frac{1}{4}$ inch drill range from about 1600 rpm for the low speed to about 2250 rpm for the high speed.

In recent years, the variable speed drill was introduced. This finally has answered the need for an infinite speed range to meet job requirements. The speed ranges listed for each of the fixed output drills generally remain as the top speed limits for the comparable sized variable speed drills. The on/off trigger, however, allows for a speed of 0 through 2250 rpm for the $\frac{1}{4}$ inch drill, 0 through 1000 rpm for the $\frac{3}{8}$ inch drill, and 0 through 550 rpm for the $\frac{1}{2}$ inch drill. To increase the speed of the drill, the trigger is squeezed. Releasing the squeezing pressure reduces the speed. Some variable speed drills are equipped with a button to lock the particular speed selected. This is provided so that speed variations due to finger movement during drilling are avoided.

Reversing drills are available in all three drill sizes. The chuck rotation in this type drill is reversed by an electrical switch generally located near the trigger. There are basically two reasons for drill reversing. First, in drilling metal or hard materials, the drill, on breaking through the workpiece, sometimes jams in the work. By reversing the drill rotation, the drill will unscrew itself from the workpiece. The second reason is that drills with appropriate driving bits can either screw or unscrew fasteners rapidly.

Most drills have what is known as a lock button. This button locks the drill on/off trigger so that finger holding of the drill switch is not necessary. The button is normally located in the trigger area. To lock the drill at a constant speed, push the locking button in at the required speed. The trigger can now be released. The drill will continue to run until a simple squeeze and release on the trigger button releases the lock and stops the drill.

When hand drilling with any drill, it is of extreme importance to maintain a good hold on the drill as well as a good footing. Injuries very often occur when a drill jams in a workpiece and the hand drill twists out of the technician's hands. When using the $\frac{1}{2}$ inch drill in hard materials, two men should control the drill at all times.

When drilling a workpiece, never attempt to drill directly to a punch mark. This will almost always result in the drill tip walking over the workpiece, causing both scratches to the workpiece and possible injury to the technician.

Predrill all center punch marks with a combined drill and countersink first (Sec. 7-3).

Hard Metal Drilling

The solution to the problem of drilling hard metal is the use of pressure and proper speed. Very often in drilling metals with fixed speed drills, the speed is too high. The drill may be jogged to remedy this situation. By momentarily starting and stopping the drill, a low drifting speed is produced that when coupled with good tool pressure, cuts into some rather hard metal workpieces.

Another step that makes drilling easier is to "predrill" large holes. In directly drilling a $\frac{1}{2}$ inch diameter hole in a metal plate, a great deal of force and time are expended by the technician. A simple and effective solution to this problem is to predrill a hole $\frac{1}{8}$ inch or so in diameter through the plate first. This cuts down the force required on the $\frac{1}{2}$ inch diameter drill and allows for a faster cutting situation. Predrilling a small hole also allows for a more accurately located finished hole.

Accessories

The drill manufacturer has introduced a line of accessories which further increase the versatility of the drill. As mentioned at the beginning of the section, these accessories allow for buffing, grinding, and driving screws. The attachments generally have shanks which are held in the drills chuck.

A polishing and sanding kit generally is a three piece set: a rubber backing disc with a shaft to fit the drill chuck, a circular disc of either emery or sandpaper, and a buffing bonnet. The rubber backing disc (Fig. 9-3) is available in several diameters. The smaller diameters have straight shanks that can be chuck held, while the larger diameter is screwed to the drill shaft after removal of the chuck.

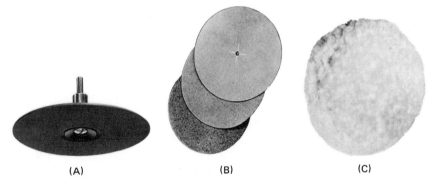

(A)   (B)   (C)

Fig. 9-3. Polishing and sanding kit: (A) rubber backing disc, (B) sanding disc, and (C) buffing bonnet. (Courtesy of Black and Decker Co.)

Attaching the abrasive sheets to the backing pad can be accomplished by two methods, either rubber gluing to the pad or by screwing the disc to the pad via a recessed screw. With this flexible backing pad, sanding can be performed on curved as well as flat surfaces.

When the need to polish or buff a job is required, the wool buffing bonnet is slipped over the face and edge of the backing pad and secured with a draw string sewn within the bonnet. If a sanding disc is already on the backing disc, it need not be removed.

Very often in the field a need to grind a hole or other surface arises. A wheel arbor and grinding wheel are used to meet this need. The arbor is designed to accept wheels with bores in mainly the $\frac{1}{2}$ inch type. The nut on the arbor is used to both lock and drive the wheel by friction only. The shank or chuck end is generally available in $\frac{1}{4}$ inch or $\frac{3}{8}$ inch diameters. This arbor also allows for the use of wire wheel brushes to be used in the removal of rust or paint from a workpiece.

WARNING

It is most important when either grinding or wire brushing that safety glasses be worn. If possible, wood or metal screens should surround the area of work to protect passing people from receiving eye or other injuries from flying particles and dust.

Variable reversing speed drills lend themselves very well to both screwing in or out all types of screws. To perform this operation, screwdriver bits in both conventional and Phillips head type are available. The Phillips head, by its design, self centers and drives that particular style screw. The conventional screwdriver, however, has a cylindrical sleeve that both aligns and keeps the slotted screw centered on the driver. These screwdriver bits work very well on large work runs where many screws must be run in or out.

Care of Drills

The modern drill motor is a rugged, dependable tool that has a few important points of care.

The gear section of the drill is greased and sealed at the factory and seldom needs to be repacked. The sleeve bearings, however, should be oiled with light machine oil as per the manufacturer's recommendations. Small holes, sometimes countersunk, are located at appropriate spots on the drill casing for this purpose. These holes are filled with felt to allow oil to pass through while straining out dirt and chips.

The frequency of oiling is dependent on the drill's use. Here again, follow the manufacturer's recommendation.

Drill motors are air cooled units. This means that air, with all its dust and other particle matter, is circulated through the housing to keep the motor cool. Over a period of time, dust starts to collect within the drill. Dust buildup causes overheating and also affects the electrical brushes on the motor. Dirty brushes can cause the motor to run at low and sometimes pulsing speeds.

WARNING

When cleaning the drill, remove the drill's electrical plug from the ac power outlet.

One method of cleaning is as follows. Obtain an air hose that delivers about 30–50 pounds of air pressure. Disconnect the drill's plug. With an on/ off valve on the air hose, allow air blasts to enter the drill casing to loosen and dislodge built-up dirt. Do this for about one to two minutes. Then run the drill for approximately $\frac{1}{2}$ minute, stop and unplug again. Repeat the air cleaning once again. Oil all oil points on the casing after the cleaning, as air cleaning removes some of the lubrication. Plug the drill in and run for approximately 30 seconds.

After long and hard use, usually after a year or more, the two electrical brushes of the drill motor must be replaced. The first sign of brush trouble is often a bright blue ring of sparks around the end of the motor. When this occurs, the drill should be stopped and new brushes installed as per the manufacturer's instructions. Also at this time, the drill casing should be opened so as to expose the armature of the motor. The armature is the section of the motor that the brushes contact. Inspect the armature for scratches or badly worn spots. If badly grooved, the armature should be machined to clean it up. Most often, only a cleaning of the armature with crocus cloth and kerosene is required. This is followed by a slight roughing with no. 400 silicon carbide paper.

After cleaning and drying the armature, the drill can be reassembled with its new brushes. Always allow the drill to "run in" for 10–15 minutes before actual use.

Removal of a drill chuck is the operation that is required to replace a badly damaged chuck or to install some accessories onto the drill. Chucks are generally screwed onto the drill shaft using right hand threads. To remove the chuck, it is necessary to apply a sharp blow to start the chuck off of the shaft thread. To perform this operation without damage to the drill or chuck, the following procedure is recommended. If possible, secure the drill in a bench vise, using wooden blocks to protect the casing. Using a strap wrench (Chap. 4) assembled to the chuck, strike a sharp blow to the wrench handle in the direction for unscrewing the chuck. After loosening, unscrew the chuck and replace with a new chuck or attachment. Never use a pipe wrench on a chuck, as damage to both chuck and wrench may occur.

9-2. Jigsaw

The jigsaw is a reciprocating saw bladed tool that lends itself well in the cutting of both thin gauge materials as well as other shapes such as bars, angles, or pipe. This tool performs well in areas where either square, circular or irregular cutouts are required.

The jigsaw is available in three speed groups: fixed, two speed, and variable. The fixed output saw has a stroke per minute (spm) rate in the range of 3000. The two speed saw has a spm rate of about 3000 for the high end and 2500 for the low end. Speed selection in this tool is controlled by a two position switch located on the handle. The variable speed saw has a speed range of from 0 to approximately 3000 spm. The speed is controlled by a trigger type switch.

Two speed and variable speed jigsaws have been developed, like the drill, to further increase the jigsaw's range of applications. The higher speeds are used in soft material applications such as wood or plastics, while the lower speed range is used in metal cutting applications.

The main items on the jigsaw are a handle with trigger switch, a blade chuck, a saw blade and a shoe plate (Fig. 9-4).

Fig. 9-4. Variable speed jigsaw. (Courtesy of Black and Decker Co.)

The cutting action of this type saw is in an upward direction toward the bottom of the shoe plate. This upward cutting action pulls the shoe plate tightly to the workpiece during cutting and relaxes its pull when the blade is moving away from the shoe plate. This up and down action of the blade imparts a high vibration factor to the saw during its operation. This vibration must be dampened by the technician by applying a firm and steady downward pressure during sawing. Failure to do so results in frequent blade breakage. The shoe plate on most jigsaws is adjustable with respect to the blade through a degree range of 0–45°. The 0° shoe setting is 90° to the blade for square edge cutting. The adjustment of the shoe, with respect to the blade, allows for bevel cutting when required. A single screw located on the angle protractor is used to adjust and lock the required angle setting.

Jigsaw Blades

Before continuing the discussion of the jigsaw and its applications, a discussion of blade varieties and uses is necessary. Jigsaw blades are available in a wide range of styles to meet specific applications. This discussion is limited to those most in keeping with this text.

Jigsaw blades are determined by basically two requirements–what material is to be cut, and what type finish should be left by the blade. The finish or type of cut is classified rough, medium, and fine.

In looking at a jigsaw blade, two designs are available: the straight shank and the canted shank (Fig. 9-5). The straight shank allows the teeth of the saw to run perpendicular to the shoe—this is the most popular type shank style. The canted shank allows the teeth of the saw to project forward at a slight angle to the shoe. The canted blade is used when sawing soft to hard woods. Its design removes saw dust quickly and keeps the blade from overheating.

(A)   (B)

Fig. 9-5. Jigsaw blade styles: (A) canted and (B) straight. (Courtesy of Black and Decker Co.)

The shank design of most blades is such that they are interchangeable between the different types of saws. Care should be taken to check this point before buying blades.

Blade lengths are found in a range of from 3 to $4\frac{1}{4}$ inches overall. The longer blade lengths usually are of the wood cutting type.

Two forms of blade teeth are available for most general work, the stagger tooth blade and the wavy tooth blade [Fig. 9-5(B)]. The stagger tooth blade is recognized by the fact that the teeth are set in an alternating pattern of right,

TABLE 9-1. JIGSAW BLADE SELECTION CHART

Material To Be Cut	Type of Cut	Speed of Cut	Teeth Per Inch	Blade Length
Soft and hard wood	Rough	Fastest	7	3″
Soft and hard wood, plywood	Medium	Medium	10	3″
Soft and hard wood, plywood	Smooth	Medium	7	3″
Plywood and veneers	Fine	Medium	10	3″
Plywood	Medium Fine	Medium Fast	7	3″
Plywood	Medium	Fast	5	3″
Non-ferrous metals	Fine	Slow	14	3″
Ferrous (iron) metals	Fine	Slow	32	3″
Leather, rubber, compositions	Smooth	Fast	Knife Edge	3″
Plywood, plastics	Medium	Fast	5	3″

left, right, left, etc. This type of blade is most often used on soft materials such as wood or plastics, and in some cases aluminum.

The wavy tooth blade generally is reserved for the finer tooth blades used in cutting ferrous metals up to $\frac{1}{4}$ inch thick, and nonferrous metals up to $\frac{1}{8}$ inch thick.

Jigsaw blades vary in number of teeth from 3 teeth per inch, used in cutting soft materials, to 32 teeth per inch, used in ferrous metal cuttings. Table 9-1 is included as an aid in selecting the correct blade for the material to be cut.

The blade chuck, used to hold the blade, is a slotted collar located at the end of the output shaft of the saw. Generally, there are two screws located on this collar, one in the front and the second on the side, 90° from the front. The sole purpose of these screws is to firmly align and lock the blade to the output shaft.

WARNING

Before inserting a blade into the collar of the jigsaw, remove the jigsaw electrical plug from the power outlet.

Blade Installation

After a proper blade for the material is selected, it should be installed in the jigsaw in the following manner. Disconnect the power plug from the outlet. Loosen both screws in the blade chuck until the shank of the blade can be inserted and bottomed. With the blade bottomed, turn the front screw in until light contact pressure with the blade is felt. Turn and tighten the side screw firmly against the blade. Return to the front screw and securely tighten it.

Metal Sawing

When there is a need to saw thin metal, the following procedure is recommended. In all cases a backup piece of either plywood or masonite should be used. This backup piece serves both to cut cleaner edges on the workpiece and to dampen vibration, which is the biggest source of blade breakage.

When sawing thicker metal pieces, such as $\frac{1}{8}$ or $\frac{1}{4}$ inch sections, ample support should be supplied. If using saw horses or some other support, always keep the area being cut as close as possible to that support to eliminate vibration.

Cutting fluids are not generally used during cutting as it would increase the need for more pressure to cut the metal and cause blade breakage. A requirement for increasing tool pressure during cutting is also the first sign of a dull blade. Jigsaw blades are expendable tools and cn dulling should be disposed of and replaced with new ones.

Jigsaw Accessories

The rip fence and circle guide has a twofold purpose: cutting a straight cut using a reference edge and cutting circles. The rip fence and circle guide inserts into the slot provided for it in the shoe of the jigsaw and can be locked at any setting. The cross bar, when set, guides the jigsaw in cutting strips parallel to a workpiece edge.

When cutting circles, the combination rip fence and circle guide is removed and reinstalled with the cross bar edge up. The required radius is then set between the pivot hole and the blade. This position is then locked. By placing a pin through the cross bar into the center of the required circle, the blade can be inserted into a predrilled hole or slot and the circle cut out, using the pin as a pivot point.

Care of Jig Saws

The same care in lubrication and cleaning as described for the drill can be extended for the jigsaw.

9-3. Hand Grinders

Hand grinders are tools that are designed to do several jobs in the field, including grinding flat surfaces, wire brushing rust or paint from a workpiece, and buffing or polishing a finish on a workpiece. Hand grinders are of two basic designs: the straight and the offset or angle type (Fig. 9-6). These two

(A)

(B)

Fig. 9-6. Hand grinders: (A) straight and (B) offset. (Courtesy of SKIL Corp.)

types answer the needs of the various field problems encountered. These tools are normally operated at a high fixed speed, generally from 5000 to 10,000 rpm. A variable speed control would create poor working conditions as well as unsafe vibrations at a low speed range.

Hand grinders are classified by the diameter of grinding wheel that they are designed to hold. Therefore, a 9 inch grinder will hold and operate a 9 inch diameter grinding wheel.

The straight or horizontal grinder (Fig. 9-6) is most often preferred when the job is a curved or circular surface, such as a hole or a shaft. Its normally held position is straight out from the technician with a firm grip on the forward or wheel end and the second hand resting on the power switch. This method of holding the grinder allows the technician to swing with a motion that will allow for curved surface grinding.

The two main features of the working or spindle end of the grinder are the wheel guard and the spindle friction clamping assembly.

WARNING

When grinding, always ensure that the wheel guard is in place.

The wheel guard is a metal shield that normally covers approximately one half of the grinding wheel or wire brush. It shields the technician from grinding particles or wire particles sprayed into the face or eyes and from the shattering of a grinding wheel. This second situation is rare, but is generally spectacular since the wheel often breaks into many jagged pieces that could cause serious injury. Always keep a grinder guard on and positioned so that your face and eyes are shielded by it. Always wear a full safety face shield when grinding or wire brushing.

The spindle clamping assembly most often consists of the spindle proper with a threaded end, a metal friction disc, and a locking nut. The spindle of the grinder is normally a flanged shaft with a threaded section at the end. The diameter of the shaft can run from $\frac{3}{8}$ to $\frac{3}{4}$ inch size in $\frac{1}{8}$ inch increments. The most popular grinder has a $\frac{1}{2}$ inch diameter shaft with a $\frac{1}{2}$—20 right hand thread on its end. This shaft accepts only wheels and brushes with a $\frac{1}{2}$ inch bore.

The function of the friction disc and nut are simple. They are used to apply pressure to clamp the wheel or brush against the flanged end of the spindle. This method of driving is preferred for a very good reason. If a grinding wheel or brush should jam in some way, the shaft assembly torque will exceed the friction clamp and continue to rotate until stopped by turning the power off. This situation is favored over a positive drive for pure safety reasons. During any grinding operation, the clamping nut and disc should be intermittently rechecked for tightness.

Grinding wheels are manufactured in a variety of shapes, grits, and hardnesses. This variety is necessary to meet the diverse job applications. A full discussion of grinding wheels is presented in Sec. 9-8.

When using straight grinders, as have just been discussed, the only shape to be used is the straight form grinding wheel. No other wheel shape should be used or substituted.

WARNING

One very important thing to remember in any type of grinding operation is never to start the grinder with the wheel in contact with the workpiece. Doing so could cause excessive jumping by the wheel that could cause injuries or a cracking of the wheel with subsequent breakage.

The angle or offset grinder (Fig. 9-6), is most often used for flat surface work. Where excess weld from plates must be removed or in some cases merely grinding off of high spots or edges is required, these job tasks are done with the offset grinder.

The construction of the offset grinder is basically the same as the straight grinder except for its mode of operation. The spindle is at 90° to the grinder housing, with a metal guard surrounding almost one-half of the grinding wheel. Securing grinding or brushing tools to the shaft is achieved in the same nature as the straight grinder previously described.

When procuring grinding wheels, discs, or wire wheels for this type grinder, they must be of the cup or recessed type. This recess is needed for the locking nut and friction disc to remain away from the grinding or brushing face of the wheel.

Accessories

To meet the wire brushing and polishing needs of the technician, several forms of accessories are available. The wheels or working attachments for the straight grinder are never used on the angle grinder.

The straight grinder, as was previously mentioned, uses a straight grinding wheel. This means that work is performed only on the outer rim of the wheel. Wire brushes for straight grinders are made to perform on their rim face also. These wire brushes are mounted on metal hubs with bores to fit the different shaft sizes.

The grades of wire wheels that are available are normally fine, medium and coarse, depending on wire size. When ordering a straight wire wheel, five requirements should be specified: brush diameter, grade or wire size (decimal diameter), shaft or arbor hole, face width, and wire material (brass, steel, stainless steel), i.e., brush 3 inch diameter, 0.020 wire size, $\frac{5}{8}$ inch arbor hole, $\frac{1}{4}$ inch face width, stainless steel.

Wire wheels are also available for the offset grinder, but these are of the cup design. The same requirements as mentioned for the straight wire wheel apply for the cup wheel except for a face width call out.

Buffing discs for both the straight and offset grinders are available to polish metals such as brass, aluminum, or stainless steels. These discs are made of either lamb's wool or sewn cotton laminates.

The most popular buffing wheel is the cotton laminate type which is made by sewing from 18 to 42 cotton discs together to form a wheel with a center hole. This laminated wheel is then affixed to an arbor with the use of two metal washers and a nut that both drive and hold the wheel to the shaft. This type wheel is only used on straight grinders or bench/pedestal grinders (described in Secs. 9-3 and 9-7, respectively).

For the angle grinder, a sanding buffing unit similar to that described for the electric drill is available. The recessed hole backing pad affords a support for both sand or emery paper discs or for attaching a wool polishing bonnet for metal finishing. A full description of metal polishing compounds is presented in Sec. 9-9.

Resinoid grinding wheels are a composition of epoxy and fiberglass with grinding grits impregnated within its structure. This type grinding wheel is used where fast removal of metal, such as weld beads, burring, or surface material removal, is required. Resinoid wheels, which are available in both the straight and recessed type, are generally run at high speeds (up to 8500 rpm). These grinding wheels are manufactured in two popular diameters—7 and 9 inches—with $\frac{7}{8}$ inch arbor holes. Adapters to allow the use of these wheels on $\frac{3}{8}$, $\frac{1}{2}$, $\frac{5}{8}$, and $\frac{13}{16}$ inch diameter shafts are available. These adapters are normally flat rings that fit the wheel bore and have a bore that fits the grinder shaft.

The wheel grit or grinding grain size is generally a coarse grain rated at 24 (grain size). A description of grain size is given in Sec. 9-8.

Care of Hand Grinders

Grinders are tools that are run at high rates of speed. This speed necessitates the use of ball bearings in both its motor and spindle sections. These bearings are made both in the sealed or unsealed variety. The sealed bearing is one that is greased and sealed at the factory and is normally replaced only when worn. The unsealed type is one which must be intermittently oiled with a light-weight oil. The manufacturer's oiling recommendations must be adhered to rigidly to prevent damage to these precision bearings. Keeping the grinder clean of grits is by far the most important fact in its life. One grinding grain in the bearing will destroy its usefulness in less that 15 seconds. All dust seals and caps should be checked prior to any grinding operation.

Cleaning procedures similar to that of the drill should be extended to the grinder but with a much more thorough approach.

Brush replacement is much more frequent in the grinder than with the drill. This is due to the higher rates of speed that are developed in the grinder. Brush replacement should be performed as per the manufacturer's recommendations.

Care of Grinding Wheels

When grinding wheels are in contact with a workpiece, rust, paint, or the material itself sometimes fills the surface of the grinding face. This process of filling in the grinding face, thereby stopping the grinding action, is called "loading" the wheel. When a wheel is loaded, a cleaning operation called dressing is used. Dressing a wheel removes both the load and the dulled grinding grits, thereby exposing a new faster cutting surface.

WARNING

When dressing a grinding wheel, always support the star wheel dresser against a bar or plate, and always wear gloves and a full face mask.

The dressing of a wheel is most often performed with a star wheel dresser. This tool consists of a series of hardened star-shaped discs mounted in a holder that allows the discs to rotate freely. The dresser is to be held with both hands and must rest against a bar or plate during wheel contact. Failure to support the dresser will cause spin-off from the rotating wheel and possible serious injury to the technician.

The grinder and wheel should be clamped securely in a vise. A support base for the dresser can then be constructed of heavy wood near the wheel face to the dressed. The height of this base should put the star dresser approximately on the centerline of the grinder shaft (Fig. 9-7). Make sure that both the grinder and the dresser base are secured. Put gloves on both hands and wear a full face mask.

After starting the grinder, take the star dresser in both hands and rest the head on the support base. Slowly advance the dresser until contact is made with the wheel. With positive straight motion, allow the dresser to traverse the loaded wheel face. This is done until the loaded face is removed and a renewed flat grinding face is formed. Don't allow grooves to be formed during dressing. After dressing, always check wheel flatness with a scale.

Wheel inspection should be performed routinely for safety of both the user and surrounding people in the grinding area. Wheels with bad chips in

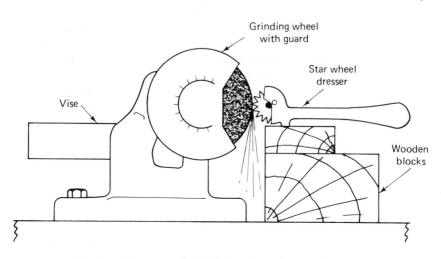

Fig. 9-7. Correct method of dressing a hand grinder wheel (grinder held in vise).

them should be either dressed if possible, or discarded. Any sign of cracks in the wheel, no matter where they are, should immediately call for the wheel to be discarded. Further use of a cracked wheel could result in its shattering, which could cause injury.

On every grinding wheel there exists an operating speed range marked in rpm's put there by the manufacturers. It should not be run in excess of that speed. In going above the recommended speed, wheel shattering could take place. This is further explained in the discussion of grinding wheels, Sec. 9-8.

9-4. Sanders

The orbital sander is mentioned here, since it plays a very important part in metal, plastic, and wood surface treatment. Used with sand or emery sheets, it is used to remove surface rust, finish a surface, deburr pieces, bevel edges, and perform a variety of other job applications.

The orbital sander (Fig. 9-8) gets its name from the motion of its grinding pad. The pad is of a rectangular shape fixed to the sander housing by four ball and socket type fittings. This pad is driven in a small oscillatory motion by way of a crankshaft drive between the motor and the floating pad. This motion, classified by oscillations per minute (opm) ranges from 4000 to 9000 opm.

To further increase the use and application of the orbital sander, a second feature is available on some models. This feature converts the orbital motion of the pad into a back-and-forth or straight-line action. A straight-line action

Fig. 9-8. Orbital sander. (Courtesy of Black and Decker Co.)

is necessary for finishing wood or metal surfaces where the surface is to have a smooth satin finish. Selection of either mode of sanding is performed by a lever located between the motor housing and the sanding pad.

The sanding pad or platen is generally lined with a rubber face to add both a firm but flexible backing to the sanding sheets during use. The sanding sheets are made in two popular sizes—$3\frac{5}{8} \times 9$ inches and $4\frac{1}{2} \times 11$ inches. The smaller dimension in each of these two sheet sizes ($3\frac{5}{8}$ and $4\frac{1}{2}$) are basically the width of the sander pad but the longer dimensions are not, since they include an extra length that is needed for clamping the sheet at each end to the sander pad.

Attaching the sanding sheet to the platen is accomplished by several methods. Some employ a spring clamping device, whereby one end of the long length of the sheet is rolled around the end of the platen and clamped. The second clamp is employed after stretching the sheet taut over the platen.

A second method involves the use of a split ratcheted cylinder located at each end of the pad. To secure a sanding sheet to the pad, the following procedure is used. Insert one end of the sanding sheet into the slotted front cylinder (opposite the handle). Draw the opposite end of the sheet across the platen and insert its end into the slotted ratchet cylinder. Using a screwdriver or supplied tool, tighten this cylinder until the sanding sheet is taut.

Abrasive sheets are made in a variety of grit sizes with the 60, 100, and 150 grits the most popular. Recently, nylon polishing cloths have been made available for use with the sander in the straight line mode. These cloths offer an extremely fine surface finish on a range of hard to semi-hard metals. The nylon pads are of random woven nylon thread with a crocus or polishing compound impregnated within and on its fibers. When using this type of pad, care should be exercised to prevent extreme heat buildup, since the nylon will begin to soften and stick to the surface being polished.

Care of Sanders

The same general care and cleaning procedures for the hand drill can be extended to the sander. Since the orbital sander runs at high output speeds, the bearings in the sander are generally of the sealed ball bearing type. With sealed bearings, there is no need to oil or grease them. Cleaning the exposed surfaces and housing of the sander should be performed on a regular basis to prevent sanding grits from entering the motor and bearing section. Always consult the manufacturer's recommendations on care and maintenance of any power hand tools. If care and maintenance manuals become lost or misplaced, the manufacturers will often supply one free of charge at your request.

BENCH AND FLOOR MOUNTED POWER TOOLS

Bench and floor mounted power tools extend the capability of the technician, making his job easier and more efficient. Mounted power tools provide more accuracy than hand power tools because of their rigid mounting. The bench and floor mounted power tools most often used by technicians are drill presses, band saws, and grinders.

9-5. Drill Presses

Possibly the most important power tool used by the technician is the drill press. Drill presses are manufactured in a range of sizes from a high-speed precision bench type to a large radial arm floor-mounted version. In our discussion of drill presses, we will limit it to the general shop bench type and the floor or pedestal type. The main differences between the bench and pedestal types are in height and mounting. Drill presses are used for a variety of job applications (other than drilling holes) where a rotary motion is needed. These applications include reaming, grinding, and buffing. Special tapping attachments also allow the drill press to tap large numbers of holes quickly and accurately. If done by hand, the tapping would involve many hours of hard labor.

Drill presses are classified by the distance between the spindle and the main supporting column. This distance is the actual capacity of the drill press. Therefore, a 15 inch drill press could drill to the center of a 30 inch diameter workpiece.

Figure 9-9 shows a typical bench model drill press. Since the working end or head of both the bench and floor model drill press are basically the same, a discussion of their main parts and functions is discussed in detail for the reader.

The drill press head consists of five basic parts: a main housing, a motor, a speed selector, a spindle, and finally a spindle feeding mechanism. The main housing is a metal casting that is designed to contain all of the essential moving

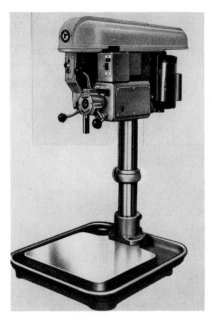

Fig. 9-9. Bench model drill press. (Courtesy of Rockwell Manufacturing Co.)

parts of the drill press. Its main function is to hold these items securely to the main drill press support column.

WARNING

Always support the drill press head before attempting to raise, lower, or turn it on its column.

There are generally two locking screws that clamp the main casting to the tubular support column. This allows for the complete drill press head to be raised, lowered, or swiveled on its support tube to meet unique job situations. The moving of a drill press head is a situation that is very rare and should not be attempted without proper support of its weight. Loosening a head clamp could send the whole head assembly crashing to the base, causing possible damage to the drill head and harm to the technician.

The electric motor of the drill press is mounted on a hinged base located at the back end of the head casting. The motor is mounted with its spindle in an upward position. To this spindle is attached either a single belt pulley or cone belt pulley.

Drill press motors are available in a wide range of horsepowers and with either a fixed or two speed output. The on/off switch for the control of this motor is generally located on the front face of the drill press head.

Speed selection on a drill press of the types under discussion is achieved by the shifting of belts or by a variable speed pulley.

When speed selection is performed by shifting a belt, stepped pulleys of four to five steps are used (Fig. 9-10). This setup gives 5 speeds from which to select. If a two speed motor is used, the speed range would increase to ten speeds (five for each speed of the two speeds of the motor).

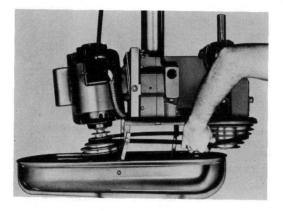

Fig. 9-10. Speed selection—cone pulley methods. (Courtesy of Rockwell Manufacturing Co.)

Since the motor base plate is both hinged and spring loaded, the motor keeps the belt taut in each of its speed ranges. To change speeds, the belt guard is removed to expose the spindle and motor pulley. Firmly grasp the motor and pull it toward the front of the drill press. This relieves the belt tension and allows for easier shifting to another speed. When changing speeds, remember that the belt must run between mating pulley steps—that is to say, if the belt is run from the second step down from the top of the pulley on the motor, it must mate or drive in the second step down from the top of the spindle pulley. Mismating causes belt damage and can cause the belt to slip off of the pulley.

A common speed range for step pulley drill presses is from 250 to 300 rpm at the low motor speed to 4500 to 5000 rpm at the high motor speed end. This speed range fluctuates as to the size of the drill press.

The variable speed drill press is unique in the fact that an infinite speed range is available.

Variable speeds are accomplished by a variable diameter pulley located between the motor and spindle pulley. The motor pulley is a fixed diameter unit. A special belt runs from this pulley to the variable pulley in the first phase in the speed reduction.

Located on the front of the machine is a speed control handle or dial. It normally is indexed in the rpm's that are available. Whenever this type of speed control is used, the one important point to be remembered is never to turn or dial a speed change while the drill press is off, as damage to the pulley and speed belt could result. Always change speeds while running. As the speed control is dialed to a higher speed, the variable pulley, which is adjustable in its groove width, opens to a larger inside belt width. This allows the special belt to bear on the pulley nearer to its center. This produces a higher speed at the variable pulley shaft. Conversely, a lower dialed speed closes the width of the variable pulley and forces the belt to run further from the center, thereby, causing a slower speed output on the shaft. At the top of this variable speed pulley shaft is a fixed diameter pulley, which uses a standard vee belt to carry the speed changes to the spindle driving pulley.

The spindle is the real workpiece of the drill press. It not only provides the rotation to the tool but also provides the up and down feeding action required in drilling, reaming, etc. The main shaft of the spindle is splined at the pulley end, while the other end is the working end. The splined end of the shaft fits into the splined bore of the spindle pulley. This allows the main shaft to travel up or down through the pulley and to be turned by it at the same time.

The main shaft and its bearings are located within the quill housing of the drill press. The quill is a nonrotating shaft that imparts the up or down feeding motion to the shaft located within. Along the back surface of the quill are a series of gear teeth; these gear teeth, coupled with a gear and handle arrangement located in the head casting, impart the up and down motion of the shaft. The working end of the drill press spindle generally has a Morse taper hole to receive both drills and attachments.

Power feeds are found on some drill presses of the smaller size. Feeding is performed by two general methods, a cone pulley belt drive or a variable power feed. These two feed motions are driven from the main motor. Both types make use of a bevel gear to drive the quill up or down. This setup allows the quill to be hand fed to any point and then by engaging the bevel gear, allows for the quill to be power fed. The feed bevel gear is most often engaged by a simple inward movement of the quill feed handle.

Some machines equipped with power feeds have depth dials that can be set to disengage the power feed at a preselected drill depth. In the problem of depth gauging of tools, several methods are available. The most simple form is merely a vernier, usually marked off in $\frac{1}{16}$ inch increments, scribed on the quill. As the quill is fed, the drill depth is checked by observing the displacement of the vernier.

A much better gauging setup is one where a threaded depth rod is attached to the quill and off to its side. As the quill moves up and down, so does the threaded rod. A flat washer with a pointer and a locking nut on either side of it is attached to the threaded rod. Affixed to the head casting and immediately adjacent to the threaded rod is a scale. By adjusting the locking nuts on the thread rod, the pointer can be referenced along any point of the scale.

Very often in jobs where holes are to be controlled to a set depth on many pieces, the nut and pointer assembly can be set and allowed to bottom on a stop. This stop is a fixed flat surface located near the bottom of the head casting. The threaded depth rod passes through this surface at 90° to it. Setting the stop is rather simple and is done with the power off. First locate the drill point to the workpiece and lock the quill with the quill lock (Fig. 9-11). If, for example, a drill depth of 4 inches (surface of the workpiece to the drill point) were required, the following procedure is recommended. Loosen both nuts holding the pointer on the threaded depth rod. Rotate the nuts until the bottom surface of the nut closest to the stop is 4 inches from it (measured with a scale or caliper). Using two wrenches, lock both the pointer and the nuts together securely. Recheck the depth. Loosen the quill lock and proceed with the drilling operation.

Fig. 9-11. Drill press depth gauge. (Courtesy of Rockwell Manufacturing Co.)

In most light to medium drill presses, a Jacobs chuck is provided on the spindle. Its capacity is keyed to the design capabilities of the drill press. In the medium to heavier duty drill press spindles, a Morse taper hole is provided to accept the various cutting or reaming tools. Jacobs chucks of all sizes can be procured with Morse taper shanks to fit these drill press spindles. A drill drift (Fig. 9-12) placed in the slot located through the spindle is used to remove a taper shank tool when its removal is required. A drill drift should be hit with either a lead or brass hammer only, as a steel hammer could glance off and damage the spindle nose and its accuracy. The flat edge side of the drift is always toward the tang of the taper shank tool.

The drill press worktable is a machined casting that is adjustable to any

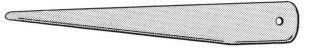

Fig. 9-12. Drill drift.

point along the main column. This is accomplished by loosening the clamping bolt located near the main column. When the need arises, this table can be rotated about the column to any position desired and locked. If a situation requires the use of the base surface table, the floating worktable can be rotated to the back of the spindle. The slots in both the floating and base tables are used to bolt jobs to the table during drilling. Some drill presses, usually in the light to medium duty range, have floating tables that also rotate through a full 90° angular rotation with respect to the spindle. This allows for holes to be drilled through a workpiece at specified angles. Whenever an adjustment in either table height or angle is performed, always recheck that the clamping locks are secured. Failure to do this results in violent chatter of both drill and the finish of the hole being drilled. Most tables have a cast hole located in their centers under the drill spindle. This is so that drilling through a workpiece can be accomplished without damage to the worktable. If a workpiece is positioned on a table out of line with the clearance hole, the job should either rest on parallels or a piece of wood. This practice protects the worktable from being accidently drilled.

WARNING

Always clamp the workpiece securely to the drill press table prior to drilling. This is discussed in the next paragraph.

Drill Press Aids

Two items of constant use in drill press work are the drill vise and vee blocks. The drill press vise (Fig. 9-13) is an aid used for securely holding a variety of shapes during drillings. The base is smooth to allow it to be slid over the worktable to any position. Workpieces are clamped by a movable

Fig. 9-13. Drill press vise. (Courtesy of Palmgren Steel Products)

flat jaw which opens or closes by a screw adjustment. Prior to further discussion, the authors would like to impress on the reader one very important rule when using a drill press. Never use hand power to hold a workpiece during drilling. Either bolt or C clamp the job or the aid holding the job to the table. Failure to obey this rule can result in severe cuts or lacerations to the technician.

The vee block is used to hold round workpieces for drill press operations. These tools are made with screw clamps to secure the job to the vee. To prevent marring of the job surface, small copper sheet pieces should be placed between the screw faces and the job. Once again, clamp this assembly prior to any drill press operation. One final suggestion on drilling procedures and safety is required. Drill presses tend to rotate workpieces clockwise if left loose. Therefore, to further secure a job, the following is recommended if a workpiece is long enough. Set the workpiece on the table and allow one end of it to rest against the left side of the main column. This prevents the workpiece from breaking loose from a setup and rotating with the drill.

9-6. Band Saw (Metal Cutting)

Possibly the second most often used large power tool used by the technician is the band saw. This tool is used for cutting a variety of materials into a variety of shapes. The most often performed operation is the cutting of circular discs or rings.

Band saws are defined by cutting capacity. For example, a 14 inch band saw is the approximate space between the blade and the saw's housing. The typical band saw (Fig. 9-14) can be broken into four basic units: motor and speed control, main frame and blade wheels, blade guides, and worktable.

In the motor and speed control unit, the same two basic types found in the drill press previously discussed are applied to the band saw. Generally a high or low speed selector lever is provided for a better range of cutting speeds to the blade. Speeds on a band saw are denoted in feet per minute units (fpm).

The band saw's main frame serves two main functions. It provides a rigid support for the two blade wheels, and it provides a support for a wheel guard door and the worktable. Located in each upper and lower half of the frame is a flat faced blade wheel. The purpose of these wheels is to provide both power to the blade and also retain the blade in its travel path.

Both wheels have rubber faces to keep the blade from slipping during cutting. Rotational power is supplied to the blade through the fixed bottom wheel. The top wheel is both an idler and tension wheel. As an idler, this wheel merely changes the direction of the blade down toward the worktable. Its real importance is in applying tension to the blade for positive power transmission. Located at the bottom of this top wheel is a blade tension

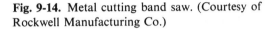

Fig. 9-14. Metal cutting band saw. (Courtesy of Rockwell Manufacturing Co.)

adjustment handle. This handle, when turned, moves the idler wheel up or down in relation to the fixed bottom driving wheel. Proper tension is generally noted on a spring scale located at the idler wheel. If a $\frac{3}{8}$ inch blade is being used, the indication must be adjusted until it aligns with the $\frac{3}{8}$ inch marked scale. When no indicator is supplied, the blade tension is arrived at by trial and error. The limiting rule is that the blade must not slip when performing its operation. Overtightening the blade could break the welded joint of the blade.

WARNING

Always have power off and wear leather gloves when making band saw blade adjustments or changes.

Normally, blades can be replaced with one adjustment (basically the tension adjusting wheel). On rare occasions, the blade will slip off of its two-wheel setup. If this occurs, the top wheel must be adjusted as to squareness with respect to the bottom fixed wheel. An adjusting screw exists at the back of the top wheel's bearing block. This screw, when adjusted, causes the wheel face to tip out of square within a small range with respect to the bottom wheel. The effect of this tipping on the blade is to cause it to run true and square to both wheels. This retruing or "tracking" of the blade is done with the power off and the saw in a neutral position. With the wheel guard door open, adjust the wheel tipping screw to slightly tip the wheel so that the top of the wheel is receding from the front of the band saw. After a small adjustment, turn the wheel and blade assembly by hand power several turns. If the blade on the top wheel starts to move to the front edge of the wheel, repeat the adjustment. For a final check after adjusting, close the guard door and engage the power to the saw. Allow the band saw to run for several minutes to make sure the adjustment is correct.

The next items of discussion are the blade guides. Two blade guide units are equipped on all band saws (Fig. 9-14). A fixed guide is located directly below the table surface while a movable (up and down) guide works off of a top support arm. The purpose of the saw guides are to both support the back of the blade during cutting and to keep the blade from bending or bowing during operation. Figure 9-15 shows a typical guide setup. A thrust bearing bears and rotates against the back edge of the blade. It takes the force due to pushing a job against the blade. The blade guides are generally adjustable metal wear plates or idler bearings located on either side of the blade, as shown in the figure. The adjustment of the blade guides is such that they both approach the side of the blade but don't really touch it. Two to three thousandths of an inch clearance between each guide and blade is sufficient. A feeler gauge is used for this adjustment. If the guides are tight against the blade, they will begin to scratch and destroy the blade sides. If this happens, remove and destroy the blade and readjust the guides on a new blade.

Fig. 9-15. Band saw blade guide (upper). (Courtesy of Rockwell Manufacturing Co.)

As previously mentioned, the top guide assembly is adjustable up or down and is locked by means of a locking nut located on the side of the top housing frame. This adjustment allows for adjusting to different thicknesses of material. Remember that this top guide assembly should be adjusted and locked to allow the material being sawed to barely pass between the top guides and the worktable. This increases blade stability and also protects the technician from excessive blade exposure.

The worktable of the band saw is a machined cast iron or steel unit. The table, by its design, is capable of allowing cutting of strips, circles, bevel cuts and compound angle cuts. The table is attached to the machine frame through a semi-circular saddle mount. This mount, when loosened, allows the table to be rotated through a total of 45° with respect to the blade.

Running from the front edge of the worktable, or in some cases from the side to the center, is a slot with a removable metal plate. The purpose of this slot is to allow replacement of the saw blades. The plate is removed by a locking nut located beneath the plate. The plate should always be in place during all sawing operations.

Two parallel machined slots, located in the top surface, are available in some tables to accept a miter gauge head. This device is used to hold and cut pieces at 90° to the blade and can also be changed through any angle up to and including 45° to the blade.

A rip fence is an adjustable metal unit that clamps over the band saw worktable and allows for the cutting of strips or for making parallel cuts. The adjustment is through a range from the blade to the edge of the work-table. The rip fence can be used on either side of the blade. Clamping of the rip fence is done by a clamp screw located at the front of the fence.

Band Saw Blades

Blade stock is bought in boxes of 250 or 450 foot lengths. This allows for the welding of specific length blade loops to meet the size requirements of the different machines. Four basic tooth forms are available to choose from to meet particular job applications: the raker set, the wavy set, the skip tooth blade, and the hook tooth.

The raker tooth is a form that, when viewed from its cutting edge, presents a repeating pattern of one tooth in line with the blade, the second tooth slanted to the right, the third tooth slanted to the left, and so on. This type tooth form is used for accurate contour sawing and for straight cut off of large solids and thick plate. The raker tooth blade is made in the following number of teeth per inch: 6, 8, 10, 12, 14, 18, and 24.

The wavy set is noticed by the slow, wavy, back and forth array of the teeth along the edge. The main purpose of this design is its use in thin metal cutting both of the mild and hard type. The range of teeth in this blade is 8, 10, 12, 14, 18, 24, and 32 teeth per inch.

The skip tooth blade, with its alternate cutting tooth array, is used where large chips are formed by the softness of the material being sawed. Nonferrous metals, plastics, and wood are best cut with this blade type. Since chip clearance between teeth must be larger to accommodate removed material, the tooth selection is normally coarse with 2, 3, 4, and 6 teeth per inch.

The hook tooth blade is basically the same as the skip tooth, but differs in one point—the tooth of this saw has a slightly curved cutting face which tends to slice or shave the material it contacts. The hook tooth blade therefore requires less feeding pressure and subsequently lasts longer than the skip tooth in certain applications. The teeth per inch range is the same as for the skip tooth blade.

Saw Blade Welding

Each band saw has a specific blade length that will run properly between the wheels. The length of the blade is the overall unwelded saw blade strip. This size is listed by the manufacturer in the saw manual.

The process of welding the ends of a blade to form a hoop is accomplished by an electric welder. This unit can be mounted within the saw's frame or mounted by itself. After cutting a length of blade, the ends should be inspected for squareness. Retrimming or grinding the blade ends should be performed if necessary. A blade length generally can be slightly over or under the actual callout, since the idler wheel on the saw can be adjusted to accept size deviations.

With the blade ends "squared," the actual process of welding can proceed. Figure 9-16 shows a typical welder setup. Five steps are necessary in welding any blade: initial clamping, weld feed setting, welding, annealing, and sizing.

Fig. 9-16. Blade welding unit. (Courtesy of Rockwell Manufacturing Co.)

Two blade clamps are located at the top of the unit. Prior to clamping, a selector arm located just below the clamps should be positioned in the clamp mode. The butted blade ends can then be clamped to the unit. Always clamp the blade in such a fashion that the back edge is resting at the back face of the clamp assembly and the butted saw ends are near the midpoint between the clamping jaws. Check that the blade is a smooth untwisted loop.

With the blade clamped, the next step is to move the clamp mode lever to a position marked or noted in some method for the blade width. This action imparts a predetermined displacement of the right clamp during actual welding. This is required to supply metal from the blade itself to form a weld bead. If the blade is a $\frac{3}{8}$ inch size, the selector should be positioned for $\frac{3}{8}$ inch.

The next step is the depressing of the weld button. The button triggers a welding current to the two clamps and melts the metal at the joint. The right hand clamp also moves slightly in toward the joint area to form the butt weld bead.

The blade is now welded, but the joint is brittle due to welding stress. A blade used in this state would break at the weld joint. To relieve the brittleness, an annealing process is performed. A button is supplied to perform this process. When depressed, a current is supplied to the weld joint which heats it to a slight red glow. When the red glow is reached, the button should be released and the blade allowed to cool. The blade weld joint is now annealed.

The final operation is to remove the welded blade and remove the excess weld from the joint. Weld removal is performed by a grinding wheel. Some welders have wheels built into them. Remove excess weld from both sides and the back edge of the blade. The weld joint should be so dressed that it will slip freely through the band saw blade guides. Many welding units have a gauge attached to check this. Excess weld left on the joint will cause damage to the blade guides when the saw blade passes through them.

Blade Installation.

Installing a blade in a band saw is a relatively simple procedure. The first step is to open the front cover doors exposing the saw wheels and remove the slot plate located in the worktable. The second step involves the loosening of the blade on its wheels. This is accomplished by loosening the tension wheel located just under the top idler wheel. The band saw blade can now be removed. Inserting the new blade requires that it be installed loosely onto the two wheels. Check that the teeth are pointing toward the floor at the table slot. Take up the slack slightly on the blade with the tension adjustment. Make sure at this point that the blade is positioned in both the top and bottom blade guides. Adjust the tension to the proper point for the blade size.

Replace and clamp the table slot plate and close the guard doors. Start the saw and allow the blade to run several seconds. This allows the blade to seat itself and also checks for defects in blade welds.

Care of a band saw is normally minimal. The manufacturer's suggestions for greasing and oiling should be adhered to rigorously. When not in use, the worktable surface should be given a light coat of oil to prevent rust from forming. If long periods of nonuse are encountered, the blade should be removed to prevent its taking a set or bend due to the wheel pressure and curvature.

9-7. Grinders

Bench and floor grinders are tools used in a variety of job applications, such as stock removal, sharpening of tool bits, shaping of special parts, removing of rust or scale, and many additional jobs.

Bench grinders and pedestal grinders (Fig. 9-17) are the same except for their method of mounting. They are defined by the diameter wheel that they accept. An 8 inch grinder accepts an 8 inch diameter wheel. Grinders are basically double ended sealed motors with a mounting shaft at each end. Motor speeds are usually of a fixed single output that is choosen for the diameter wheel used on them. Some grinders are available with a two speed switch that allows for a high and low speed selection.

Fig. 9-17. Bench grinder on pedestal. (Courtesy of Rockwell Manufacturing Co.)

The driving shafts on grinders vary through a diameter range to suit the size wheel that it is to drive. Basically the diameters are $\frac{1}{2}$, $\frac{5}{8}$, $\frac{7}{8}$, $1\frac{1}{4}$, and $1\frac{1}{2}$ inches.

The shaft ends are threaded to provide for clamping of the grinding wheels. To prevent loosening during operation, the thread on the right side of the shaft is a right hand thread while the left shaft end has a left hand thread. This arrangement keeps the clamping nuts from vibrating loose during a grinding operation. Clamping a wheel is performed along the same lines mentioned for the hand grinders in Sec. 9-3.

Attached to the motor casing and surrounding most of the wheel is a shatter guard. Its purpose is to protect both the technician and other personnel in the area from flying grinding grit or the shattering of a wheel. The end of this guard is covered with a removable end plate. This plate is removed when the need to replace a wheel is encountered.

To further protect the face of the technician during grinding, a safety glass shield should be attached to the wheel guard housing. This shield can be moved by hand pressure to cover any possibility of large chips of either a workpiece or wheel from hitting the technician.

The work or tool rest is the support plate located in front of each wheel. This metal plate is notched so as to extend on either side of the wheel. By loosening the clamping screw to this assembly, the support plate can be adjusted in toward the wheel and can also tip through an angular range with respect to the wheel. This tipping ability allows for the grinding of cutting tool bits when the angle is critical.

The purpose of this support plate is to allow a workpiece to rest on it during a grinding operation. During any operation, the gap between it and the wheel face should be as close as possible. If the gap is left large, a small workpiece could slip and lodge between the rotating wheel and the support plate, causing the wheel to shatter.

Many grinders have a tray or cup to hold water for grinding. During all grinding operations, heat buildup becomes very noticeable. To protect both the tool and the technician's hands from burning, frequent water quenching of the workpiece is recommended.

Grinders in their usual setup have a roughing wheel and a finish wheel. When forming a job, the roughing wheel is used to remove the large excess amounts of stock, while the finish wheel is used to produce both a sharp and a highly smooth, finished surface. When approaching a grinder, the general rule of thumb is that the coarsest grit wheel is the roughing wheel. This can be ascertained by visual inspection before starting.

When forming a job or tool on a grinder, the cardinal rule is to make use of the whole face of the wheel, whether the workpiece is smaller or larger than the width of the wheel. The recommended method is to move the workpiece back and forth slowly across the wheel. Two important benefits result: one, the abrasive cutting action is faster and cooler; second, the wheel face remains flat.

The normal tendency on first grinding encounters is to grind at one spot. This grooves the wheel and causes poor grinding on future jobs. Although the practice is sometimes performed, the use of the side of the wheel to grind a job is never recommended. This practice can and does cause a weakening of the wheel's structure. When a wheel face becomes grooved or loaded with material, it must be redressed. Using the star dresser, the wheel should be retrued in accordance with the procedures in Sec. 9-3.

After dressing the grinding wheel, the nuts should be inspected for tightness. The tool rest support plate should then be adjusted in toward the wheel to again keep the wheel face and tool rest gap at a minimum.

Tool Grinding Hints

The task most often performed on the grinder is the resharpening of cutting tools such as drills, tool bits, chisels, etc. These jobs require the grinding of sharp points or edges.

The following rule must be remembered in all sharpening operations. Always grind away from the cutting edge. Adherence to this rule will produce tools with the sharpest cutting edges and will also eliminate wire edging. Wire edging is a term used to describe a thin flexible burr on a cutting edge that prevents the tool from cutting properly.

When sharpening tools, the first process is to grind one face both true and smooth. The second step is to turn the tool over and grind the adjacent face until a sharp edge is produced. During the grinding of the second side, a simple observation tells the technician when the required cutting edge has been sharpened—sparks will be observed flying over the edge being ground. This is an indication of a sharp ground edge.

GRINDING WHEELS AND POLISHING COMPOUNDS

Grinding wheels and polishing compounds are used on the hand, bench, and pedestal mounted grinders or drill accessories. This section describes the wheels and compounds used.

9-8. Grinding Wheels

Grinding and grinding polishing wheels are tools that employ hard abrasive grains to remove or finish materials. These grits are bonded together by various methods and are then made into various shapes and sizes. Figure 9-18 shows some of the most commonly used wheel forms available and their tool uses. As noted, these wheels have a recommended grinding face.

In 1958 an American standard marking system was adopted for use on all grinding wheels except those employing diamond grit and sharpening stones used in cutlery and associated fields. This system is basically a six step callout. This notation as well as the wheel's maximum operating speed is located on the wheel either on a paper label or printed on the wheel itself. The six basic steps or notations are *abrasive type, grain size, grade* (soft to hard), *structure* (dense to open), *bond type*, and *manufacturer's record*. Table 9-2 shows a sample callout with a group range listed under each of the six basic notations. To fully understand these markings and their meaning a brief discussion will follow.

Abrasive Type

Two types of abrasives of most commonly use are aluminum oxide, designated by the letter (A), and silicon carbide, designated by the letter (C). Many

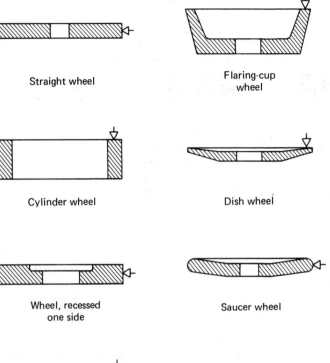

Straight wheel

Flaring-cup
wheel

Cylinder wheel

Dish wheel

Wheel, recessed
one side

Saucer wheel

Straight-cup
wheel

Note: Arrows ◁— indicate grinding surface

Fig. 9-18. Various shapes of grinding wheels.

manufacturers use their own trade names for these abrasives; for example, aluminum oxide is sometimes referred to as aloxite, alundum, and borolon, while silicon carbide (C) is often called out as carborundum, crystolon, and electrolon.

Aluminum oxide is by far the most commonly used abrasive grit for general grinding. In application on very hard metals or tool bits, silicon carbide is the recommended type.

Grain size (grit size) is another important factor in wheel selection. Large grain sizes are used in areas where fast material removal is needed or where soft nonferrous material (aluminum) is to be worked on. Finer grains are used for producing smooth finishes. Abrasive grain sizes are cataloged into four groups; coarse, medium, fine, and very fine. Table 9-2 shows the

TABLE 9-2. GRINDING WHEEL STANDARD NOTATIONS

Position:	1	2	3	4	5	6
	Abrasive	Grain Size	Grade	Structure	Bond Type	Manufacturer's Record

RA   46-J   6-V   8

Abrasive	Coarse	Medium	Fine	Very Fine	Dense to Open		Bond Symbols	
A Borolon				Very Fine			Bond Symbols	
NA Borolon								
JA Borolon	10	30	70	220	1	9	1	9
TA Borolon	12	36	80	240	2	10	2	98
ZA Borolon	14	46	90	280	3	12	3	99
YA Borolon	16	54	100	320	4	14		C
BA Borolon	20	60	120		5	16		F
RA Borolon	24		150		6		64	F52
								F53
							64	FN
SA Borolon			180		7	V—Vitrified		FXX
WA Borolon					8	B—Resinoid	8	IL
						R—Rubber		ILF
C Electrolon				Grade Scale		E—Shellac		
GC Electrolon								

Very Soft	Soft	Medium	Hard	Very Hard
D E F G	H I J K	L M N O	P Q R S	T U V W - X Y Z

grouping. The general rule is that the larger the number indicated on the wheel, the finer the grain.

The term grade, as applied to grinding wheels, is a rating of grain or grit bond strength. The rating runs from the letter A (the softest) through to and including the letter Z (the hardest)—refer to Table 9-2. This grade range is broken into subunits: very soft, soft, medium, hard, and very hard. When grinding soft materials such as brass, there are times when it is important to keep a clean cutting surface. A very soft wheel is available to solve the cleanliness problem. When using this type of wheel, note that the material holding the grains tends to loosen and allow the surface grains to fall away. This falling away performs two important functions: it carries away impregnated material (retards loading), and it continually renews its abrasive cutting action. This type of wheel naturally wears away very fast. In applications on metals in the hard to very hard range, a wheel grade selection in the soft to very soft range is recommended.

A grinding wheel's structure is defined as either dense or open-type grain. A dense wheel has grains very tightly packed and presents a somewhat smooth appearance. An open grain wheel is one with large pockets or holes between grains. The dense or closed face wheel usually is used to produce smooth polished finishes on hard metal, while the open grain is used for both soft or rapid stock removal. As shown in Table 9-2, the dense to open face range goes from 1 to 16 with 1 being the densest and 16 the most open face structure.

Bond Type

The bond type refers to how the grinding grains are held together to produce the wheel. There are four main bond types available: vitrified (V), resinoid (B), rubber (R), and shellac (E). The vitrified wheel, which is the most common type currently in use, uses a ceramic-type material to bond the wheel together. These wheels are porous and are not affected by water, oil, acids or use in hot or cold environments.

The resinoid bond (B) uses an epoxy resin bond to produce the finished wheel. This type bond is recommended in uses where high speed and rapid stock removal are required.

Rubber bonded wheels are used basically in imparting high luster finishes to metal surfaces. This type bond is also used to produce thin wheels of $\frac{1}{32}$ inch or less used to cut off metal stock. This wheel bond also lends itself well to deburring wheels used on both small and large pieces.

Shellac bond wheels use a shellac based bond to form the wheel. The most important use of this type of wheel is in producing extremely high luster finishes on steel rolls and other items requiring excellent surface finish. This bond is also used in producing thin cut off wheels, but due to its make-up it cannot stand any heat buildup. This bond type is not recommended in heavy grinding applications.

Each wheel contains a manufacturer's record mark. Its position is optional on the wheel and it is used for the private record keeping of the manufacturer. If at anytime a wheel seems to be defective in bond or any other area, the manufacturer should be informed by referring to this number.

Wheel Speed.

The manufacturers of all wheels, whatever their type, list a very important extra item—maximum speed. This number, usually in rpm, is the safe operating speed for that particular wheel. Whenever installing a wheel to a particular power tool, always check that the spindle speed of the tool is not greater than the speed noted on the grinding wheel.

Using a higher than recommended speed on a grinding wheel can cause it to explode into pieces causing very serious injury to the operator and surrounding personnel.

In summary, several points should be made. Aluminum oxide is recommended for the grinding of steel and most of its alloys. For grinding cast iron, nonferrous and nonmetallic materials, silicon carbide is recommended. Grain selection should be based on the hardness of materials; a fine grain size is used on hard materials and a coarse grain is used on soft, ductile material. Selection of grade of hardness is as follows: a soft wheel is used on hard materials and hard wheels are used on soft materials. In some applications, such as brass, a soft wheel is used to prevent loading. A closed grain wheel is used on hard and brittle material while an open face wheel is used on soft, ductile material.

9-9. Polishing Compounds

High luster finishes are produced on a wide range of materials by means of a process called buffing. This process makes use of either wool or cotton wheels charged with a fine polishing compound. Charging is a term used to describe the transfer of a compound to the buffing wheel.

With the charged buffer rotating at a high speed, the workpiece to be polished is brought in gentle contact with the wheel. Frequent recharging is necessary for a good finish.

Polishing compounds are generally made from a variety of abrasive materials and have a very fine grain makeup. A wide range of compounds are made to meet the full range of polishing requirements. Some manufacturers use a color code to call out the compound's uses. Table 9-3 gives a range of applications for some of the standard compounds.

TABLE 9-3. COMPOUNDS AND THEIR USES

Compound	Uses
Crocus (red)	Steel—brass
Tripoli	Brass, copper, gold, silver
Rottenstone	Brass, steel
Rouge	Gold, silver
Pumice	Metal, wood
Putty	Glass
Emery	Steel, stone

SUMMARY

Power tools extend man's working capability and versatility by increasing power (torque) and by decreasing the time required to complete a job. Power tools are driven by electrical or pneumatic power, but the electrically powered tools are the more popular because of the availability of a power source. Portable tools are used for applications where less accuracy is required and for field use. Bench tools are used for accurate work in the shop.

REVIEW QUESTIONS

1. During use of an electric hand tool, blue sparks coupled with a pulsing speed are noticed. What should be your reaction?

2. List two reasons for a "reversing" type drill.

3. In drilling large diameter holes by hand through metal, considerable pushing force is required. What could you do to reduce the force in drilling?

4. What is the most common cause of blade breakage on a jigsaw?

5. Explain the purpose of a straight and a canted shank jigsaw blade.

6. What type jobs are straight grinders used on? Offset grinders?

7. It is observed that removing the guard on a straight grinder would allow for a faster grinding job. What would you do?

8. When should you dress a grinding wheel?

9. How do you change from an orbital to a straight line action on an orbital sander?

10. What are the five basic parts of a drill press head?

11. On a variable speed drill press do you change speed while running the spindle or when stopped?

12. What is the function of the quill on the drill press?

13. What is the purpose of the upper wheel on a band saw?

14. How is the tension set on a newly installed blade?

15. It is noticed that a blade while running tends to creep out of its guides. How is this corrected and what is the procedure called?

16. Name the four basic tooth forms in a band saw blade?

17. List the five steps used in welding a band saw blade.

18. How is the size of a grinder determined?

19. What is the rule of thumb concerning a roughing wheel when grit alone is considered?

20. Why should the tool rest on a grinder be properly adjusted to the wheel for a minimum gap?

21. What are the six basic notations found on a grinding wheel?

22. Describe the process of charging in reference to buffing wheels.

23. Select a polishing compound(s) that can be used on silver; on glass.

10

FASTENERS

Fasteners are devices that are used to locate and secure two or more workpieces to each other. They are generally classed as threaded, fixed, or aligning fasteners. Threaded fasteners, such as machine screws, bolts, and nuts, are designed for use in applications where the removal and reinstallations of workpieces may be required. Fixed fasteners, such as rivets, are used to permanently secure workpieces—there is never any intention of detaching the workpieces. Rather than turning out the fixed fastener, as is done with the threaded fastener, the fixed fastener must be drilled out or ground off when removal is required.

Aligning fasteners, such as the common pin of either the hardened and ground dowel, tapered, or spring type, do not exert a compressive force like a screw or rivet, but instead retain workpieces only in the plane that is perpendicular to the pin. This fastener merely prevents the slippage of one workpiece with respect to another.

The first section of this chapter provides information on screw thread types. The chapter then describes the characteristics, uses, and applications of the following types of common fasteners: machine screws, nylon fasteners, bolts, nuts, washers, setscrews, sheet metal/self tapping screws, rivets, and pins. A thorough knowledge of this chapter will enable the technician to choose the proper fastener for his particular application.

SCREW THREAD TYPES

All commercial bolt and screw hardware in the United States is generally of the American Standard thread form (Chapter 8). Listed among the total range of screw hardware are some sizes which are of the unified system. This merely means that these particular sizes can be used and interchanged throughout the United States, the United Kingdom, and Canada.

A threaded fastener consists of a partially or fully threaded rod with one end enlarged or headed in such a way so as to provide for the application of a tool to turn or torque it.

Screws or bolts are available in both a left-handed thread or a right-handed thread. Since left-handed threaded products are not in common demand, some difficulty may be experienced in obtaining them. When a situation arises where left-handed threaded screws must be used and are not commercially available, most manufacturers will produce the required thread for an additional "special set up charge." The ability to obtain special lengths, material, etc. is also readily available from most screw products manufacturers on a "set up charge" basis.

MACHINE SCREWS

Machine screws are a group of threaded fasteners that are hand driven by three basic tools: conventional screwdriver, Phillips head screwdriver and key setscrew (Allen) or splined setscrew wrenches (Secs. 4-1 and 4-5). When using either of the two types of screwdrivers to drive a fastener—the normal maximum diameter fastener available is a $\frac{3}{4}$ inch diameter screw. When using a key setscrew or spline setscrew wrench to drive machine screws, the maximum diameter fastener available is normally a $1\frac{1}{2}$ inch diameter piece. The reason for these ranges of diameters is coupled with the torquing or screwing ability of the particular hand powered tool. It can be seen that tightening a $1\frac{1}{2}$ inch diameter screw with a screwdriver would present a difficult torquing or tightening problem; but this torquing is not difficult with a key or spline setscrew wrench. Machine screws usually have the thread running along their full length. Their head designs vary and can be broken into two categories, slotted or Phillips head type and hex (Allen) or splined key type.

10-1. Screwdriver Machine Screw Heads

There are seven types of heads in common usage: fillister, round, flat, oval, pan, binding, and truss (Fig. 10-1). These seven head forms are available in both the slotted—for conventional screwdrivers—and the Phillips head—

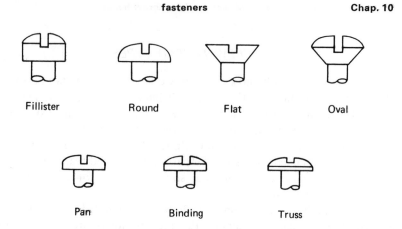

Fillister   Round   Flat   Oval

Pan   Binding   Truss

Fig. 10-1. Screwdriver type screw head forms.

for the Phillips head screwdriver. Each design is unique and is used for a specific job requirement.

The fillister headed machine screw, by its design, allows for large torques to be applied to it. This is achieved by the fact that the head shape approaches an almost cylindrical configuration. The fillister head screw also lends itself well when used in counterbored holes.

The round head machine screw, which is probably one of the oldest used head forms, was originally made to be used on external surfaces where they would neither snag nor catch softer materials. In areas where the presence of a screw head above the surface is objectionable, the flat head screw is employed. When using flat head screws to join two or more workpieces, it is extremely important that the countersunk head holes in the one workpiece all align very accurately to the tap or clearance holes in the mating workpiece. Any hole misalignment between the two workpieces will cause the head of the screw to bear heavily on one side of the countersunk hole and in some cases will not allow the screw head to flush mount to the workpiece.

The oval head machine screw is used mainly for mounting panels. In areas where both the critical location of a panel and heavy torquing pressure are required, the oval head form is the solution. The included angle of the oval head screw is 80 to 82°. The head is basically a "rounded off" flat head form. It is this extra rounding that allows for heavier torquing of the screw while still allowing a round, non-snagging outer face.

The pan head machine screw is used to answer the need of a low profile screw head in areas where a fillister head would be too thick. The pan head screw has another distinct advantage over the fillister head screw in that the diameter of the head is larger than the fillister and it is, therefore, used in situations where extra head clamping area is needed (e.g., thin sheet metal pieces to large sections).

The binding head machine screw is very similar to the pan head but the basic head diameter is slightly larger. The one unique difference of this type screw head is its undercut bearing face. This undercut face allows the head to bear on only the outer edge of the screw head when tightened to a workpiece. This bearing on only the outer edge imparts a stronger screw head-to-workpiece bond that is very effective in units where light to medium vibration is encountered.

Truss head screws are used in areas where relatively flexible workpiece sections must be fastened. The head diameter of this screw is the largest diameter available in any of the head styles. This extra large diameter allows it to be used in securing thin as well as soft or crushable workpieces.

To aid the technician in future use of machine screws, Tables 10-1 to 10-7 list the head dimensions of the seven types of screwdriver machine screw heads.

TABLE 10-1. FILLISTER HEAD SCREW HEAD DIMENSIONS

| Nom. Size | Head Diameter | | Height of Head | | | | Width of Slot | | Depth of Slot | | Dimensions of Recess | | | | Phillips Driver Size |
| | | | Side Height | | Total Height | | | | | | Diameter | | Depth Width | | |
	Max.	Min.	Max.	Min.	Max.	Min.	Max.	Min.	Max.	Min.	Max.	Min.	Max.	Min.	
2	0.140	0.124	0.062	0.053	0.083	0.066	0.031	0.023	0.037	0.025	0.104	0.091	0.059	0.017	1
3	0.161	0.145	0.070	0.061	0.095	0.077	0.035	0.027	0.043	0.030	0.112	0.099	0.068	0.019	1
4	0.183	0.166	0.079	0.069	0.107	0.088	0.039	0.031	0.048	0.035	0.122	0.109	0.078	0.019	1
5	0.205	0.187	0.088	0.078	0.120	0.100	0.043	0.035	0.054	0.040	0.148	0.135	0.067	0.027	2
6	0.226	0.208	0.096	0.086	0.132	0.111	0.048	0.039	0.060	0.045	0.166	0.153	0.091	0.028	2
7	0.248	0.229	0.105	0.094	0.144	0.122	0.048	0.039	0.065	0.049	0.176	0.163	0.100	0.029	2
8	0.270	0.250	0.113	0.102	0.156	0.133	0.054	0.045	0.071	0.054	0.182	0.169	0.108	0.030	2
10	0.313	0.292	0.130	0.118	0.180	0.156	0.060	0.050	0.083	0.064	0.199	0.186	0.124	0.031	2
12	0.357	0.334	0.148	0.134	0.205	0.178	0.067	0.056	0.094	0.074	0.259	0.246	0.141	0.034	3
14	0.400	0.376	0.165	0.151	0.230	0.201	0.075	0.064	0.105	0.084	0.281	0.268	0.161	0.036	3
1/4	0.414	0.389	0.170	0.155	0.237	0.207	0.075	0.064	0.109	0.087	0.281	0.268	0.161	0.036	3
5/16	0.518	0.490	0.211	0.194	0.295	0.262	0.084	0.072	0.137	0.110	0.322	0.309	0.235	0.042	3
3/8	0.622	0.590	0.253	0.233	0.355	0.315	0.094	0.081	0.164	0.133	0.393	0.380	0.233	0.065	4

TABLE 10-2. ROUND HEAD SCREW HEAD DIMENSIONS

| Nom. Size | Head Diameter | | Height of Head | | Width of Slot | | Depth of Slot | | Dimensions of Recess | | | | Phillips Driver Size |
| | | | | | | | | | Diameter | | Depth Width | | |
	Max.	Min.	Max.	Min.	Max.	Min.	Max.	Min.	Max.	Min.	Max.	Min.	
2	0.162	0.146	0.069	0.059	0.031	0.023	0.048	0.037	0.097	0.087	0.071	0.017	1
3	0.187	0.169	0.078	0.067	0.035	0.027	0.053	0.040	0.106	0.096	0.080	0.018	1
4	0.211	0.193	0.086	0.075	0.039	0.031	0.058	0.044	0.115	0.105	0.090	0.017	1
5	0.236	0.217	0.095	0.083	0.043	0.035	0.063	0.047	0.151	0.141	0.104	0.027	2
6	0.260	0.240	0.103	0.091	0.048	0.039	0.068	0.051	0.159	0.149	0.112	0.027	2
7	0.285	0.264	0.111	0.099	0.048	0.039	0.072	0.055	0.167	0.157	0.120	0.029	2
8	0.309	0.287	0.120	0.107	0.054	0.045	0.077	0.058	0.175	0.165	0.128	0.030	2
10	0.359	0.334	0.137	0.123	0.060	0.050	0.087	0.065	0.192	0.182	0.145	0.031	2
12	0.408	0.382	0.153	0.139	0.067	0.056	0.096	0.072	0.246	0.236	0.165	0.036	3
14	0.457	0.429	0.170	0.155	0.075	0.064	0.106	0.080	0.262	0.252	0.182	0.037	3
1/4	0.472	0.443	0.175	0.160	0.075	0.064	0.109	0.082	0.265	0.255	0.187	0.037	3
5/16	0.590	0.557	0.216	0.198	0.084	0.072	0.132	0.099	0.305	0.295	0.227	0.043	3
3/8	0.708	0.670	0.256	0.237	0.094	0.081	0.155	0.117	0.384	0.374	0.281	0.064	4

TABLE 10-3. FLAT HEAD SCREW HEAD DIMENSIONS

Nom. Size	Head Diameter				Height of Head		Width of Slot		Depth of Slot		Dimensions of Recess				
	Sharp		Modified								Diameter		Depth	Width	Phillips Driver Size
	Max.	Min.	Abs. Min.	Min. Flat	Max.	Min.	Max.	Min.	Max.	Min.	Max.	Min.	Max.	Min.	
2	0.172	0.156	0.150	0.003	0.051	0.040	0.031	0.023	0.023	0.015	0.099	0.089	0.060	0.017	1
3	0.199	0.181	0.175	0.004	0.059	0.048	0.035	0.027	0.027	0.017	0.104	0.094	0.065	0.018	1
4	0.225	0.207	0.200	0.004	0.067	0.055	0.039	0.031	0.030	0.020	0.125	0.115	0.086	0.018	1
5	0.252	0.232	0.225	0.005	0.075	0.062	0.043	0.035	0.034	0.022	0.151	0.141	0.083	0.027	2
6	0.279	0.257	0.249	0.005	0.083	0.069	0.048	0.039	0.038	0.024	0.171	0.161	0.103	0.029	2
7	0.305	0.283	0.274	0.005	0.091	0.076	0.048	0.039	0.041	0.027	0.182	0.169	0.114	0.030	2
8	0.332	0.308	0.300	0.006	0.100	0.084	0.054	0.045	0.045	0.029	0.186	0.176	0.118	0.031	2
10	0.385	0.359	0.348	0.007	0.116	0.098	0.060	0.050	0.053	0.034	0.201	0.191	0.133	0.032	2
12	0.438	0.410	0.397	0.008	0.132	0.112	0.067	0.056	0.060	0.039	0.265	0.255	0.153	0.038	3
14	0.491	0.461	0.447	0.009	0.148	0.127	0.075	0.064	0.068	0.044	0.280	0.270	0.168	0.039	3
$\frac{1}{4}$	0.507	0.477	0.462	0.009	0.153	0.131	0.075	0.064	0.070	0.046	0.280	0.270	0.168	0.039	3
$\frac{5}{16}$	0.635	0.600	0.581	0.011	0.191	0.165	0.084	0.072	0.088	0.058	0.362	0.352	0.213	0.061	4
$\frac{3}{8}$	0.762	0.722	0.700	0.013	0.230	0.200	0.094	0.081	0.106	0.070	0.390	0.380	0.242	0.065	4

TABLE 10-4. OVAL HEAD SCREW HEAD DIMENSIONS

Nom. Size	Head Diameter — Sharp Max.	Sharp Min.	Modified Abs. Min.	Modified Min. Flat	Height of Head — Side Height Max.	Side Height Min.	Total Height Max.	Total Height Min.	Width of Slot Max.	Width of Slot Min.	Depth of Slot Max.	Depth of Slot Min.	Recess Diameter Max.	Recess Diameter Min.	Recess Depth	Recess Width	Phillips Driver Size
2	0.172	0.156	0.150	0.003	0.051	0.040	0.080	0.063	0.031	0.023	0.045	0.037	0.109	0.099	0.078	0.018	1
3	0.199	0.181	0.175	0.004	0.059	0.048	0.092	0.073	0.035	0.027	0.052	0.043	0.121	0.111	0.090	0.019	1
4	0.225	0.207	0.200	0.004	0.067	0.055	0.104	0.084	0.039	0.031	0.059	0.049	0.133	0.123	0.103	0.019	1
5	0.252	0.232	0.225	0.005	0.075	0.062	0.116	0.095	0.043	0.035	0.067	0.055	0.155	0.145	0.098	0.028	2
6	0.279	0.257	0.249	0.005	0.083	0.069	0.128	0.105	0.048	0.039	0.074	0.060	0.175	0.165	0.119	0.030	2
7	0.305	0.283	0.274	0.005	0.091	0.076	0.140	0.116	0.048	0.039	0.081	0.066	0.183	0.170	0.124	0.030	2
8	0.332	0.308	0.300	0.006	0.100	0.084	0.152	0.126	0.054	0.045	0.088	0.072	0.189	0.179	0.133	0.031	2
10	0.385	0.359	0.348	0.007	0.116	0.098	0.176	0.148	0.060	0.050	0.103	0.084	0.206	0.196	0.151	0.033	2
12	0.438	0.410	0.397	0.008	0.132	0.112	0.200	0.169	0.067	0.056	0.117	0.096	0.267	0.257	0.174	0.038	3
14	0.491	0.461	0.447	0.009	0.148	0.127	0.224	0.190	0.075	0.064	0.132	0.108	0.288	0.275	0.193	0.039	3
1/4	0.507	0.477	0.462	0.009	0.153	0.131	0.232	0.197	0.075	0.064	0.136	0.112	0.287	0.277	0.194	0.040	3
5/16	0.635	0.600	0.581	0.011	0.191	0.165	0.290	0.249	0.084	0.072	0.171	0.141	0.387	0.377	0.266	0.065	4
3/8	0.762	0.722	0.700	0.013	0.230	0.200	0.347	0.300	0.094	0.081	0.206	0.170	0.407	0.397	0.285	0.068	4

TABLE 10-5. PAN HEAD SCREW HEAD DIMENSIONS

Nom. Size	Head Diameter		Height of Head Slotted		Height of Head Recessed		Width of Slot		Depth of Slot		Radius	Dimensions of Recess Diameter		Depth	Width	Phillips Driver Size
	Max.	Min.	Max.	Min.	Max.	Min.	Max.	Min.	Max.	Min.	Nom.	Max.	Min.	Max.	Min.	Size
2	0.167	0.155	0.053	0.045	0.062	0.053	0.031	0.023	0.033	0.023	0.035	0.101	0.091	0.071	0.017	1
3	0.193	0.180	0.060	0.051	0.071	0.062	0.035	0.027	0.037	0.027	0.037	0.109	0.099	0.080	0.018	1
4	0.219	0.205	0.068	0.058	0.080	0.070	0.039	0.031	0.041	0.030	0.042	0.119	0.109	0.090	0.018	1
5	0.245	0.231	0.075	0.065	0.089	0.079	0.043	0.035	0.045	0.032	0.044	0.155	0.145	0.103	0.028	2
6	0.270	0.256	0.082	0.072	0.097	0.087	0.048	0.039	0.050	0.038	0.046	0.163	0.153	0.111	0.028	2
7	0.296	0.281	0.089	0.079	0.106	0.096	0.048	0.039	0.054	0.041	0.049	0.173	0.163	0.119	0.028	2
8	0.322	0.306	0.096	0.085	0.115	0.105	0.054	0.045	0.058	0.043	0.052	0.179	0.169	0.127	0.029	2
10	0.373	0.357	0.110	0.099	0.133	0.122	0.060	0.050	0.067	0.050	0.061	0.196	0.186	0.145	0.030	2
12	0.425	0.407	0.125	0.112	0.151	0.139	0.067	0.056	0.077	0.060	0.078	0.256	0.246	0.171	0.031	3
14	0.476	0.457	0.139	0.126	0.169	0.156	0.075	0.064	0.085	0.068	0.087	0.278	0.268	0.192	0.037	3
1/4"	0.492	0.473	0.144	0.130	0.175	0.162	0.075	0.064	0.087	0.070	0.087	0.278	0.268	0.192	0.039	3
5/16"	0.615	0.594	0.178	0.162	0.218	0.203	0.084	0.072	0.109	0.092	0.099	0.347	0.337	0.227	0.059	4
3/8"	0.740	0.716	0.212	0.195	0.261	0.244	0.094	0.081	0.130	0.113	0.143	0.390	0.380	0.266	0.065	4

TABLE 10-6. BINDING HEAD SCREW HEAD DIMENSIONS

| Nom. Size | Head Dia. | | Height of Head | | | | Wdth of Slot | | Dpth of Slot | | Undercut Dimensions | | | | Dimensions of Recess | | | | | Phillips Driver Size |
| | | | Hgt. of Oval | | Total Hgt. | | | | | | Diameter | | Depth | | Diameter | | Dpth | Wdth | | |
	Max.	Min.	Max.	Min.	Max.	Min.	Max.	Min.	Max.	Min.	Max.	Min.	Max.	Min.	Max.	Min.	Max.	Min.	
2	0.181	0.171	0.018	0.013	0.046	0.041	0.031	0.023	0.030	0.024	0.141	0.124	0.010	0.005	0.097	0.087	0.063	0.017	1
3	0.208	0.197	0.022	0.016	0.054	0.048	0.035	0.027	0.036	0.029	0.162	0.143	0.011	0.006	0.107	0.097	0.073	0.017	1
4	0.235	0.223	0.025	0.018	0.063	0.056	0.039	0.031	0.042	0.034	0.184	0.161	0.012	0.007	0.115	0.105	0.081	0.017	1
5	0.263	0.249	0.029	0.021	0.071	0.064	0.043	0.035	0.048	0.039	0.205	0.180	0.014	0.009	0.145	0.135	0.085	0.018	2
6	0.290	0.275	0.032	0.024	0.080	0.071	0.048	0.039	0.053	0.044	0.226	0.199	0.015	0.010	0.157	0.147	0.098	0.026	2
8	0.344	0.326	0.039	0.029	0.097	0.087	0.054	0.045	0.065	0.054	0.269	0.236	0.017	0.012	0.183	0.173	0.125	0.028	2
10	0.399	0.378	0.045	0.034	0.114	0.102	0.060	0.050	0.077	0.064	0.312	0.274	0.020	0.015	0.202	0.192	0.145	0.029	2
12	0.454	0.430	0.052	0.039	0.130	0.117	0.067	0.056	0.089	0.074	0.354	0.311	0.023	0.018	0.264	0.254	0.169	0.032	3
1/4	0.513	0.488	0.061	0.046	0.153	0.138	0.075	0.064	0.105	0.088	0.410	0.360	0.026	0.021	0.278	0.268	0.185	0.046	3
5/16	0.641	0.609	0.077	0.059	0.193	0.174	0.084	0.072	0.134	0.112	0.513	0.450	0.032	0.027	0.347	0.337	0.221	0.068	4
3/8	0.769	0.731	0.094	0.071	0.234	0.211	0.094	0.081	0.163	0.136	0.615	0.540	0.039	0.034	0.397	0.387	0.271	0.076	4

TABLE 10-7. TRUSS HEAD SCREW HEAD DIMENSIONS

Nom. Size	Head Diameter		Height of Head		Width of Slot		Depth of Slot		Radius	Dimensions of Recess				Phillips Driver Size
										Diameter		Depth Width		
	Max.	Min.	Max.	Min.	Max.	Min.	Max.	Min.	Max.	Max.	Min.	Max.	Min.	
2	0.194	0.180	0.053	0.044	0.031	0.023	0.031	0.022	0.129	0.104	0.091	0.059	0.017	1
3	0.226	0.211	0.061	0.051	0.035	0.027	0.036	0.026	0.151	0.110	0.097	0.066	0.017	1
4	0.257	0.241	0.069	0.059	0.039	0.031	0.040	0.030	0.169	0.112	0.099	0.069	0.017	1
5	0.289	0.272	0.078	0.066	0.043	0.035	0.045	0.034	0.191	0.128	0.115	0.085	0.018	1
6	0.321	0.303	0.086	0.074	0.048	0.039	0.050	0.037	0.211	0.158	0.145	0.084	0.027	2
7	0.352	0.333	0.094	0.081	0.048	0.039	0.054	0.041	0.231	0.165	0.152	0.091	0.028	2
8	0.384	0.364	0.102	0.088	0.054	0.045	0.058	0.045	0.254	0.173	0.160	0.099	0.029	2
10	0 448	0.425	0.118	0.103	0.060	0.050	0.068	0.053	0.283	0.188	0.175	0.115	0.030	2
12	0.511	0.487	0.134	0.118	0.067	0.056	0.077	0.061	0.336	0.248	0.235	0.128	0.032	3
14	0.557	0.530	0.146	0.129	0.075	0.064	0.085	0.068	0.375	0.263	0.250	0.143	0.033	3
$\frac{1}{4}$	0.573	0.546	0.150	0.133	0.075	0.064	0.087	0.070	0.375	0.263	0.250	0.143	0.033	3
$\frac{5}{16}$	0.698	0.666	0.183	0.162	0.084	0.072	0.106	0.085	0.457	0.352	0.339	0.193	0.059	4
$\frac{3}{8}$	0.823	0.787	0.215	0.191	0.094	0.081	0.124	0.100	0.538	0.388	0.375	0.239	0.065	4

10-2. Machine Screw Lengths

The commercial screw manufacturers have increased the range of the lengths of machine screws through the years to meet the general growing work needs. Some of these machine screws, particularly in the more popular sizes, have length increments in units of $\frac{1}{32}$ inch. Generally, the most common increments are increments of $\frac{1}{16}$ and $\frac{1}{8}$ inch.

Measuring a screw to determine its length is sometimes confusing. This is particularly true for flat head and oval head screws. When using a machine screw or a bolt with a flat or oval head design, the length of the fastener is determined by measuring the length from the point where the flat head bevel meets the face of the screw to the end of the thread (Fig. 10-2). All other screws or bolts are measured by the length under the head to the end of the threaded section.

Lengths of flat and oval head screws

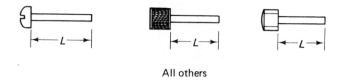

All others

Fig. 10-2. Various types of machine screw lengths are measured as shown by the dimension *L*.

Since it is extremely difficult to accurately define the full range of commercial machine screw lengths, Table 10-8 is included for the technician's use.

TABLE 10-8. SLOTTED SCREW LENGTHS

Size (no.)	Manufactured Length Range (inches)	Available Increments Shown in Parentheses (inches) for Lengths
0	$\frac{1}{16}$ to 1	$\frac{1}{32}$ ($\frac{1}{16}$ to $\frac{3}{16}$)
1	$\frac{1}{8}$ to $\frac{3}{4}$	$\frac{1}{16}$ ($\frac{1}{8}$ to $\frac{1}{2}$)
		$\frac{1}{8}$ ($\frac{1}{2}$ to $\frac{3}{4}$)
2	$\frac{3}{32}$ to 1$\frac{1}{2}$	$\frac{1}{32}$ ($\frac{3}{32}$ to $\frac{5}{16}$)
		$\frac{1}{16}$ ($\frac{5}{16}$ to $\frac{3}{4}$)
		$\frac{1}{8}$ ($\frac{3}{4}$ to 1$\frac{1}{2}$)
3	$\frac{1}{8}$ to 1	$\frac{1}{16}$ ($\frac{1}{8}$ to $\frac{5}{8}$)
		$\frac{1}{8}$ ($\frac{5}{8}$ to 1)
4	$\frac{1}{8}$ to 2$\frac{1}{2}$ (coarse thrd)	$\frac{1}{32}$ ($\frac{1}{8}$ to $\frac{3}{8}$)
	$\frac{1}{8}$ to $\frac{3}{4}$ (fine thrd)	$\frac{1}{16}$ ($\frac{3}{8}$ to $\frac{3}{4}$)
5	$\frac{1}{8}$ to 2	$\frac{1}{16}$ ($\frac{1}{8}$ to $\frac{1}{2}$)
		$\frac{1}{8}$ ($\frac{1}{2}$ to 2)
6	$\frac{1}{8}$ to 3	$\frac{1}{16}$ ($\frac{1}{8}$ to $\frac{7}{8}$)
		$\frac{1}{8}$ ($\frac{7}{8}$ to 1$\frac{3}{4}$)
		$\frac{1}{4}$ (1$\frac{3}{4}$ to 3)
8	$\frac{1}{8}$ to 4 (UNC)	$\frac{1}{16}$ ($\frac{1}{8}$ to $\frac{1}{2}$)
	$\frac{1}{4}$ to 1 (UNF)	$\frac{1}{8}$ ($\frac{1}{2}$ to 1$\frac{3}{4}$)
		$\frac{1}{4}$ (1$\frac{3}{4}$ to 3)
		$\frac{1}{2}$ (3 to 4)
10	$\frac{3}{16}$ to 6 (UNC)	$\frac{1}{16}$ ($\frac{3}{16}$ to $\frac{5}{8}$)
	$\frac{3}{16}$ to 3 (UNF)	$\frac{1}{8}$ ($\frac{5}{8}$ to 1$\frac{3}{4}$)
		$\frac{1}{4}$ (1$\frac{3}{4}$ to 2$\frac{1}{2}$)
		$\frac{1}{2}$ (2$\frac{1}{2}$ to 5)
		1 (5 to 6)
12	$\frac{1}{4}$ to 2$\frac{1}{4}$	$\frac{1}{16}$ ($\frac{1}{4}$ to $\frac{1}{2}$)
		$\frac{1}{8}$ ($\frac{1}{2}$ to 1)
		$\frac{1}{4}$ (1 to 2$\frac{1}{4}$)

NOTE: Slotted Phillips head machine screws are not made in the no. 0 or no. 1 screw gauge size. Available lengths run generally shorter than the popular slotted head type listed above.

10-3. Hex (Allen) and Splined Key Head Screws

When using socket head screws, four basic head forms are available: the standard socket head cap screw, the socket flat head cap screw, the socket

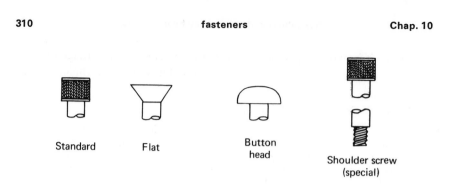

Fig. 10-3. Socket drive head styles of screw products.

button head cap screw and the standard hexagon socket head shoulder screw (Fig. 10-3).

Socket head screws, except for the hexagon socket head shoulder screw, are available in either the more popular hexagon shaped socket or the splined shaped socket. The main advantage of this type of head over the slotted or Phillips type head is the fact that it can be tightened with greater force through the use of a hardened driving key (key setscrew or splined setscrew wrench, Sec. 4-5).

Three of the four types of socket screws have basically the same applications as mentioned for the slotted head screws. The standard socket cap screw can be used in place of the fillister head; the flat socket head can be used in place of the slotted flat; and finally the button socket head can be used in place of the round slotted head screw.

The fourth type of socket head screw, the socket head shoulder screw, has a unique use. Its main function is its use as a fixed shaft about which a workpiece can rotate. This is accomplished by the unthreaded body of the screw having a machined section whose diameter is normally several thousandths of an inch under a fractional size. For example, a body diameter of 0.248 inch is intended for use in a 0.250 inch hole. This allows a total clearance of 0.002 inch between the workpiece hole and the body of the screw. The threaded section of this screw is always smaller in relation to the rest of the screw's diameter. The second feature is the length of the body fit; it is such that it will encompass specific fractional lengths. For example, a square bar ($\frac{1}{2}$ inch) is to be pivoted about a $\frac{1}{4}$ inch diameter shoulder bolt with approximately 0.002 inch clearance between the head and the bar. The best choice for a body length is 0.500 or $\frac{1}{2}$ inch. To increase the length by 0.002 inch to 0.502 inch one simply installs a 0.002 inch thick shim between the end face of the body fit and the tapped hole. This method of adding shims, either under the head or at the threaded end, has the effect of decreasing or increasing the body length to suit a particular need. Table 10-9 lists the available commercial lengths of socket head screws.

TABLE 10-9. SOCKET HEAD CAP SCREWS (COMMERCIAL LENGTHS)

Fractional sizes ($\frac{1}{4}$—$\frac{3}{4}$) UNC, UNF threads

Size	Length Range (inches)	Increments by (inches)
$\frac{1}{4}$ inches	$\frac{3}{8}$ to 1	$\frac{1}{8}$
	1 to 3	$\frac{1}{4}$
$\frac{5}{16}$	$\frac{3}{8}$ to 1	$\frac{1}{8}$
	1 to $3\frac{1}{2}$	$\frac{1}{4}$
$\frac{3}{8}$	$\frac{1}{2}$ to 1	$\frac{1}{8}$
	1 to $3\frac{1}{2}$	$\frac{1}{4}$
$\frac{7}{16}$	$\frac{1}{2}$ to 1	$\frac{1}{8}$
	1 to $3\frac{1}{2}$	$\frac{1}{4}$
	$3\frac{1}{2}$ to 4	$\frac{1}{2}$
$\frac{1}{2}$	$\frac{1}{2}$ to 1	$\frac{1}{8}$
	1 to $3\frac{1}{2}$	$\frac{1}{4}$
	$3\frac{1}{2}$ to 7	$\frac{1}{2}$
	7 to 8	1
$\frac{5}{8}$	1 to $3\frac{1}{2}$	$\frac{1}{4}$
	$3\frac{1}{2}$ to 7	$\frac{1}{2}$
	7 to 8	1
$\frac{3}{4}$	$1\frac{1}{4}$ to $3\frac{1}{2}$	$\frac{1}{4}$
	$3\frac{1}{2}$ to 7	$\frac{1}{2}$
	7 to 8	1

Screw gauge sizes (UNC, UNF threads)

Size (no.)	Length Range (inches)	Increments by (inches)
0	$\frac{1}{8}$ to $\frac{3}{8}$	$\frac{1}{16}$
1	$\frac{1}{8}$ to $\frac{3}{8}$	$\frac{1}{16}$
2	$\frac{3}{16}$ to $\frac{1}{2}$	$\frac{1}{16}$
3	$\frac{3}{16}$ to $\frac{1}{2}$	$\frac{1}{16}$
4	$\frac{1}{4}$ to $\frac{3}{4}$	$\frac{1}{8}$
5	$\frac{1}{4}$ to $\frac{3}{4}$	$\frac{1}{8}$
6	$\frac{1}{4}$ to 1	$\frac{1}{8}$
8	$\frac{1}{4}$ to 1	$\frac{1}{8}$
	1 to $1\frac{1}{2}$	$\frac{1}{4}$
10	$\frac{3}{8}$ to 1	$\frac{1}{8}$
	1 to 2	$\frac{1}{4}$

NOTE: Button head socket screws are available in the same lengths as noted above except in the range of no. 5 through no. 10 where they have lengths slightly shorter than specified above.

10-4. Machine Screw Materials

Machine screws are manufactured in a wide range of materials to meet most needs. The selection of fastener material is governed by three main factors: strength, environment, and cost.

To answer the needs of screw strength, two materials are outstanding in the field—stainless steel, sometimes noted as 18-8, and heat treated carbon steels. In applications where extreme strength is required, the heat treated carbon steel type screw is used most often. This material is used quite extensively in the manufacture of the socket head type screws. It affords both a hard non-denting exterior coupled with a high working strength. This material is quite resistant to rust since it has a dull blue-to-black oxide coating covering its entire surface. This coating is imparted to the screw in the heat treating process. If the coating becomes scraped or burred, rusting results at that point.

Stainless steel (18-8) screws have both strength and resistance to rusting and corrosive environments. In applications where either heat or corrosives, such as acids, salt water, etc., are encountered, 18-8 is most often chosen. This material, while meeting most of the requirements of strength and corrosion resistance, does not take the shock and shear pressure as well as the heat treated steels. Deformation, such as burring and bending, of the screw sometimes takes place. The entire line of screw products, as previously mentioned, can be procured in 18-8 stainless steel.

By far the most common material used in the manufacture of machine screws is low carbon content steel. This material answers the need for both economical and normal strength fastener uses. Screw products made from this material have a poor rust or corrosion factor and should not be used if this is a critical design requirement. Of the total range of fastener material, low carbon steel screws are the most economical in price and most common in use.

For appearance and rust resistance, steel screws are available with either a nickel or cadmium plate at a slight increase in cost. Low carbon steel has one drawback in use—if excessive pressure is applied to the head to drive it, it will sometimes burr and deform the head, since the metal is somewhat soft in relation to the stainless and heat treated steel.

Brass is a material that is being replaced in many applications by the availability of 18-8 stainless products. Brass hardware has some corrosive resistance properties but, in general, is very brittle and quite susceptible to thread stripping or shearing during hard use. Its most used application is in areas where a decorative appearance is required or where electrical connections must be secured to buss bars, terminal strips etc. Brass products generally are high in cost due to the initial cost of the material itself. Socket head screws are not made of brass on a commercial basis.

NYLON FASTENERS

10-5. Nylon Fasteners

Within recent years, nylon has become an essential material in the fastener trade. This material exhibits two very important characteristics: its anticorrosive qualities (for the chemical and food fields) and its nonelectrical conducting properties (for the electrical fields).

Structurally and thermally, nylon is very poor. It should not be used in areas where structural fastening is required or heat above 200°F is encountered. It should be used in those areas where only light work sections are assembled.

To further enhance its strength, some manufacturers produce a steel-cored fastener. The core is cast within the nylon and is not exposed to the exterior surface in any way. Nylon fasteners are comparable in price to the low carbon steel fasteners. The steel-cored nylon fastener is much higher in cost due to the extra cost involved to produce this product. Nylon fasteners are generally made in the slotted or Phillips head type screw only.

In summary, there are many other materials of which threaded fasteners are made. Some of these materials are: phosphor bronze, monel, aluminum, copper and titanium. The use and need for these materials are generally dictated by unique and specific job needs and are not the general day-to-day variety.

BOLTS

At the beginning of this chapter, we defined machine screws as being threaded fasteners that are torqued by use of a standard screwdriver, a Phillips head screwdriver, or a splined or key setscrew wrench. In this section, we define a bolt as a threaded fastener which has either a square or hexagonal shaped head. Bolts, unlike machine screws, can have threaded lengths that do not run the full length. Bolts are torqued by open end, box, combination, or adjustable wrenches.

10-6. Bolt Sizes Available

Bolts are commercially available from the $\frac{1}{4}$ inch—20 UNC size to the 3 inch diameter—4 UNC size. Square headed bolts normally are available in the coarse thread series only, with a size range of $\frac{1}{4}$ to $1\frac{5}{8}$ inches. The hex head bolt range is from $\frac{1}{4}$ to a 3 inch diameter.

The square head bolt is commercially available in only one grade. The hexagon (hex) bolt head however, can be procured in three grades: regular, semi-finished, and finished. The difference between the three grades lies in the undercut washer effect located under the head and the chamfering of the head itself. The regular bolt head is devoid of any washer effect and generally has the outer head surface chamfered. The semi-finished bolt has an outer head chamfer and a washer effect cut under the head. The purpose of this washer effect is to allow the head to bear squarely on the workpiece.

The fully finished hex head bolt incorporates, along with the washer effect, an extra chamfer cut about the bottom of the head. The purpose of this bottom chamfer is to allow the proper seating of the bolt head by eliminating the possibility of burr contact at the hex points. The term "cap screw" is sometimes used to describe bolts with sizes running from $\frac{1}{4}$ to $1\frac{1}{2}$ inches in diameter that have the full finished hex bolt configuration.

Tables 10-10 and 10-11 detail the dimensions of regular square bolts and of American standard finished hexagon bolts and American standard hexagon head cap screws.

10-7. Bolt Grading

In grading bolts for strength, two standards are frequently used, ASTM and SAE. The American Society for Testing Materials (ASTM) and the Society of Automotive Engineers (SAE) have developed a method of grading bolts as to their strength. This allows for the proper selection of a bolt to

TABLE 10-10. SQUARE BOLTS

Nominal Size (dia.)	Body Diam., Max.	Width Across Flats		Width Across Corners		Height of Head	
		Max.	Min.	Max.	Min.	Max.	Min.
$\frac{1}{4}$	0.260	$\frac{3}{8}$	0.362	0.530	0.498	0.188	0.156
$\frac{5}{16}$	0.324	$\frac{1}{2}$	0.484	0.707	0.665	0.220	0.186
$\frac{3}{8}$	0.388	$\frac{9}{16}$	0.544	0.795	0.747	0.268	0.232
$\frac{7}{16}$	0.452	$\frac{5}{8}$	0.603	0.884	0.828	0.316	0.278
$\frac{1}{2}$	0.515	$\frac{3}{4}$	0.725	1.061	0.995	0.348	0.308
$\frac{5}{8}$	0.642	$\frac{15}{16}$	0.906	1.326	1.244	0.444	0.400
$\frac{3}{4}$	0.768	$1\frac{1}{8}$	1.088	1.591	1.494	0.524	0.476
$\frac{7}{8}$	0.895	$1\frac{5}{16}$	1.269	1.856	1.742	0.620	0.568
1	1.022	$1\frac{1}{2}$	1.450	2.121	1.991	0.684	0.628
$1\frac{1}{8}$	1.149	$1\frac{11}{16}$	1.631	2.386	2.239	0.780	0.720
$1\frac{1}{4}$	1.277	$1\frac{7}{8}$	1.812	2.652	2.489	0.876	0.812
$1\frac{3}{8}$	1.404	$2\frac{1}{16}$	1.994	2.917	2.738	0.940	0.872
$1\frac{1}{2}$	1.531	$2\frac{1}{4}$	2.175	3.182	2.986	1.036	0.964
$1\frac{5}{8}$	1.658	$2\frac{7}{16}$	2.356	3.447	3.235	1.132	1.056

TABLE 10-11. AMERICAN STANDARD FINISHED HEXAGON BOLTS AND AMERICAN STANDARD HEXAGON HEAD CAP SCREWS

Nominal Size (dia.)	Body Diam., Min.	Width Across Flats		Width Across Corners		Height of Head	
		Max.	Min.	Max.	Min.	Max.	Min.
FINISHED HEXAGON BOLTS AND HEXAGON HEAD CAP SCREWS							
$\frac{1}{4}$	0.2450	$\frac{7}{16}$	0.428	0.505	0.488	0.163	0.150
$\frac{5}{16}$	0.3065	$\frac{1}{2}$	0.489	0.577	0.557	0.211	0.195
$\frac{3}{8}$	0.3690	$\frac{9}{16}$	0.551	0.650	0.628	0.243	0.226
$\frac{7}{16}$	0.4305	$\frac{5}{8}$	0.612	0.722	0.698	0.291	0.272
$\frac{1}{2}$	0.4930	$\frac{3}{4}$	0.736	0.866	0.840	0.323	0.302
$\frac{9}{16}$	0.5545	$\frac{13}{16}$	0.798	0.938	0.910	0.371	0.348
$\frac{5}{8}$	0.6170	$\frac{15}{16}$	0.922	1.083	1.051	0.403	0.378
$\frac{3}{4}$	0.7410	$1\frac{1}{8}$	1.100	1.299	1.254	0.483	0.455
$\frac{7}{8}$	0.8660	$1\frac{5}{16}$	1.285	1.516	1.465	0.563	0.531
1	0.9900	$1\frac{1}{2}$	1.469	1.732	1.675	0.627	0.591
$1\frac{1}{8}$	1.1140	$1\frac{11}{16}$	1.631	1.949	1.859	0.718	0.658
$1\frac{1}{4}$	1.2390	$1\frac{7}{8}$	1.812	2.165	2.066	0.813	0.749
$1\frac{3}{8}$	1.3630	$2\frac{1}{16}$	1.994	2.382	2.273	0.878	0.810
$1\frac{1}{2}$	1.4880	$2\frac{1}{4}$	2.175	2.598	2.480	0.974	0.902
FINISHED HEXAGON BOLTS ONLY							
$1\frac{3}{4}$	1.7380	$2\frac{5}{8}$	2.538	3.031	2.893	1.134	1.054
2	1.9880	3	2.900	3.464	3.306	1.263	1.175
$2\frac{1}{4}$	2.2380	$3\frac{3}{8}$	3.262	3.897	3.719	1.423	1.327
$2\frac{1}{2}$	2.4880	$3\frac{3}{4}$	3.625	4.330	4.133	1.583	1.479
$2\frac{3}{4}$	2.7380	$4\frac{1}{8}$	3.988	4.763	4.546	1.744	1.632
3	2.9880	$4\frac{1}{2}$	4.350	5.196	4.959	1.935	1.815

meet a particular work load. Naturally, high strength rated bolts are more expensive and should be used only when required.

Manufacturers permanently mark a bolt by stamping marks or notations on the face of the bolt head. Table 10-12 defines the grade markings and load strength for a size range and material of which the bolt is produced.

10-8. Bolt/Cap Screw Lengths

Bolt manufacturers offer a wide range of popular lengths in both the hex and square head configurations. These lengths have been increased over the years to meet demands in the machine and construction building trades.

Tables 10-13 and 10-14 list fastener lengths in the most popular size ranges. Square head bolts and hex head cap screws are also listed.

TABLE 10-12. ASTM AND SAE GRADE MARKINGS FOR STEEL BOLTS AND SCREWS

Grade Marking	Specification	Material	Bolt and Screw Size, inches	Proof Load, psi	Tensile Strength min., psi
No mark	SAE—Grade 1		$\frac{1}{4}$ thru $1\frac{1}{2}$	33,000	60,000
	ASTM—A307	Low Carbon Steel	$\frac{1}{4}$ thru $1\frac{1}{2}$	33,000	60,000
			Over $1\frac{1}{2}$ thru 4	......	55,000
	SAE—Grade 2	Low Carbon Steel	$\frac{1}{4}$ thru $\frac{3}{4}$	55,000	74,000
			Over $\frac{3}{4}$ thru $1\frac{1}{2}$	33,000	60,000
	SAE—Grade 3	Medium Carbon Steel, Cold Worked	$\frac{1}{4}$ thru $\frac{1}{2}$	85,000	110,000
			Over $\frac{1}{2}$ thru $\frac{5}{8}$	80,000	100,000
	SAE—Grade 5	Medium Carbon Steel, Quenched and Tempered	$\frac{1}{4}$ thru 1	85,000	120,000
			Over 1 thru $1\frac{1}{2}$	74,000	105,000
	ASTM—A449		$\frac{1}{4}$ thru 1	85,000	120,000
			Over 1 thru $1\frac{1}{2}$	74,000	105,000
			Over $1\frac{1}{2}$ thru 3	55,000	90,000
A325	ASTM—A325	Medium Carbon Steel, Quenched and Tempered	$\frac{1}{2}$, $\frac{5}{8}$, $\frac{3}{4}$	85,000	120,000
			$\frac{7}{8}$, 1	78,000	115,000
			$1\frac{1}{8}$ thru $1\frac{1}{2}$	74,000	105,000
BB	ASTM—A354 Grade BB	Low Alloy Steel, Quenched and Tempered	$\frac{1}{4}$ thru $2\frac{1}{2}$	80,000	105,000
			Over $2\frac{1}{2}$ thru 4	75,000	100,000

TABLE 10-12. (CONTINUED)

	Specification	Steel	Size		
(BC)	ASTM—A354 Grade BC	Low Alloy Steel, Quenched and Tempered	$\frac{1}{4}$ thru $2\frac{1}{2}$ Over $2\frac{1}{2}$ thru 4	105,000 95,000	125,000 115,000
	SAE—Grade 5.1	Low or Medium Carbon Steel, Quenched and Tempered with Assembled Lock Washer	Up to $\frac{3}{8}$ incl.	85,000	120,000
	SAE—Grade 7	Medium Carbon Alloy Steel, Quenched and Tempered, Roll Thread after heat treatment	$\frac{1}{4}$ thru $1\frac{1}{2}$	105,000	133,000
	SAE—Grade 8	Medium Carbon Alloy Steel, Quenched and Tempered	$\frac{1}{4}$ thru $1\frac{1}{2}$	120,000	150,000
	ASTM—A354 Grade BD	Alloy Steel, Quenched and Tempered			
(A490)	ASTM—A490	Alloy Steel, Quenched and Tempered	$\frac{1}{2}$ thru $2\frac{1}{2}$ Over $2\frac{1}{2}$ thru 4	120,000 105,000	150,000 140,000

ASTM Specifications

A307—Low Carbon Steel Externally and Internally Threaded Standard Fasteners.

A325—High Strength Steel Bolts for Structural Steel Joints, Including Suitable Nuts and Plain Hardened Washers.

A449—Quenched and Tempered Steel Bolts and Studs.

A354—Quenched and Tempered Alloy Steel Bolts and Studs with Suitable Nuts.

TABLE 10-13. HEX HEAD CAP SCREWS

Diameter (inches)	Series	Length (inches)	By Increments of Inches
$\frac{1}{4}$	NC/NF 20 28	$\frac{1}{2}$ to 1	$\frac{1}{8}$
		1 to 4	$\frac{1}{4}$
		4 to 8	$\frac{1}{2}$
$\frac{5}{16}$	NC/NF	4 to 8	$\frac{1}{2}$
$\frac{3}{8}$	NC/NF	4 to 8	$\frac{1}{2}$
		4 to 12	$\frac{1}{2}$
$\frac{1}{2}$	NC/NF	$\frac{3}{4}$ to 1	$\frac{1}{8}$
		1 to 4	$\frac{1}{4}$
		4 to 12	$\frac{1}{2}$
$\frac{9}{16}$	NC/NF	1 to 4	$\frac{1}{4}$
		4 to 12	$\frac{1}{2}$
$\frac{5}{8}$	NC/NF	1 to 4	$\frac{1}{4}$
		4 to 12	$\frac{1}{2}$
$\frac{3}{4}$	NC/NF	1 to 4	$\frac{1}{4}$
		4 to 12	$\frac{1}{2}$
$\frac{7}{8}$	NC/NF	$1\frac{1}{2}$ to 4	$\frac{1}{4}$
		4 to 12	$\frac{1}{2}$
1	NC/NF	$1\frac{1}{2}$ to 4	$\frac{1}{4}$
		4 to 12	$\frac{1}{2}$

NUTS

Nuts are internally threaded metal pieces that are used on a screw, bolt, or other threaded product of the same thread configuration to both tighten and clamp a workpiece or pieces together. Clamping is achieved as the nut advances toward the head of the screw. As the nut advances, tremendous compressive pressure is created on the material located between the nut and the screw or bolt head. Nuts are manufactured in two distinct shapes—square and hexagonal (hex).

10-9. Square Nuts

The square nut is manufactured in two forms—the regular square and the heavy duty square. The real difference in these two forms is that the heavy duty square is larger across the flat dimension and in the head thickness than the regular form. The heavy duty square is used in areas where both sturdy clamping and strength are required.

The square nut is an economical item and costs much less than its hex

TABLE 10-14. SQUARE HEAD BOLTS (COARSE THREADS)

Diameter Size (inches)	Series	Length (inches)	By Increments of Inches
$\frac{1}{4}$	coarse	$\frac{1}{2}$ to 1	$\frac{1}{8}$
		1 to 5	$\frac{1}{4}$
		5 to 12	$\frac{1}{2}$
		12 to 20	1
$\frac{5}{16}$	coarse	$\frac{1}{2}$ to 1	$\frac{1}{8}$
		1 to 5	$\frac{1}{4}$
		5 to 12	$\frac{1}{2}$
		12 to 20	1
$\frac{3}{8}$	coarse	$\frac{3}{4}$ to 5	$\frac{1}{4}$
		5 to 12	$\frac{1}{2}$
		12 to 25	1
$\frac{7}{16}$	coarse	1 to 5	$\frac{1}{4}$
		5 to 12	$\frac{1}{2}$
		12 to 25	1
$\frac{1}{2}$	coarse	1 to 5	$\frac{1}{4}$
		5 to 12	$\frac{1}{2}$
		12 to 25	1
$\frac{5}{8}$	coarse	1 to 5	$\frac{1}{4}$
		5 to 12	$\frac{1}{2}$
		12 to 25	1
$\frac{3}{4}$	coarse	1 to 5	$\frac{1}{4}$
		5 to 12	$\frac{1}{2}$
		12 to 30	1
$\frac{7}{8}$	coarse	$1\frac{1}{2}$ to 5	$\frac{1}{4}$
		5 to 12	$\frac{1}{2}$
		12 to 30	1
1	coarse	$1\frac{1}{2}$ to 5	$\frac{1}{4}$
		5 to 12	$\frac{1}{2}$
		12 to 30	1

shaped counterpart. The full range of threads of the square nut is broken into two groups, machine screw square nuts and fractional size square nuts. Square nuts are normally available in the coarse thread range only. In the machine screw size, nuts are available between no. 0 and no. 12. Fractional square nuts have a size range of $\frac{1}{4}$ to $1\frac{1}{2}$ inches.

Most smaller size square nuts are formed by punching plate steel or brass. Steel nuts $\frac{1}{4}$ inch and up have a somewhat rough outer finish which at times is dark blue to black in appearance. This dark coating prevents rust from forming on the nut.

The square nut is not normally used in machine building but has its real use in the construction fields.

10-10. Hex Nuts

The hex nut is by far the most widely used form of nut in industry. Its shape lends itself well in a variety of work applications. The hex shape allows for tightening in cramped corners, since it can be turned using any two of its parallel flats. There are five basic hex nut forms in common use: regular nut, semi-finished nut, finished nut, slotted nut, and finished castle nut.

Hex nuts are made in both the coarse and fine series of threads. The full range of machine screw nuts from no. 0 through no. 12 is readily available. The fractional size range of nuts is from $\frac{1}{4}$ to 4 inches in the regular, semi-finished, and finished hex nuts.

Further grading of nuts leads into the categories of regular, heavy, and jam grade. The regular grade is used in normal applications and the heavy grade is used in areas where heavy work loads are encountered. The jam nut is a low profile nut used to either backup and lock another nut or to mount threaded pieces to a thin panel. The main difference between the three grades is in their thickness.

The regular hex nut is the least expensive type in use. It is characterized by the fact that its bearing face is machined flat while the other face is crowned or beveled at approximately a 30° angle. This nut has one setback; burrs can develop on the points of the hex at the bearing face and impair proper seating. This nut type, however, does lend itself well in use with a flat washer (discussed in Sec. 10-11).

The semi-finished nut is characterized by a washer effect that exists on the one bearing face of the nut. This nut also has the 30° bevel on its outer face edge. The washer effect adds a distinct advantage to the nut; it allows for a more solid and flat bearing surface when the nut comes in contact with the material it is securing.

The finished nut incorporates both the washer effect on the bearing face coupled with the chamfer on each face of the nut. It should be noted that the word "finished" does not necessarily mean that the entire surface of the nut is machined.

Of the three previously mentioned nuts, regular, semi-finished, and finished, the finished nut is the more expensive due to the extra machining involved in its manufacture.

Slotted nuts, as shown in Fig. 10-4, are manufactured to meet a special need. When rotating assemblies such as flanged shafts, etc., are bolted together, it is very important that nuts or bolts do not become loose and fly off. To prevent this problem, the slotted nut is used in conjunction with a bolt or screw that has a hole through its body perpendicular to the thread.

When a slotted nut is used with a drilled screw or bolt, the nut is first tightened to its required torque value. While tightening, try to keep in mind

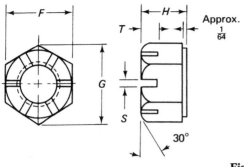

H = Thickness
F = Across the flats dimension
G = Across the diagonal size
S = Width of slot
T = Depth of slot

Fig. 10-4. Slotted hex nut.

that one slot in the nut is to align with the drilled hole in the screw or bolt. When the slot and hole are aligned, a cotter pin or soft steel wire is inserted through the slotted nut and bolt assembly and the ends of the pin or wire are bent. The pin or wire keeps the nut from unscrewing from the thread during use. On flange assemblies having several slotted nuts, a single wire can be fed through each nut-bolt assembly and the ends of the wire twisted. This single wire securely retains any number of nut-to-bolt assemblies. The most preferred method, however, is the use of cotter pins on each nut-bolt assembly.

Occasions arise when the technician must drill a hole in the screw or bolt to mate the slot in the nut. The preferred method is to mark the location and then drill the hole in a drill press. It is bad practice to hand drill the hole in place using the nut slot as a guide. This method generally proves to be poor because when the drill breaks through the other side of the thread, it generally does not line up with the slot in the nut.

Castellated nuts are easily recognized by their buttressed or crowned tops which are slotted (Fig. 10-5). This type nut is not in general use, but it has many applications in the automotive field. Many automobile manufacturers make use of the castellated nut in the bearing caps of piston connecting rods and front wheel bearings. Castellated nuts are considered finished as far as grade and design. The effect of the thinned buttress top is that it tends to spring into and hold onto its mating thread. This spring action coupled with

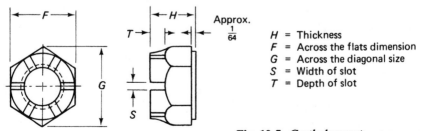
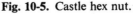

H = Thickness
F = Across the flats dimension
G = Across the diagonal size
S = Width of slot
T = Depth of slot

Fig. 10-5. Castle hex nut.

its cotter pin lock, makes this nut an excellent choice for rotating or reciprocating assemblies.

The jam nut is a thin nut available in either the regular, the semi-finished, or the finished style of nut. Its purpose is to back up or jam a regular nut, preventing the loosening of the main nut. This locking or jamming is performed by restraining the main nut with a wrench while the jam nut is tightened firmly against it. The second application of the jam or thin nut is in its use in mounting threaded components to thin panels. The nut's low profile allows for the close mounting of dials or knobs on shafts extending from the end of the threaded component.

WASHERS

Washers are metal rings or discs of various designs which are used between a nut/bolt head and workpiece. The basic types of washers are flat washers and lock washers.

10-11. Flat Washers

Flat washers are flat metal discs with an appropriate clearance hole in the center to accept a particular size screw or bolt. The functions of a flat washer are: (1) to allow a nut and screw to be used in a larger than normal clearance hole assembly, particularly where the head or nut would otherwise pass through the hole; (2) to more evenly spread the compressive loading of the nut, thereby creating a more rigid assembly; (3) to prevent marring of the workpiece surface by allowing the nut/bolt head to rotate and scratch only against the washer and (4) to aid in removing a corroded or rusted nut/bolt assembly.

Flat washers can be commercially procured in two patterns, the large pattern and the small pattern. The difference in pattern lies in both the diameter and the thickness. Selection of a pattern is governed by its application. If the need is to spread the clamping load over a large or soft material area or to cover an extra large clearance hole, the large pattern is used. In assemblies where the need is mainly to prevent a nut from marring or jamming a work surface, the thinner small pattern is used.

Flat washers are available throughout the whole range of screw sizes to a 3 inch size. The size of a washer is that size which will pass a particular threaded fastener through its bore. For example, a no. 5, small or large pattern washer, allows all no. 5 threaded fasteners to pass through it. When ordering flat washers, three things must be specified: the size of the fastener, the pattern, and the material.

Flat washers are available commercially in low carbon steel, stainless steel (18-8), and brass. In the larger sizes, some washers are made from hot rolled steel which has a somewhat scaly and dull surface finish. The majority of flat washers are made by punching them from flat sheet stock of the desired thickness.

Brass or stainless steel washers are often used with steel nut and bolt assemblies where rust is a problem. The use of washers made from these materials keeps rust from locking the nut to a workpiece. This enables easy disassembly under corroded conditions.

10-12. Lock Washers

Lock washers are metal rings or discs used to "lock" a nut or bolt head to a workpiece. This locking action prevents fastener hardware from loosening and eventually falling away from assemblies where there is either a rotational or a vibrational motion. There are two main types of lock washers: the spring lock and the toothed lock (Fig. 10-6).

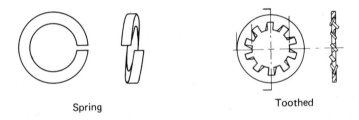

Spring   Toothed

Fig. 10-6. Lock washer types.

The spring lock washer is a rectangular cross sectioned ring which is split at one point. One end of the split is bent slightly from the split joint. This lock washer has a spring action that closes to a flattened condition when it is tightened between a nut and a workpiece. Under the flat compressed condition, the washer split ends tend to spring out, dig into, and keep the nut from unscrewing.

There are four basic grades to the spring type lock washer: light, medium, heavy, and extra heavy. This grading is basically related to the amount of spring due to the difference of the thickness and width of the washers. The size range is normally available in a full range from a no. 2 machine screw through to a $1\frac{1}{2}$ inch screw size.

Spring lock washers are made from alloy heat treated steel, stainless steel, silicon bronze, and phosphor bronze. Alloy heat treated steel is the least expensive of the materials.

Toothed lock washers are made in internal tooth and external tooth styles. Each style washer is further categorized into type A or type B. In the type A lock washer, only one trailing edge of each tooth is bent, all in one direction; on the type B lock washer, the trailing edge of each tooth is bent in one direction and the leading edges are bent in the opposite direction. Where a more secure locking washer is required, the type B is preferred since the better formed teeth dig into the nut and workpiece much more than the type A.

External toothed lock washers leave sharp burrs exposed on assemblies. These burrs can snag clothing or the hands. The internal tooth lock washer teeth are completely hidden in an assembly. This makes a much safer assembly.

Besides the standard flat tooth lock washers, a countersunk lock washer for use under the head of 82° flat head screws is available. Both the internal and external tooth types are available.

The size range of the toothed lock washer runs from a no. 2 through a $1\frac{1}{4}$ inch size. When ordering this type of hardware, the size, style, (internal/external) and type (A or B) must be specified and whether it is flat or countersunk. Toothed washers are available in carbon steel and stainless steel.

SETSCREWS

Setscrews are threaded fasteners used to lock one workpiece part to another. The most common application of a setscrew is its use in securing a belt pulley or gear to a shaft. In this application, the setscrew is turned through a threaded hole in the pulley or gear collar. The pressure of the point of the setscrew against the shaft locks the pulley or gear to the shaft.

There are two types of setscrews in common use: the hexagon socket setscrew and the square head setscrew.

10-13. Hex Socket Setscrews

The hex socket setscrew is the most popular in use today. This type of setscrew is devoid of any external head and has threads that run its entire length. Its design allows it to be inserted into a threaded hole and completely screwed into the hole by the use of a key setscrew wrench. This means that no part of a properly selected setscrew is exposed from the assembled parts to catch or jam either adjacent assemblies or snag a technician's hand.

Hex socket setscrews are available in six point styles: flat point, cone point, oval point, cup point, full dog point, and half dog point (Fig. 10-7). The flat point setscrew is used in applications where the bearing surface for the point is flat. Often a driving shaft has a small flat surface machined on it. This is particularly true in small electric motors. To fit a pulley or gear on this type shaft, the flat point setscrew is used. On larger shaft assemblies, a square metal key between the shaft and pulley is used to transmit extra

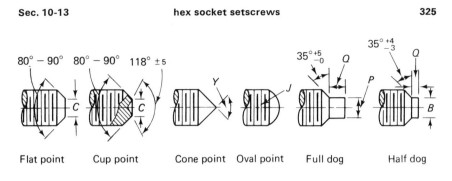

Fig. 10-7. Point styles on set screws.

torque. Generally, a tapped hole for a flat setscrew is located over the pulley or gear keyway. When inserted, the setscrew locates and clamps the entire assembly.

The cone point and oval point setscrews are discussed together because of their close parallelism. Both setscrews, but particularly the cone point, cause a permanent indentation on a shaft during assembly. This deformation or pocket, provides two definite advantages. The first advantage is that it lessens the possibility of slippage between shaft and workpiece. The second advantage is that it repositions a pulley or gear to its original location. Both setscrew point styles realign into their original indentations.

The cup point setscrew is used on assemblies where a self-locking effect is required. The cup point setscrew has a series of knurled edges around the edge of the cup. When used in assemblies, the knurled edges dig into the shaft and also push the shaft metal out and into the knurled edges. This loading of the edges acts as a lock to keep the setscrew from loosening during use.

The full and half dog point hex socket setscrews are identified by an unthreaded flat cylindrical end point. The length of the half dog point is, as expected, one half the length of the full dog point. The diameter and length of this point vary according to the setscrew size.

The dog point setscrew is used when a gear or pulley must often be removed and repositioned to its original location. By using a drilled hole (the diameter of the dog point in the shaft), the dog point can then both position and drive the gear/pulley and shaft through a large number of cycles. The dog point hex socket setscrew is also used in a situation where a keyway exists in a shaft, but not in the pulley or gear. The dog point can be used to extend into and to the bottom of the shaft keyway affording both clamping and torque transmission to the entire assembly.

The hex socket setscrew is available in a size range of a no. 0 through a 2 inch diameter. A wide range of lengths in each size is provided to meet most needs.

These type setscrews are produced in either a hardened high alloy steel or 18-8 stainless steel. When high strength and performance are required, the hardened alloy steel variety is far more superior than the stainless type.

10-14. Square Head Setscrews

The square head setscrew, of which there are two head styles, is generally being replaced by the socket head type in many applications. The square head setscrew is available in only the coarse thread series, generally starting with the $\frac{1}{4}$ inch diameter size and terminating with a $1\frac{1}{2}$ inch diameter size. Square head setscrews are available in the same point styles as the hex socket type but commercially are found only in the cup point. Requests for other point styles are very often considered a special order.

Applications of point styles for the square head setscrew are the same as previously discussed for the hex socket setscrews. Manufacturers of this type setscrew supply them normally in the hardened alloy steel material only.

SHEET METAL AND SELF TAPPING SCREWS

10-15. Introduction to Sheet Metal, Self Tapping Screws

Sheet metal and self tapping screws are fasteners which were developed to meet three specific needs: securing thin sheet metal sections together by deforming the metal into threads, securing a thin material to a thicker material by a self thread cutting action, and, finally, as a means of fabricating mass produced consumer items economically and rapidly.

To use these type screws the only requirement is that a correctly sized hole for the specific screw be available in the workpieces.

Sheet metal and self tapping screws are made in two basic types: screws with coarse high twist threads for use in thin gauge metals and screws whose size and thread are noted by the standard machine screw sizes.

Table 10-15 shows the ASA types of sheet metal/self tapping screws in general use. Types A, B, BP, and C form threads in the metal by a displacing action. These are the basic types used in thin gauge metal fastening. Types D, F, G, T, BF, and BT form threads by a cutting action when turned into a hole. Type U is used in both metal and plastic applications and is pressed into the workpiece hole. The U fastener is intended for permanent or tamper-proof applications.

The head styles available on these screws, for both slotted and Phillips head drives, are the same as previously discussed for the machine screws in Sec. 10-1. The U style, however, has no slots whatsoever. Two more head styles are also available: the hex head and the combination hex washer head styles. These hex head forms are used where extra turning power is required to install the fastener.

TABLE 10-15. SHEET METAL/SELF TAPPING SCREWS

Type	ASA	Manuf.
	AB	AB
	A	A
	B	B
	BP	BP
	C	C
	D	1
	F	F
	G	G
	T	23
	BF	BF
	BT	25
	U	U

10-16. Type Selection and Application

This section is included to aid the technician in his fastener selection as defined by the need or application. Refer to Table 10-16 for sizes.

Type A   —used on thin gauge metal, plywood, and asbestos composition materials

Type B   —used on thin to heavy sheet metal, plastics, plywood asbestos composition and non-ferrous metals.

Type BP —same application as type B, but because of cone point, it is used in misaligned hole situations.

Type C   —this screw is basically the same diameter and thread series as the machine screw. Used in heavy metal sections, this type is a self tapping screw.

Types F, —thread cutting types also. These types can be used in aluminum,
G, D,   brass, cast iron, plastics, die castings, and steel
and T

Types BF—thread cutting types for use in plastics, die castings, resin type
and BT   plywoods, and asbestos

Type U   —this type fastener with its high helix design is pressed into place in both metals and plastics

TABLE 10-16. SIZE RANGES OF THE DIFFERENT TYPES OF SCREWS

Type	Size Range	Available Length Range (inches)
A	4, 6, 7, 8, 10, 12, 14	$\frac{1}{4}$ to 2
B	4, 6, 7, 8, 10, 12, $\frac{1}{4}$	$\frac{3}{16}$ to 2
BP	Same as B	$\frac{1}{4}$ to 2
C	4–40, 6–32, 8–32, 10–24, 10–32, 12–24, $\frac{1}{4}$–20, $\frac{1}{4}$–28, $\frac{5}{16}$–18	$\frac{1}{4}$ to $1\frac{1}{2}$
F, G, D, & T	Same as C but with 2–56, 3–48, $\frac{3}{8}$–16, $\frac{3}{8}$–24	$\frac{1}{8}$ to 1 (small sizes) $\frac{1}{4}$ to $1\frac{1}{2}$ (large sizes)
BF and BT	2–32, 3–28, 4–24, 5–20 6–20, 8–18, 10–16, 12–14, $\frac{1}{4}$–14, $\frac{5}{16}$–12, $\frac{3}{8}$–12	$\frac{1}{8}$ to 1 (small sizes) $\frac{1}{4}$ to $1\frac{1}{2}$ (large sizes)
U	00, 0, 2, 4, 6, 7	$\frac{1}{8}$ to $\frac{5}{16}$

Since the action of these fasteners either incorporates a deforming or cutting action, it is imperative that the fastener be a hard, durable material. These fasteners are, therefore, made of high alloy steel and stainless steel that is heat treated for strength. The size range of each type varies.

The proper use of the sheet metal and self tapping screw is closely tied to the drilling of the proper diameter hole for the insertion of the fastener.

Tables 10-17 through 10-22 are provided to help in the selection of hole to screw type situations. It should be pointed out here that it may be necessary for the technician to "adjust" the particular hole size to allow the fastener to be both easily and adequately installed into the workpiece.

RIVETS

For many years, the rivet has been the method used to permanently secure both large and small assemblies. With the advance of newer techniques in design and welding processes, the use of the rivet has steadily declined. This section briefly describes the standard rivet and a newer type of rivet known as the expanding rivet.

TABLE 10-17. SELF TAPPING SCREWS—TYPE A

Approximate Hole Sizes for Type A Steel Thread Forming Screws
In Steel, Stainless Steel, Monel Metal, Brass, and Aluminum Sheet Metal

Screw Size	Metal Thickness	Hole Required Drilled or Clean Punched	Drill Size	Screw Size	Metal Thickness	Hole Required Drilled or Clean Punched	Drill Size
4	.015	.086	44	8	.024	.113	33
	.018	.086	44		.030	.116	32
	.024	.093	42		.036	.120	31
	.030	.093	42		.048	.128	30
	.036	.098	40	10	.018	.128	30
6	.015	.099	39		.024	.128	30
	.018	.099	39		.030	.128	30
	.024	.099	39		.036	.136	29
	.030	.101	38		.048	.149	25
	.036	.106	36	12	.024	.147	26
7	.015	.104	37		.030	.149	25
	.018	.104	37		.036	.152	24
	.024	.110	35		.048	.157	22
	.030	.113	33	14	.024	.180	15
	.036	.116	32		.030	.189	12
	.048	.120	31		.036	.191	11
8	.018	.113	33		.048	.196	9

TABLE 10-18. SELF TAPPING SCREWS—TYPE B

Approximate Drilled Hole Sizes for Type B Thread Forming Screws

Screw Size	Metal Thickness	Hole Required	Drill Size	Screw Size	Metal Thickness	Hole Required	Drill Size	Screw Size	Metal Thickness	Hole Required	Drill Size
				In Steel, Stainless Steel, Monel Metal, and Brass Sheet Metal							
2	.015	.063	52		.018	.113	33	10	.125	.169	18
	.018	.063	52		.024	.113	33		.135	.169	18
	.024	.067	51		.030	.116	32		.164	.173	17
	.030	.070	50	7	.036	.116	32		.024	.166	19
	.036	.073	49		.048	.120	31		.030	.166	19
	.048	.073	49		.060	.128	30		.036	.166	19
	.060	.076	48		.075	.136	29	12	.048	.169	18
4	.015	.086	44		.105	.140	28		.060	.177	16
	.018	.086	44		.024	.116	32		.075	.182	14
	.024	.089	43		.030	.120	31		.105	.185	13
	.030	.093	42		.036	.120	31		.125	.196	9
	.036	.093	42	8	.048	.128	30		.135	.196	9
	.048	.096	41		.060	.136	29		.164	.201	7
	.060	.099	39		.075	.140	28		.030	.185	13
	.075	.101	38		.105	.149	25		.036	.185	13
6	.015	.104	37		.125	.149	25		.048	.191	11
	.018	.104	37		.135	.152	24		.060	.199	8
	.024	.106	36		.024	.144	27	$\frac{1}{4}$	.075	.204	6
	.030	.106	36		.030	.144	27		.105	.209	4
	.036	.110	35	10	.036	.147	26		.125	.228	1
	.048	.111	34		.048	.152	24		.135	.228	1
	.060	.116	32		.060	.152	24		.164	.234	$\frac{15}{64}$
	.075	.120	31		.075	.157	22		.187	.234	$\frac{15}{64}$
	.105	.128	30		.105	.161	20		.194	.234	$\frac{15}{64}$

TABLE 10-19. SELF TAPPING SCREWS—TYPE C

Approximate Hole Sizes for Type C Steel Thread Forming Screws

Screw Size	Metal Thickness	Hole Required	Drill Size	Screw Size	Metal Thickness	Hole Required	Drill Size	Screw Size	Metal Thickness	Hole Required	Drill Size
				In Sheet Steel							
	.037	.093	42		.037	.154	23		.037	.221	2
	.048	.093	42		.048	.161	20		.048	.221	2
4–40	.062	.096	41	10–24	.062	.166	19	$\frac{1}{4}$–20	.062	.228	1
	.075	.0995	39		.075	.1695	18		.075	.234	A
	.105	.101	38		.105	.173	17		.105	.234	A
	.134	.101	38		.134	.177	16		.134	.236	6mm
	.037	.113	33		.037	.1695	18		.037	.224	5.7mm
	.048	.116	32		.048	.1695	18		.048	.228	1
6–32	.062	.116	32	10–32	.062	.1695	18	$\frac{1}{4}$–28	.062	.232	5.9mm
	.075	.122	3.1mm		.075	.173	17		.075	.234	A
	.105	.125	$\frac{1}{8}$		.105	.177	16		.105	.238	B
	.134	.125	$\frac{1}{8}$		.134	.177	16		.134	.238	B
	.037	.136	29		.037	.189	12		.037	.290	L
	.048	.144	27		.048	.1935	10		.048	.290	L
8–32	.062	.144	27	12–24	.062	.1935	10	$\frac{5}{16}$–18	.062	.290	L
	.075	.147	26		.075	.199	8		.075	.295	M
	.105	.1495	25		.105	.199	8		.105	.295	M
	.134	.1495	25		.134	.199	8		.134	.295	M

TABLE 10-20. SELF TAPPING SCREWS—TYPES D, F, G, AND T

Approximate Hole Sizes for Types D, F, G, and T Steel Thread Cutting Screws

Screw Size	.050	.060	.083	.109	.125	.140	$\frac{3}{16}$	$\frac{1}{4}$	$\frac{5}{16}$	$\frac{3}{8}$	$\frac{1}{2}$
								Hole Sizes in Steel			
2–56	.0730	.0730	.0730	.0730	.0760	.0760					
3–48	.0810	.0810	.0820	.0860	.0860	.0860	.0890				
4–40	.0890	.0890	.0935	.0960	.0980	.0980	.1015				
5–40	.1060	.1060	.1060	.1065	.1094	.1100	.1160	.1160			
6–32	.1100	.1130	.1160	.1160	.1160	.1200	.1250	.1250			
8–32	.1360	.1405	.1405	.1440	.1440	.1470	.1495	.1495	.1495		
10–24	.1520	.1540	.1610	.1610	.1660	.1695	.1730	.1730	.1730	.1730	
10–32	.1590	.1660	.1660	.1695	.1695	.1695	.1770	.1770	.1770	.1770	
12–24		.1800	.1820	.1875	.1910	.1910	.1990	.1990	.1990	.1990	.1990
$\frac{1}{4}$–20			.2130	.2188	.2210	.2210	.2280	.2280	.2280	.2280	.2280
$\frac{1}{4}$–28			.2210	.2280	.2280	.2340	.2344	.2344	.2344	.2344	.2344
$\frac{5}{16}$–18				.2770	.2770	.2813	.2900	.2900	.2900	.2900	.2900
$\frac{5}{16}$–24				.2900	.2900	.2900	.2950	.2950	.2950	.2950	.2950
$\frac{3}{8}$–16					.3390	.3390	.3480	.3580	.3580	.3580	.3580
$\frac{3}{8}$–24					.3480	.3480	.3580	.3580	.3580	.3580	.3580

TABLE 10-21. SELF TAPPING SCREWS—TYPES BF, BG, AND BT

Approximate Hole Sizes for Types BF, BG, and BT Steel Thread Cutting Screws

Stock Thickness	2–32	3–28	4–24	5–20	6–20	8–18	10–16	12–14	$\frac{1}{4}$–14	$\frac{5}{16}$–12	$\frac{3}{8}$–12
					Hole Sizes in Zinc and Aluminum Die Castings						
.060	.0730	.0860									
.083	.0730	.0860									
.109	.0760	.0860	.0980	.1110							
.125	.0760	.0860	.0995	.1110	.1200	.1490	.1660	.1910	.2210	.2810	.3438
.140	.0760	.0890	.0995	.1130	.1200	.1490	.1660	.1910	.2210	.2810	.3438
$\frac{3}{16}$		.0890	.0995	.1130	.1200	.1490	.1660	.1910	.2210	.2810	.3438
$\frac{1}{4}$			.1015	.1160	.1250	.1520	.1695	.1960	.2280	.2810	.3438
$\frac{5}{16}$					.1250	.1520	.1719	.1960	.2280	.2900	.3438
$\frac{3}{8}$							.1719	.1960	.2280	.2900	.3438

TABLE 10-22. DRIVE SCREWS—TYPE U

Approximate Hole Sizes for Type U Hardened Steel Metallic Drive Screws

In Ferrous and Non-Ferrous Castings, Sheet Metals, Plastics,
Plywood (Resin-Impregnated) and Fiber

Screw Size	Hole Size	Drill Size	Screw Size	Hole Size	Drill Size
00	.052	55	8	.144	27
0	.067	51	10	.161	20
2	.086	44	12	.191	11
4	.104	37	14	.221	2
6	.120	31	$\frac{5}{16}$	.295	M
7	.136	29	$\frac{3}{8}$	.358	T

10-17. Description and Application of the Standard Rivet

A rivet is very similar to an "unthreaded" machine screw. It has a head and a round body; both head designs and lengths vary. Large rivets, having a body diameter of $\frac{1}{2}$ inch or larger, are available in six styles of head designs: button head, high button, cone head, pan head, flat top countersunk head, and round top countersunk head. These rivets are used in heavy construction units and range from a $\frac{1}{2}$ to a $1\frac{3}{4}$ inch diameter body size.

In the smaller sized rivets ($\frac{1}{16}$ to $\frac{7}{16}$ inch diameter body) four basic head designs are used: flat head—sometimes called tinners rivets, countersunk head, button head, and pan head (Fig. 10-8). Small rivets are made in a variety of soft malleable materials such as monel, copper, aluminum, and steel. Monel is a widely used choice because of its forming ability and its corrosive resistance.

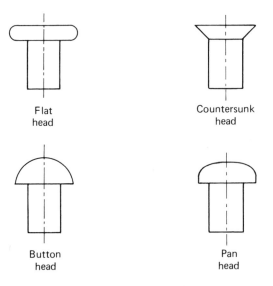

Flat
head

Countersunk
head

Button
head

Pan
head

Fig. 10-8. Basic head designs of small rivets.

In the small rivet range, there exists another group of rivets whose body diameter range runs from 0.081 to 0.347 inch diameter. This group is specified by a size number which is in either ounces or pounds. The size number is the average weight of 1000 of that particular size rivet.

The application of a rivet is as follows: the rivet is inserted into a hole connecting two or more pieces to be joined. With the head rigidly backed up (bucked) with a steel bar or hammer, the other end (which protrudes from the workpiece) is peened over or set until a second "head" is formed tightly against the workpiece. Very often, a formed tool is used to produce the second head. This action permanently secures the workpiece between the two heads of the rivet.

The forming tool is usually struck with a hammer and is designed to produce a rounded or button head rivet form.

10-18. Description and Application of the Expanding Rivet

Within recent years, a unique type of rivet design has been made available. Its design was prompted by a need to produce rivet joints quickly and to also rivet blind holes. A blind hole is the name given to any hole whose back side is not accessible for riveting (forming a head).

The expanding rivet is a two part design that incorporates a hollow body rivet with a steel mandrel or pin built into it. To use this type rivet, a special hand riveting tool is used (Fig. 10-9). The mandrel or pin end of the rivet is inserted into the gripper nose of the tool which grasps the mandrel only. The held rivet is then inserted into a properly drilled hole. To form the rivet, the technician merely squeezes the handles of the tool together until the mandrel breaks from the rivet. If the mandrel doesn't break on the first squeeze, the handles are released and then resqueezed until the mandrel breaks. What happens in this operation is quite simple. A head on the mandrel, located within the rivet, causes the body to flare out when withdrawn by the tool and this in turn forms a rivet head. When a breaking pressure, defined by the strength of the mandrel, is reached, the mandrel breaks and leaves a riveted joint.

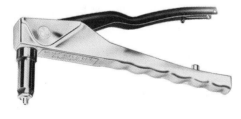

Fig. 10-9. Hand "POP" riveter. (Courtesy of USM Corp., "POP" Rivet Div.)

Expanding rivets are available in several head forms with body diameters of $\frac{3}{32}$ to a $\frac{1}{4}$ inch. In applications where the rivet is used in pressure work, a special closed end expanding rivet is available. It is produced in monel, aluminum, steel, and stainless steel.

PINS

Pins that are used in assembly work perform two important functions: pins fix the position of one piece relative to another and they preserve critical alignments. Pins, unlike threaded fasteners, do not exert a compressive force but rather constrain translation of workpieces in a plane perpendicular to the pin. Three basic dowel pin types are in common use: the straight hardened and ground pin, the tapered pin, and the split pin. Applications and job requirements dictate the use of these important items.

10-19. Straight, Hardened, and Ground Dowel Pins

The straight hardened, and ground dowel pin is most often preferred in tool or gauge building. Often a gauge is built of several high precision parts that must fit and stay together very accurately. Reamed holes are placed into such an assembly and the unit is then pinned with hardened dowels. These dowels maintain the assembly pieces in location and take up shearing shock between the pieces. Most often, threaded fasteners are used in such an assembly to supply the clamping or compressive loading needed to retain the assembly. Reamed holes, which were discussed in Chap. 7, are used whenever a straight hardened and ground dowel pin or tapered pin is to be used. Never use a drilled hole as the finished dowel hole, as neither the diameter nor the finish is compatible with the dowel.

The straight hardened and ground dowel pin is available in a range of diameters from $\frac{1}{8}$ (0.125) to 1 inch, usually in $\frac{1}{16}$ inch increments. The range of lengths for this type pin runs from $\frac{1}{2}$ to 3 inch for diameters up to $\frac{7}{16}$ inch and $\frac{3}{4}$ to 6 inch for the $\frac{1}{2}$ to 1 inch diameter sizes.

Diameters are made in two standard sizes, 0.0002 inch oversize and 0.001 inch oversize. The 0.0002 inch oversized pin is made to produce a better fit when using a standard 0.250 inch ream. As was mentioned in Chap. 7, a range of over and undersized reamers are available to produce pin fits from a "slide" fit to a press fit to meet the required application.

Before moving to our next discussion, one final important point is made. Whenever using straight, hardened and ground dowel pins, always be sure that its reamed hole is through an assembly. The use of a straight through hole is suggested to allow the removal of a broken pin from an assembly. It is extremely difficult to remove broken pins from a blind hole.

10-20. Tapered Pins

Tapered pins are alloy or stainless steel pins made to be used in workpieces where frequent removal and reassembly of the unit takes place. If a straight, hardened and ground pin was used in this situation instead of a tapered pin, the hole would soon become worn and reinsertion of a straight dowel pin would not accurately relocate the workpiece. The tapered pin with its $\frac{1}{4}$ inch to the foot taper, allows for a pin to be installed and removed through many cycles. Wear that may develop in the tapered hole merely lets the pin enter the hole until it seats itself. The tapered pin is also used to secure items such as small gears or collars to rotating assemblies since the pin does not loosen easily.

A thorough explanation in producing tapered ream holes has been covered in Chap. 7. Table 10-23 lists the specifications of tapered pins.

TABLE 10-23. AMERICAN STANDARD TAPER PINS

No. of Taper Pin	Diam. Large End	Approx. Size	Range of Lengths	No. of Taper Pin	Diam. Large End	Approx. Size	Range of Lengths
7/0	0.0625	$\frac{1}{16}$	$\frac{3}{8}$ to $\frac{5}{8}$	3	0.219	$\frac{7}{32}$	$\frac{3}{4}$ to $1\frac{3}{4}$
6/0	0.078	$\frac{5}{64}$	$\frac{3}{8}$ to $\frac{3}{4}$	4	0.250	$\frac{1}{4}$	$\frac{3}{4}$ to 2
5/0	0.094	$\frac{3}{32}$	$\frac{1}{2}$ to 1	5	0.289	$\frac{19}{64}$	1 to $2\frac{1}{4}$
4/0	0.109	$\frac{7}{64}$	$\frac{1}{2}$ to 1	6	0.341	$\frac{11}{32}$	$1\frac{1}{4}$ to 3
3/0	0.125	$\frac{1}{8}$	$\frac{1}{2}$ to 1	7	0.409	$\frac{13}{32}$	2 to $3\frac{3}{4}$
2/0	0.141	$\frac{9}{64}$	$\frac{1}{2}$ to $1\frac{1}{4}$	8	0.492	$\frac{1}{2}$	2 to $4\frac{1}{2}$
0	0.156	$\frac{5}{32}$	$\frac{1}{2}$ to $1\frac{1}{4}$	9	0.591	$\frac{19}{32}$	$2\frac{3}{4}$ to $5\frac{1}{4}$
1	0.172	$\frac{11}{64}$	$\frac{5}{8}$ to $1\frac{1}{4}$	10	0.706	$\frac{45}{64}$	$3\frac{1}{2}$ to 6
2	0.193	$\frac{3}{16}$	$\frac{3}{4}$ to $1\frac{1}{2}$				

10-21. Split Dowel Pins

Split dowel pins (also called spring or roll pins) are hollow hardened metal tubes that are split along one side (Fig. 10-10). The purpose of the split is to allow the pin to be driven into a rough drilled hole and to lock itself into the hole with its spring temper. Split dowel pins are used in production work where the need to secure assemblies with speed and economy are important.

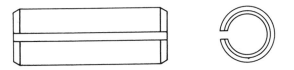

Fig. 10-10. Split or spring type dowel pin.

The installation of the split dowel pin requires only a drilled hole of the proper size. Hole reaming is eliminated. To insert the pin, either press it in or drive it in with a hammer.

The split dowel pin is excellent in assemblies where impact, shock, or vibration is encountered. Split dowel pins can be used over and over since they take up the variation and wear that may develop in the hole.

Split dowel pins are removed by the use of a pin punch and hammer. This variety of pin is available in carbon steel, stainless steel, and beryllium copper. The range of available diameters is from $\frac{1}{16}$ nominal diameter to a $\frac{1}{2}$ inch diameter pin. Table 10-24 lists the recommended hole diameter for a particular size spring pin and also lists the range of available lengths.

TABLE 10-24. SPRING PINS

Nominal Pin Dia.	Maximum Dia.	Minimum Dia.	Max.	Wall Thickness		Recommended Hole Size		Commercial Length Range (inches)
				Nom.	Standard	Min.	Max.	
0.062	0.069	0.066	0.059	0.011	0.012	0.062	0.065	$\frac{3}{16} \longrightarrow 1$
0.078	0.086	0.083	0.075	0.014	0.018	0.078	0.081	$\frac{3}{16} \longrightarrow 1\frac{1}{2}$
0.094	0.103	0.099	0.091	0.018	0.022	0.094	0.097	$\frac{1}{4} \longrightarrow 1\frac{1}{2}$
0.125	0.135	0.131	0.122	0.024	0.028	0.125	0.129	$\frac{1}{4} \longrightarrow 2$
0.156	0.167	0.162	0.151	0.028	0.032	0.156	0.160	$\frac{7}{16} \longrightarrow 2\frac{1}{2}$
0.187	0.199	0.194	0.182	0.036	0.042	0.187	0.192	$\frac{3}{8} \longrightarrow 2\frac{1}{2}$
0.219	0.232	0.226	0.214	0.042	0.050	0.219	0.224	$\frac{9}{16} \longrightarrow 3$
0.250	0.264	0.258	0.245	0.042	0.050	0.250	0.256	$\frac{1}{2} \longrightarrow 3\frac{1}{2}$
0.312	0.328	0.321	0.306	0.060	0.062	0.312	0.318	$\frac{3}{4} \longrightarrow 4$
0.375	0.392	0.385	0.368	0.060	0.077	0.375	0.382	$\frac{3}{4} \longrightarrow 4$
0.437	0.456	0.448	0.430	0.060	0.077	0.437	0.445	$1 \longrightarrow 4$
0.500	0.521	0.513	0.485	0.060	0.094	0.500	0.510	$1\frac{1}{4} \longrightarrow 4$

SUMMARY

This chapter has provided the technician with an insight into the types of fasteners available for attaching workpieces together. Three major classifications of fasteners were presented: threaded, for applications where removal and reinstallation of workpieces may be required; fixed, for permanently attaching workpieces together; and aligning, for keeping workpieces only in a plane that is perpendicular to the pin.

REVIEW QUESTIONS

1. List the two methods of producing threaded fasteners.
2. Name the seven head forms used in screw products.
3. What is the chief advantage of Allen socket head hardware over the screwdriver head hardware?
4. Show an application of a socket head shoulder screw.
5. List some applications of nylon hardware over metal hardware.
6. Define cap screws.
7. What is the common preference in shop assemblies—square or hex head screws? Why?
8. How can you tell a graded bolt from an ungraded type?
9. What is the purpose of a jam nut?
10. When would you use a slotted nut and why?
11. How does a spring lock washer work?
12. What are the six point styles available in hex socket setscrews?
13. List the fastening actions performed by the sheet metal and self tapping screws.
14. How do self tapping fasteners work?
15. When would you select a U type fastener for a particular application?
16. Large rivets have six head forms. List them.
17. Describe the fastening action of a rivet.
18. What is an expanding rivet and how is it installed?
19. Give two functions that a pin fulfills in an assembly.
20. What is the purpose of an oversize and an undersize straight dowel pin?
21. List some advantages of a taper over a straight dowel pin.
22. What is an split dowel pin and where is it most often used?

APPENDIX

Eight tables are included to aid the technician in making conversions within systems and between systems of measures and weights. The tables are:

TABLE A-1. INCH-MILLIMETER EQUIVALENTS OF DECIMAL AND COMMON FRACTIONS

Inch	½'s	¼'s	8ths	16ths	32nds	64ths	Millimeters	Decimals of an Inch[a]
						1	0.397	0.015 625
					1	2	0.794	0.031 25
						3	1.191	0.046 875
				1	2	4	1.588	0.062 5
						5	1.984	0.078 125
					3	6	2.381	0.093 75
						7	2.778	0.109 375
			1	2	4	8	3.175[a]	0.125 0
						9	3.572	0.140 625
					5	10	3.969	0.156 25
						11	4.366	0.171 875
				3	6	12	4.762	0.187 5
						13	5.159	0.203 125
					7	14	5.556	0.218 75
						15	5.953	0.234 375
		1	2	4	8	16	6.350[a]	0.250 0
						17	6.747	0.265 625
					9	18	7.144	0.281 25
						19	7.541	0.296 875
				5	10	20	7.938	0.312 5
						21	8.334	0.328 125
					11	22	8.731	0.343 75
						23	9.128	0.359 375
			3	6	12	24	9.525[a]	0.375 0
						25	9.922	0.390 625
					13	26	10.319	0.406 25
						27	10.716	0.421 875
				7	14	28	11.112	0.437 5
						29	11.509	0.453 125
					15	30	11.906	0.468 75
						31	12.303	0.484 375
	1	2	4	8	16	32	12.700[a]	0.500 0
						33	13.097	0.515 625
					17	34	13.494	0.531 25
						35	13.891	0.546 875
				9	18	36	14.288	0.562 5
						37	14.684	0.578 125
					19	38	15.081	0.593 75
						39	15.478	0.609 375
			5	10	20	40	15.875[a]	0.625 0

TABLF A-1. (CONT'D.)

Inch	½'s	¼'s	8ths	16ths	32nds	64ths	Millimeters	Decimals of an Inch[a]
						41	16.272	0.640 625
					21	42	16.669	0.656 25
						43	17.066	0.671 875
				11	22	44	17.462	0.687 5
						45	17.859	0.703 125
					23	46	18.256	0.718 75
						47	18.653	0.734 375
		3	6	12	24	48	19.050[a]	0.750 0
						49	19.447	0.765 625
					25	50	19.844	0.781 25
						51	20.241	0.796 875
				13	26	52	20.638	0.812 5
						53	21.034	0.828 125
					27	54	21.431	0.843 75
						55	21.828	0.859 375
			7	14	28	56	22.225[a]	0.875 0
						57	22.622	0.890 625
					29	58	23.019	0.906 25
						59	23.416	0.921 875
				15	30	60	23.812	0.937 5
						61	24.209	0.953 125
					31	62	24.606	0.968 75
						63	25.003	0.984 375
1	2	4	8	16	32	64	25.400[a]	1.000 0

[a] Exact.

TABLE A-2. DECIMAL EQUIVALENTS OF 6THS, 12THS, AND 24THS

6th	12th	24th	Decimal	6th	12th	24th	Decimal
		1	0.041667	3			0.5
	1		0.083333			13	0.541666
		3	0.125		7		0.583333
1			0.166666			15	0.625
		5	0.208333	4			0.666666
	3		0.25			17	0.708333
		7	0.291666		9		0.75
2			0.333333			19	0.791666
		9	0.375				
				5			0.833333
	5		0.416666				
						21	0.875
		11	0.458333				
					11		0.916666
						23	0.958333

TABLE A-3. DECIMAL EQUIVALENTS OF 7THS, 14THS, AND 28THS

7th	14th	28th	Decimal	7th	14th	28th	Decimal
		1	0.035714			15	0.535714
	1		0.071429	4			0.571429
		3	0.107143			17	0.607143
1			0.142857		9		0.642867
		5	0.178571			19	0.678571
	3		0.214286	5			0.714286
		7	0.25			21	0.75
2			0.285714		11		0.785714
		9	0.321429			23	0.821429
	5		0.357143	6			0.857143
		11	0.392857			25	0.892857
3			0.428571		13		0.928571
		13	0.464286			27	0.964286
	7		0.5				

TABLE A-4. DECIMAL EQUIVALENTS OF MILLIMETERS
(0.01 to 100 mm)

mm.	Inches	mm.	Inches	mm.	Inches	mm.	Inches	mm.	Inches
0.01	0.00039	0.41	0.01614	0.81	0.03189	21	0.82677	61	2.40157
0.02	0.00079	0.42	0.01654	0.82	0.03228	22	0.86614	62	2.44094
0.03	0.00118	0.43	0.01693	0.83	0.03268	23	0.90551	63	2.48031
0.04	0.00157	0.44	0.01732	0.84	0.03307	24	0.94488	64	2.51968
0.05	0.00197	0.45	0.01772	0.85	0.03346	25	0.98425	65	2.55905
0.06	0.00236	0.46	0.01811	0.86	0.03386	26	1.02362	66	2.59842
0.07	0.00276	0.47	0.01850	0.87	0.03425	27	1.06299	67	2.63779
0.08	0.00315	0.48	0.01890	0.88	0.03465	28	1.10236	68	2.67716
0.09	0.00354	0.49	0.01929	0.89	0.03504	29	1.14173	69	2.71653
0.10	0.00394	0.50	0.01969	0.90	0.03543	30	1.18110	70	2.75590
0.11	0.00433	0.51	0.02008	0.91	0.03583	31	1.22047	71	2.79527
0.12	0.00472	0.52	0.02047	0.92	0.03622	32	1.25984	72	2.83464
0.13	0.00512	0.53	0.02087	0.93	0.03661	33	1.29921	73	2.87401
0.14	0.00551	0.54	0.02126	0.94	0.03701	34	1.33858	74	2.91338
0.15	0.00591	0.55	0.02165	0.95	0.03740	35	1.37795	75	2.95275
0.16	0.00630	0.56	0.02205	0.96	0.03780	36	1.41732	76	2.99212
0.17	0.00669	0.57	0.02244	0.97	0.03819	37	1.45669	77	3.03149
0.18	0.00709	0.58	0.02283	0.98	0.03858	38	1.49606	78	3.07086
0.19	0.00748	0.59	0.02323	0.99	0.03898	39	1.53543	79	3.11023
0.20	0.00787	0.60	0.02362	1.00	0.03937	40	1.57480	80	3.14960
0.21	0.00827	0.61	0.02402	1	0.03937	41	1.61417	81	3.18897
0.22	0.00866	0.62	0.02441	2	0.07874	42	1.65354	82	3.22834
0.23	0.00906	0.63	0.02480	3	0.11811	43	1.69291	83	3.26771
0.24	0.00945	0.64	0.02520	4	0.15748	44	1.73228	84	3.30708
0.25	0.00984	0.65	0.02559	5	0.19685	45	1.77165	85	3.34645
0.26	0.01024	0.66	0.02598	6	0.23622	46	1.81102	86	3.38582
0.27	0.01063	0.67	0.02638	7	0.27559	47	1.85039	87	3.42519
0.28	0.01102	0.68	0.02677	8	0.31496	48	1.88976	88	3.46456
0.29	0.01142	0.69	0.02717	9	0.35433	49	1.92913	89	3.50393
0.30	0.01181	0.70	0.02756	10	0.39370	50	1.96850	90	3.54330
0.31	0.01220	0.71	0.02795	11	0.43307	51	2.00787	91	3.58267
0.32	0.01260	0.72	0.02835	12	0.47244	52	2.04724	92	3.62204
0.33	0.01299	0.73	0.02874	13	0.51181	53	2.08661	93	3.66141
0.34	0.01339	0.74	0.02913	14	0.55118	54	2.12598	94	3.70078
0.35	0.01378	0.75	0.02953	15	0.59055	55	2.16535	95	3.74015
0.36	0.01417	0.76	0.02992	16	0.62992	56	2.20472	96	3.77952
0.37	0.01457	0.77	0.03032	17	0.66929	57	2.24409	97	3.81889
0.38	0.01496	0.78	0.03071	18	0.70866	58	2.28346	98	3.85826
0.39	0.01535	0.79	0.03110	19	0.74803	59	2.32283	99	3.89763
0.40	0.01575	0.80	0.03150	20	0.78740	60	2.36220	100	3.93700

TABLE A-5. ENGLISH SYSTEM OF WEIGHTS AND MEASURES

Linear Measure (Length)

1000 mils = 1 inch (in.)
12 inches = 1 foot (ft.)
3 feet = 1 yard (yd.)
5280 feet = 1 mile

Square Measure (Area)

144 square inches (sq.in.) = 1 square foot (sq.ft.)
9 square feet = 1 square yard (sq.yd.)

Cubic Measure (Volume)

1728 cubic inches (cu.in.) = 1 cubic foot (cu.ft.)
27 cubic feet = 1 cubic yard (cu.yd.)
231 cubic inches = 1 U.S. gallon (gal.)
277.27 cubic inches = 1 British imperial gallon (i.gal.)

Liquid Measure (Capacity)

4 fluid ounces (fl.oz.) = 1 gill (gi.)
2 pints = 1 quart (qt.)
4 quarts = 1 gallon

Dry Measure (Capacity)

2 pints = 1 quart
8 quarts = 1 peck (pk.)

Weight (Avoirdupois)

27.3438 grains = 1 dram (dr.)
16 drams = 1 ounce (oz.)
16 ounces = 1 pound (lb.)
100 pounds = 1 hundredweight (cwt.)
112 pounds = 1 long hundredweight (l.cwt.)
2000 pounds = 1 short ton (S.T.)
2240 pounds = 1 long ton (L.T.)

Weight (Troy)

24 grains = 1 pennyweight (dwt.)
20 pennyweights = 1 ounce (oz.t.)
12 ounces = 1 pound (lb.t.)

Angular or Circular Measure

60 seconds = 1 minute
60 minutes = 1 degree
57.2958 degrees = 1 radian
90 degrees = 1 quadrant or right angle
360 degrees = 1 circle or circumference

TABLE A-6. METRIC SYSTEM OF WEIGHTS AND MEASURES

Linear Measure (Length)

1/10 meter = 1 decimeter (dm)
1/10 decimeter = 1 centimeter (cm)
1/10 centimeter = 1 millimeter (mm)
1/1000 millimeter = 1 micron (μ)
1/1000 micron = 1 millimicron (mμ)
10 meters = 1 dekameter (dkm)
10 dekameters = 1 hectometer (hm)
10 hectometers = 1 kilometer (km)
10 kilometers = 1 myriameter

Square Measure (Area)

1 are = 1 square dekameter (dkm^2)
1 centare = 1 square meter (m^2)
1 hectare = 1 square hectometer (hm^2)

Cubic Measure (Volume)

1 stere = 1 cubic meter (m^3)
1 decistere = 1 cubic decimeter (dm^3)
1 centistere = 1 cubic centimeter (cm^3)
1 dekastere = 1 cubic dekameter (dkm^3)

Capacity

1/10 liter = 1 deciliter (dl)
1/10 deciliter = 1 centiliter (cl)
1/10 centiliter = 1 milliliter (ml)
10 liters = 1 dekaliter (dkl)
100 liters = 1 hectoliter (hl)
1000 liters = 1 kiloliter (kl)
1 kiloliter = 1 stere (s)

Weight

1/10 gram = 1 decigram (dg)
1/10 decigram = 1 centigram (cg)
1/10 centigram = 1 milligram (mg)
10 grams = 1 dekagram (dkg)
100 grams = 1 hectogram (hg)
1000 grams = 1 kilogram (kg)
10,000 grams = 1 myriagram
100,000 grams = 1 quintal (q)
1,000,000 grams = 1 metric ton (t)

TABLE A-7. CONVERSION BETWEEN ENGLISH AND METRIC UNITS

English to Metric   *Metric to English*

Units of Length

1 millimeter = 0.03937 inch or about 1/25 inch

1 inch = 2.540 centimeters	1 centimeter = 0.3937 inch
1 foot = 0.3048 meter	1 decimeter = 3.937 inches
1 yard = 0.9144 meter	1 meter = 39.37 inches
1 mile = 1.6093 kilometers	= 3.281 feet
	= 1.094 yards
	1 kilometer = 0.62137 mile

Units of Area

1 sq. inch = 6.4516 sq. centimeters	1 sq. centimeter = 0.1549997 sq. inch
1 sq. foot = 0.0929 sq. meter	1 sq. meter = 10.764 sq. feet
1 sq. mile = 2.590 sq. kilometers	1 sq. kilometer = 0.3861 sq. mile
1 acre = 0.4047 hectare	1 hectare = 2.471 acres

Units of Volume

| 1 cu. inch = 16.387 cu. centimeters | 1 cu. centimeter = 0.061023 cu. inch |
| 1 cu. foot = 0.028317 cu. meter | 1 cu. meter = 35.31445 cu. feet |

Capacity (Liquid)

1 gill = 0.11829 liter	1 liter = 8.4537 gills
1 pint = 0.4732 liter	1 liter = 2.1134 pints
1 quart = 0.9463 liter	1 liter = 1.0567 quarts

Capacity (Dry)

1 pint = 0.5506 liter	1 liter = 1.816 pints
1 quart = 1.1012 liters	1 liter = 0.908 quart
1 peck = 8.8096 liters	1 liter = 0.1135 peck
1 bushel = 3.52383 dekaliters	1 dekaliter = 0.28378 bushel

Units of Mass

1 grain = 0.0648 gram	1 gram = 15.432 grains
1 ounce (avdp.) = 28.3495 grams	1 kilogram = 35.274 oz. avdp.
1 pound (avdp.) = 0.45359 kilogram	1 kilogram = 2.2046 lbs. avdp.
1 short ton (2000 lb.) = 0.9072 metric ton	1 metric ton = 1.1023 short tons
1 long ton (2240 lb.) = 1.016 metric tons	1 metric ton = 0.9842 long ton

TABLE A-8. METRIC CONVERSION TABLE

Millimeters	×	0.03937	= Inches
Millimeters	=	25.400	× Inches
Meters	×	3.2809	= Feet
Meters	=	0.3048	× Feet
Kilometers	×	0.621377	= Miles
Kilometers	=	1.6093	× Miles
Square centimeters	×	0.15500	= Square inches
Square centimeters	=	6.4515	× Square inches
Square meters	×	10.76410	= Square feet
Square meters	=	0.09290	× Square feet
Cubic centimeters	×	0.061025	= Cubic inches
Cubic centimeters	=	16.3866	× Cubic inches
Cubic meters	×	35.3156	= Cubic feet
Cubic meters	=	0.02832	× Cubic feet
Cubic meters	×	1.308	= Cubic yards
Cubic meters	=	0.765	× Cubic yards
Liters	×	61.023	= Cubic inches
Liters	=	0.01639	× Cubic inches
Liters	×	0.26418	= U.S. gallons
Liters	=	3.7854	× U.S. gallons
Grams	×	15.4324	= Grains
Grams	=	0.0648	× Grains
Grams	×	0.03527	= Ounces, avoirdupois
Grams	=	28.3495	× Ounces, avoirdupois
Kilograms	×	2.2046	= Pounds
Kilograms	=	0.4536	× Pounds
Kilograms per square centimeter	×	14.2231	= Pounds per square inch
Kilograms per square centimeter	=	0.0703	× Pounds per square inch
Kilograms per cubic meter	×	0.06243	= Pounds per cubic foot
Kilograms per cubic meter	=	16.01890	× Pounds per cubic foot
Metric tons (1,000 kilograms)	×	1.1023	= Tons (2,000 pounds)
Metric tons	=	0.9072	× Tons (2,000 pounds)
Calories	×	3.9683	= B.T. units
Calories	=	0.2520	× B.T. units

GLOSSARY

Throughout this text, the authors have attempted to define specific words or expressions as used in their proper location. The following glossary is presented as a quick reference to some of the more commonly used words.

alloy
A base metal that has one or more other metals mixed with it to produce a desired end product.

ampere
A unit of electric current—the constant current which, if maintained in two straight parallel conductors of infinite length, negligible circular cross sections, and placed one meter apart in a vacuum, will produce between these conductors a force equal to 2×10^{-7} newton per meter of length.

angle plates
L-shaped metal castings whose outer surfaces are both machined and perpendicular to each other. They are used to hold workpieces perpendicular to a work surface.

annealing
The act of removing the structural hardness of a workpiece usually by use of heat.

backoff

The process of reversing a tap intermittently during tapping. This keeps large chips from developing and jamming the tap.

bevel

The act of removing sharp, dangerous edges on workpieces. This is performed very often by the use of a hand file or grinder.

bore

The process of machining a hole to exacting dimensions, usually performed on a lathe, milling machine, boring mill, e.g., bore the hole to $1.000^{+.005}_{-.000}$ dia.

bottom tap

A tap so named because of its ability to cut threads close to the bottom of a blind drilled hole.

buffing

The process of imparting a smooth shiny surface to a workpiece by using a soft cloth and fine polishing compounds.

burnish

The act of polishing or producing a shine.

burr

A jagged piece of material formed in all machine operations where cutting tools enter or exit a workpiece.

caliper

The determination of an outside or inside dimension of a workpiece by use of a spring type or vernier type caliper tool.

candela

A unit of such a value that the luminance of a full (blackbody) radiator at the freezing temperature of platinum is 60 candelas per square centimeter.

C-clamps

Cast or forged metal C-shaped pieces with an adjustable clamping screw located normal to the opening. These tools are used to apply clamping pressure to workpieces.

charging

The act of impregnating a material onto a buffing wheel to buff or polish a workpiece.

chase

To chase a thread is an expression that defines the cutting of a thread in a lathe.

chip

A general term applied to any material that is removed from a workpiece by a tool.

clearance

The amount of space that is allowed between two or more workpieces so that assembly of them is easily accomplished.

countersink

A tool used to produce a conical shaped hole at the entrance to an existing hole. This is done to allow flush mounting of screw heads or rivets.

counterbore

Describes a process of producing a flat bottomed hole which is in line and bigger in diameter than an existing hole. The tool performing this operation is called a counterbore.

degree Kelvin

The unit of temperature determined by the Carnot cycle with the triple-point temperature of water defined as exactly $273.16°K$.

die

A hardened metal tool with internal formed threads used to produce threads on round stock.

die cast

The term used to describe a part that is initially cast to exacting dimensions in a mold. Further machining may or may not be necessary.

dope

Pipe dope is a general term used to describe a material put on the threads of an assembly to seal and lock their union. This is done mainly in pipe thread work.

dress

The act of renewing a cutting surface that has become dull or worn.

drunken thread

A thread in which part of the thread helix at some point either increases or decreases with respect to the true helix. This causes the thread to jam when used with its mating threaded part.

ductile

Having the characteristic of being hammered or drawn into thinner sections (gold, copper).

expendable tools

Those tools that, on becoming dull or broken, are discarded (sawblades, taps, thread dies).

gall

A term used to describe the process of two materials that destructively rub and seize at their common contact faces.

gauges

Precision metal pieces which are made to exacting shapes, sizes, and forms. Gauges are used to compare a workpiece or part for accuracy.

jig

A general term of a tool that aids in the production of a part. This tool can be used to aid drilling, reaming, bending, etc. and is intended for use by unskilled people. It is custom built to suit particular job needs.

job

A term used in the machine trade to describe in general terms a workpiece being operated on.

layout

The process of scribing or transferring a shape or location on a workpiece to aid in the machining/fabrication of it into a useful end product.

loaded

A term used to describe the impregnation of a tool with the material it is cutting. This prevents proper removal of the material and increases heat buildup.

meter

The basic unit of measure in the metric system, equal to 39.37 inches. This is a wavelength (krypton 86 orange-red) of 6057.802×10^{-10} m.

mic (mike)

A term used to describe the dimensional sizing of a workpiece by use of a micrometer measuring tool. Also used when referring to a micrometer.

parallels

Metal bars that have the same dimensions of thickness used to rest workpiece on during layout or machining operations—usually two per set.

pin

The process of inserting either a straight or taper bodied pin into a workpiece to accurately align the pieces for assembly and reassembly purposes.

pitch

The distance, in inches or millimeters, that exists between any two successive threads on a threaded product.

plug tap

The second tap in a tap set. Used after the starting tap.

radian

The unit of measure of a plane angle with its vertex at the center of a circle and subtended by an arc equal in length to the radius.

ream

The process of producing extremely accurate and smooth holes in workpieces by a tool called a reamer.

scrape
> The process of developing a random finished surface on a machined part. This produces a non-galling surface, as well as removing high spots that could affect the accuracy of the mating part. Scraping is done on special parts of machines that have rotary or sliding action involved. To scrape a sleeve bearing in means to custom fit the bearing to match the shaft that rides within it.

square
> To square a workpiece means to bring into a perpendicular position one edge or side with respect to a chosen edge or side.

stake
> A word used to describe the procedure of retaining one workpiece to another. This is done by upsetting the material with a punch in such a fashion that the one workpiece is locked to the other.

start tap
> A tap so named because its design allows it to easily enter and start to tap a hole.

steradian
> The unit of measure of a solid angle with its vertex at the center of a sphere and enclosing an area of the spherical surface equal to that of a square with sides equal in length to the radius.

stock
> A metal bar that is used to hold and drive dies for producing threads. Stocks are driven with hand power.

tap
> This word is used to describe the process of forming internal threads by use of a tap.

thread
> The act of forming external threads on a round workpiece. This is done either by machine or hand dies.

tolerance
> The size range that a particular part may be made in and still be useful.

T-tapper
> A "tee" shaped wrench that is used basically to hold and drive a tap, or hand reamer.

turning
> A term used to define any form of round or conical shaped workpiece finished by the use of a lathe.

workpiece
> The name given to any item that is to be machined, formed, welded, or otherwise have some operation performed on it.

INDEX